FORESTS IN A MARKET ECONOMY

FORESTRY SCIENCES

Volume 72

The titles published in this series are listed at the end of this volume.

Forests in a Market Economy

edited by

Erin O. Sills
North Carolina State University, U.S.A.

and

Karen Lee Abt
USDA Forest Service, U.S.A.

KLUWER ACADEMIC PUBLISHERS
DORDRECHT / BOSTON / LONDON

A C.I.P. Catalogue record for this book is available from the Library of Congress.

ISBN 1-4020-1028-1

Published by Kluwer Academic Publishers,
P.O. Box 17, 3300 AA Dordrecht, The Netherlands.

Sold and distributed in North, Central and South America
by Kluwer Academic Publishers,
101 Philip Drive, Norwell, MA 02061, U.S.A.

In all other countries, sold and distributed
by Kluwer Academic Publishers,
P.O. Box 322, 3300 AH Dordrecht, The Netherlands.

Cover illustration: Subhrendu K. Pattanayak

Printed on acid-free paper

All Rights Reserved
© 2003 Kluwer Academic Publishers
No part of this work may be reproduced, stored in a retrieval system, or transmitted
in any form or by any means, electronic, mechanical, photocopying, microfilming, recording
or otherwise, without written permission from the Publisher, with the exception
of any material supplied specifically for the purpose of being entered
and executed on a computer system, for exclusive use by the purchaser of the work.

Printed in the Netherlands.

Contents

Contributors	ix
Preface	xiii
1. Introduction ERIN O. SILLS AND KAREN LEE ABT	1
2. Global Forests JACEK P. SIRY AND FREDERICK W. CUBBAGE	9
3. Private Forests FREDERICK W. CUBBAGE, ANTHONY G. SNIDER, KAREN LEE ABT, AND ROBERT J. MOULTON	23
SECTION I: TIMBER PRODUCTION AND MARKETS	39
4. Optimal Stand Management KAREN LEE ABT AND JEFFREY P. PRESTEMON	41
5. Forest Production JACEK P. SIRY, FREDERICK W. CUBBAGE, AND ERIN O. SILLS	59
6. Financial Analysis of Timber Investments F. CHRISTIAN ZINKHAN AND FREDERICK W. CUBBAGE	77
7. Timber Production Efficiency Analysis DOUGLAS R. CARTER AND JACEK P. SIRY	97

8. Aggregate Timber Supply 117
   DAVID N. WEAR AND SUBHRENDU K. PATTANAYAK

9. Timber Demand 133
   ROBERT C. ABT AND SOEUN AHN

10. Structure And Efficiency Of Timber Markets 153
    BRIAN C. MURRAY AND JEFFREY P. PRESTEMON

11. International Trade In Forest Products 177
    JEFFREY P. PRESTEMON, JOSEPH BUONGIORNO,
    DAVID N. WEAR, AND JACEK P. SIRY

SECTION II: MULTIPLE PRODUCTS FROM FORESTS 201

12. Public Timber Supply under Multiple-Use Management 203
    DAVID N. WEAR

13. Economics of Forest Carbon Sequestration 221
    BRIAN C. MURRAY

14. Timber and Amenities on Nonindustrial Private Forest Land 243
    SUBHRENDU K. PATTANAYAK, KAREN LEE ABT,
    AND THOMAS P. HOLMES

15. Nontimber Forest Products in the Rural Household Economy 259
    ERIN O. SILLS, SHARACHCHANDRA LELE,
    THOMAS P. HOLMES, AND SUBHRENDU K. PATTANAYAK

16. Agroforestry Adoption By Smallholders 283
    D. EVAN MERCER AND SUBHRENDU K. PATTANAYAK

SECTION III: NON-MARKET VALUATION 301

17. Contingent Valuation of Forest Ecosystem Protection 303
    RANDALL A. KRAMER, THOMAS P. HOLMES,
    AND MICHELLE HAEFELE

18. Stated Preference Methods for Valuation of Forest Attributes 321
    THOMAS P. HOLMES AND KEVIN J. BOYLE

19. Estimating Forest Recreation Demand Using Count Data Models   341
    JEFFREY E. ENGLIN, THOMAS P. HOLMES, AND ERIN O. SILLS

20. Forest Ecosystem Services As Production Inputs   361
    SUBHRENDU K. PATTANAYAK AND DAVID T. BUTRY

Contributors

Karen Lee Abt, USDA Forest Service
  Research Economist, Southern Research Station
  Research Triangle Park, North Carolina
Robert C. Abt, North Carolina State University
  Professor, Department of Forestry
  Raleigh, North Carolina
SoEun Ahn, North Carolina State University
  Research Assistant Professor, Department of Forestry
  Raleigh, North Carolina
Kevin J. Boyle, University of Maine
  Professor, Department of Resource Economics and Policy
  Orono, Maine
Joseph Buongiorno, University of Wisconsin
  Professor, Department of Forest Ecology and Management
  Madison, Wisconsin
David T. Butry, USDA Forest Service
  Research Economist, Southern Research Station
  Research Triangle Park, North Carolina
Douglas R. Carter, University of Florida
  Associate Professor, School of Forest Resources and Conservation
  Gainesville, Florida
Frederick W. Cubbage, North Carolina State University
  Professor and Head, Department of Forestry
  Raleigh, North Carolina

Jeffrey E. Englin, University of Nevada
Professor and Chair, Department of Applied Economics and Statistics
Reno, Nevada

Michele A. Haefele, Colorado State University
Post Doctoral Research Associate, Department of Economics
Fort Collins, Colorado

Thomas P. Holmes, USDA Forest Service
Research Forester, Southern Research Station
Research Triangle Park, North Carolina

Randall A. Kramer, Duke University
Professor, Nicholas School of the Environment and Earth Sciences
Durham, North Carolina

Sharachchandra Lélé, Centre for Interdisciplinary Studies in
Environment and Development
Senior Fellow and Coordinator
Bangalore, India

D. Evan Mercer, USDA Forest Service
Research Economist, Southern Research Station
Research Triangle Park, North Carolina

Robert J. Moulton, USDA Forest Service
Retired, State and Private Forestry
Research Triangle Park, North Carolina

Brian C. Murray, Research Triangle Institute
Director, Environment and Natural Resource Economics Program
Research Triangle Park, North Carolina

Subhrendu K. Pattanayak, Research Triangle Institute
Senior Economist,
Environment and Natural Resource Economics Program
Research Triangle Park, North Carolina
And
North Carolina State University
Visiting Assistant Professor, Department of Forestry
Raleigh, North Carolina

Jeffrey P. Prestemon, USDA Forest Service
Research Forester, Southern Research Station
Research Triangle Park, North Carolina

Erin O. Sills, North Carolina State University
Assistant Professor, Department of Forestry
Raleigh, North Carolina

Jacek P. Siry, University of Georgia
  Assistant Professor, Warnell School of Forest Resources
  Athens, Georgia
  *At North Carolina State University during work on this book
Anthony G. Snider, University of Minnesota
  Assistant Professor, Department of Forest Resources
  St. Paul, Minnesota
  *At North Carolina State University during work on this book
David N. Wear, USDA Forest Service
  Project Leader, Southern Research Station
  Research Triangle Park, North Carolina
F. Christian Zinkhan, The Forestland Group, LLC
  Managing Director
  Chapel Hill, North Carolina
    And
  Campbell University
  Professor, Lundy-Fetterman School of Business
  Buies Creek, North Carolina

Preface

This book draws together contributions from forest economists in the Research Triangle of North Carolina, with co-authors from institutions around the world. It represents our common belief that rigorous empirical analysis in an economic framework can inform forest policy. We intend the book as a guide to the empirical methods that we have found most useful for addressing both traditional and modern areas of concern in forest policy, including timber production and markets, multiple use forestry, and valuation of non-market benefits.

The book editors and most chapter authors are affiliated with three institutions in the Research Triangle: the Southern Research Station of the USDA Forest Service (K. Abt, Butry, Holmes, Mercer, Moulton, Prestemon, Wear), the Department of Forestry at North Carolina State University (R. Abt, Ahn, Cubbage, Sills), and the Environmental and Natural Resource Economics Program of Research Triangle Institute (Murray, Pattanayak). Two other Triangle institutions are also represented among the book authors: Duke University (Kramer) and the Forestland Group (Zinkhan). In addition to our primary affiliations, many of us are adjunct faculty and/or graduates of Triangle universities. Many of our co-authors also graduated from or were previously affiliated with Triangle institutions. Thus, the selection of topics, methods, and case studies reflects the work of this particular network of economists, and to some degree, our location in the southeastern United States. However, our work and the chapters encompass other regions of the United States and the world, including Latin America and Asia.

All of the chapters in this volume were subject to rigorous peer review by at least one other contributing author and at least one external reviewer. We held two workshops to discuss the internal reviews and share suggestions for improvements. Additional feedback on draft chapters and coordination among authors were faciliated by the section editors (Holmes, Murray,

Pattanayak, and Prestemon). External reviews were coordinated by the book editors or by Kluwer. We greatly appreciate the careful reviews provided by Vic Adamowicz, Janaki Alavalapati, Ralph Alig, Greg Amacher, Peter Boxall, Diane Burton, Don Dennis, George Dutrow, Donald Grebner, Peter Holmgren, Bill Hyde, Hunter Jenkins, Jan Laarman, Ian Munn, David Newman, Peter Parks, Matthew Pelkki, Daniel Phaneuf, Dixie Reaves, Kim Rollins, Roger Sedjo, R. David Simpson, Jeffrey C. Stier, Steve Swallow, and Roger H. von Haefen.

We owe thanks to our respective institutions for in-kind and direct support of this project including the USDA Forest Service for contract editing (Sara Jenkins) and technical assistance (Renee Boozer and John Pye), Research Triangle Institute for support of Murray and Pattanayak with a Professional Development Award, and NC State University for technical formatting (Judy Rogers) and graduate student research assistance (Stibniati Atmadja, Nevin Dawson, and Shubhayu Saha). Finally, we have benefited from extremely patient editors at Kluwer Academic Publishers, and we extend our thanks to Noeline Gibson, Helen Buitenkamp, Mary Kelly, and Ursula Hertling.

Chapter 1

Introduction

Erin O. Sills and Karen Lee Abt
North Carolina State University and USDA Forest Service

This book demonstrates how economic principles can be used to analyze forest policy issues across existing and developing market economies. The majority of the chapters address timber production and timber markets, primarily from private forest lands. However, policy makers and forest owners are increasingly concerned with a wide range of forest outputs, including ecosystem services, amenities, recreation, and fuelwood, as well as timber. While many of these outputs are not traded in formal markets, the chapters in this book demonstrate that the market paradigm is a useful framework for examining the behavior and values of forest owners and users. Market concepts can be applied broadly to improve our understanding of public policy in the contentious arena of forest management.

Forest economics addresses the significance of forests to the economy, the impact of the economy on forests, and the means by which government and landowners achieve forest management goals. There are several factors that distinguish forest economics as a separate applied field of economics. First, the diversity of forest landowners, both by groups (public, and private industrial and nonindustrial) and within groups, leads to a diversity of preferences, expectations, and constraints. Second, the long time frames involved in forest production give rise to the classical problem of choosing optimal rotation lengths, capital budgeting, and modern financial analysis. A third complicating factor is that forests jointly produce multiple outputs, some extracted and some valued *in situ*, some traded in the market and some not, and some accruing to forest owners and some to the public. Those not traded in the market, whether consumed by landowners or by the public, have no market price signals to predict behavior or guide allocation. Fourth, the immobility of forests lends greater importance to the issue of market power and to travel costs as necessary inputs to forest use. These and other

Sills and Abt (eds.), Forests in a Market Economy, 1–7. ©Kluwer Academic Publishers. Printed in The Netherlands.

aspects of forest economics are reviewed in many textbooks, such as Buongiorno and Gilless (2003), Gregory (1987), Johansson and Lofgren (1985), Klemperer (1996), Nautiyal (1988), and Pearse (1990).

1. BACKGROUND AND PURPOSE

This book is designed as a handbook of applied, empirical forest economics for practitioners, policy analysts, and graduate students. The reader of this book is assumed to be familiar with basic market concepts, including marginal analysis of production and consumption decisions under profit and utility maximization. The chapters are designed for readers with a background in quantitative microeconomics (Varian 1999), introductory calculus (partial derivatives), and statistics including multivariate ordinary least squares regression (OLS) and common maximum likelihood estimation techniques such as probit and logit. All of the chapters provide references for the reader wishing to understand the methods in greater depth. Econometric textbooks such as Greene (2002) and Gujarati (1998) are recommended as general references. A more intuitive presentation of many of the methods can be found in Kennedy (1998).

The chapters encompass traditional and modern areas of concern in forest policy, explaining and illustrating how to apply a range of empirical analytical methods (table 1.1). The two chapters following this introduction summarize the status of the world's forests (chapter 2) and the state of research on private forest management (chapter 3). The rest of the book is divided into three sections. The first focuses on timber production, primarily from private US forest lands, and markets. The second addresses multiple use management and considers a diversity of forest owners and outputs. The third section focuses on the valuation of non-market benefits from forests, including stated and revealed preference methods. In each chapter, the goal is to demonstrate rigorous, policy-relevant, empirical analysis in a manner accessible to readers with a background in intermediate microeconomic theory and statistics.

2. ORGANIZATION OF THE BOOK

Traditional forest economics is concerned with producers who are assumed to maximize profits subject to production technology and exogenous prices. The chapters in the first section follow in that tradition, extending the basic theory (chapter 5) in several directions to more accurately model forest landowners' objectives and constraints by

Introduction 3

Table 1.1. Chapter themes, data and methods

No.	Theme	Data and Location	Methods
2	Status of the world's forests	Aggregate Worldwide	Summary
3	Private forest management	Aggregate and micro US	Summary
4	Neotraditional optimal rotation	Micro North Carolina	Faustmann (logit)
5	Timber production and harvesting	Micro US South	Cost and production functions (OLS^a)
6	Forests and land as investments	Aggregate US	Modern portfolio theory - CAPM^b, efficient frontier, and option pricing
7	Efficient production frontiers	Micro US South	Stochastic frontier analysis (Math. programming, OLS^a)
8	Modeling aggregate timber supply	Aggregate US South	Timber supply from profit function (3-stage least squares)
9	Modeling aggregate timber demand	Aggregate US South	Derived demand from cost function (Seemingly unrelated regression)
10	Efficiency of timber markets	Aggregate US South	Law of One Price (OLS^a, ARIMA^c, Dickey-Fuller)
11	Trade in forest products	Aggregate Worldwide	Partial equilibrium trade models (simulation)
12	Timber harvests from public lands	Parameters from literature US	Engineering supply (simulation, smoothed by OLS^a)
13	Carbon sequestration	Parameters from literature US	Faustmann, land rent theory (simulation)
14	Timber and amenity as joint outputs	Aggregate and micro North Carolina	Household production (OLS^a and probit)
15	Nontimber forest products	Household survey India and Brazil	Household production (OLS^a, Tobit, and neg. binomial)
16	Adoption of agroforestry	Household survey Mexico, Philippines	Household production, adoption choice (probit and logit)
17	Demand for forest ecosystem health	Household survey US South	Contingent valuation (bivariate probit)
18	Preferences for forest management	Household survey Maine	Stated preference, attribute based (multinomial logit)
19	Demand for forest recreation	Recreationist survey Brazil	Travel cost (poisson and negative binomial)
20	Ecosystem services as production inputs	Household survey Indonesia	Weak complementarity, derived demand (OLS^a)

^{a}OLS=ordinary least squares ^{b}CAPM = Capital Asset Pricing Model
^{c}ARIMA = autoregressive integrated moving average

incorporating risk (chapter 6), the possibility of inefficient production (chapter 7), and market power (chapter 10). The cumulative impacts of individual decisions are observed in markets, and the link between landowner decisions and market analysis is one theme of the first section. The section addresses markets for wood products, including aggregate supply (chapter 8), derived demand from domestic industry (chapter 9), and international trade (chapter 11). Recommended texts for background in production economics are Antle and Capalbo (1988) and Chambers (1988).

One key decision for forest landowners is rotation length, or when to harvest the timber from a given forest stand. The underlying theory of optimal rotations was developed by Faustmann (1849) and extended by Hartman (1976). This theoretical framework is introduced in the first section (chapter 4) and also underlies the first two chapters of the second section (chapters 12 and 13). The Faustmann solution provides the optimal harvest age for the deterministic case with positive financial income from timber harvests. Hartman expanded the analysis to examine tradeoffs between timber returns and other outputs such as amenities and ecosystem services. This and other issues of multiple-use are central to public forest land management in the US, which is addressed in chapter 12 of this book and in references such as Bowes and Krutilla (1989) and Loomis (1993).

The remaining chapters in the second section rely on the household production framework, in which economic agents are modelled as both producers and consumers of forest outputs (chapters 14, 15 and 16). This framework is appropriate for non-industrial private forest landowners in the US and for agricultural households in developing countries. These households use public forests and/or plant trees to obtain a variety of benefits. Where markets are complete, their production decisions can be modelled in the standard profit maximization framework, as demonstrated by the last chapter of the book (chapter 20). Key references on household production theory are Singh et al. (1986) and Sadoulet and deJanvry (1995). The other chapters in section three focus on demands for forest outputs and thus rely on consumer theory, or utility maximization and welfare estimation (chapters 17, 18, and 19). The standard reference for non-market valuation is Freeman (2003), with relevant econometric methods discussed in greater detail in Haab et al. (2002).

3. STUDY SITES AND DATA

Many of the chapters address forestry issues in the US South. Forest land in the South is a market driven, ecologically and culturally significant part of the regional landscape. It provides an excellent laboratory for examining market influence on a diversity of landowners across a diverse landscape. The South has active markets for timber, and an increasing population with increasing demands for other forest outputs such as recreation and ecosystem services (Wear and Greis 2002). Several chapters consider other regions in the US (e.g., Maine), address the US as a whole, or are not specific to a region. Other chapters focus on developing countries (Brazil, India, Indonesia, Mexico, and the Phillipines). The literature reviewed in the most of the chapters is international, including many examples from Europe. Finally, the international trade chapter addresses the US at a national level in concert with other players in international markets.

The empirical examples in the chapters draw on data from various sources, including secondary data from the US and the South in particular. Examples include the Forest Inventory and Analysis of the USDA Forest Service, the Timber Mart-South price series (Norris Foundation), the Total Timberland Index of NCREIF (National Council of Real Estate Investment Fiduciaries), surveys of logging firms by the American Pulpwood Association, sector-specific producer price indices from the Bureau of Economic Analysis, and US Census of Population and of Manufacturers. Some chapters draw on other literature for parameters to use in simulations. The third type of data is from surveys of households or individuals conducted by the authors. By definition, survey data is required for stated preference valuation methods (chapters 17 and 18). All of the international applications (chapters 15, 16, 19, and 20) analyze household survey data, perhaps in part due to the lower cost of collecting data in developing countries. Deaton (1997) and Mukherjee et al. (1998) are excellent references for analysis of such household survey data.

4. SUMMARY

Most of the chapters present general theory and methodology relevant to a set of forest policy or management questions, review the findings of previous literature, and derive key testable hypotheses. These hypotheses are then tested in the context of case studies, using the variety of data sources and econometric or other quantitative methods described above. Thus, the results are both methodological and policy-related, and both specific to the case studies and generalizable. Table 1.2 lists selected findings from the chapters. Many of the chapters suggest areas for further work, either in testing hypotheses or advancing the methodology to address other issues and other regions of the world.

Table 1.2. Summary of key findings

No.	Theme	Findings
2	Status of the world's forests	The world has 3.9 billion ha of forests, 87% publicly owned, 5% in plantations, 43% with management plans.
3	Private forest management	Landowner and timber characteristics, market and policy variables determine private forest management.
4	Neotraditional optimal rotation	Neotraditional models are consistent with nonindustrial, traditional with industrial. Future rotations matter to both.
5	Timber production and harvesting	Estimated cost and production functions indicate only limited structural change within logging technology class, 1979-1987.
6	Forests and land as investments	Timberland provides portfolio diversification benefits and is attracting new capital from institutional investors.
7	Efficient production frontiers	Average technical efficiency approximately 60%, similar between years, but increasing over time.
8	Modeling aggregate timber supply	Structure and amount of forest inventory capital affect timber supply estimates; pulp & sawtimber supply respond differently.
9	Modeling aggregate timber demand	Heterogeneity of firms, from processing, input definition, or output definition, may drive observed aggregate relationships.
10	Efficiency of timber markets	Timber prices are temporally efficient. Some oligopsony exists in the pulpwood market.
11	Trade in forest products	Rapid growth in trade has been encouraged by trade agreements (NAFTA, WTO), but barriers still exist.
12	Timber harvests from public lands	Public intervention through harvesting increases public supply, reduces private supply, lowers price, increases price volatility.
13	Carbon sequestration	Prices for carbon sequestration influence optimal rotation age and land use allocation.
14	Timber and amenity as joint outputs	Joint production depends on plot, market and landowner factors and affects timber supply elasticities.
15	Nontimber forest products	NTFP collection driven by functional relationship to other household activities, preferences and assets.
16	Adoption of agroforestry	Market incentives, biophysical conditions, preferences, risk/uncertainty and resource endowments influence adoption.
17	Demand for forest ecosystem health	Forest condition is an economic good, and existence value is largest component of public value of spruce-fir forests.
18	Preferences for forest management	Maine general public prefers balance of harvest and protection and is WTP to protect more forest area.
19	Demand for forest recreation	Count data models adjusted to reflect characteristics of the trip decision demonstrate value of infrequent forest recreation.
20	Ecosystem services as production inputs	Micro theory helps identify data-efficient methods that show substantive contributions of forest ecosystem services.

5. LITERATURE CITED

ANTLE, J.M., AND S.M. CAPALBO (EDS). 1988. Agricultural Productivity Measurement and Explanation. Resources for the Future. Washington DC.

BOWES, M. AND J. KRUTILLA. 1989. Multiple Use Management: The Economics Of Public Forestlands. Resources For The Future, Washington DC: 357 p.

BUONGIORNO, J. AND J. K. GILLESS. 2003. Decision methods for Forest Managers. Academic Press, London, UK. 400 p.

CHAMBERS, R. 1988. Applied Production Analysis: A Dual Approach. Cambridge University Press. Cambridge, U.K. 331 p.

DEATON, A. 1997. The analysis of household surveys: A microeconometric approach to development policy. Johns Hopkins University Press, Baltimore, MD. 479 p.

FAUSTMANN, M. 1849. On the determination of the value which forest land and immature stands possess for forestry. Institute Paper 42 (1968), M. Gane, ed. Oxford: Commonwealth Forestry Institute, Oxford University.

FREEMAN, A.M. 2003. The Measurement of Environmental and Resource Values: Theory and Methods. RFF Press, Washington DC. 496 p.

GREENE, W.H. 2002. Econometric Analysis. Prentice Hall, New Jersey. 1056 p.

GREGORY, G.R. 1987. Resources Economics for Foresters. John Wiley & Sons, New York. 477 p.

GUJARATI, D. 1998. Essentials of Econometrics. McGraw Hill, Boston. 534 p.

HAAB, T.C., K.E. MCCONNELL, AND P.E. EARL. 2002. Valuing Environmental and Natural Resources: the Econometrics of NonMarket Values. Edward Elgar Publishing, UK. 326 p.

HARTMAN, R. 1976. The harvesting decision when a standing forest has value. Econ. Inq. 14:52-58.

JOHANSSON, P.O. AND K. LOFGREN. 1985. The Economics of Forestry and Natural Resources. Basil Blackwell, Oxford 292 p.

KENNEDY, P. A. 1998. Guide to Econometrics. The MIT Press, Cambridge, MA. 468 p.

KLEMPERER, W. DAVID. 1996. Forest Resource Economics and Finance. McGraw-Hill Inc., New York. 551 p.

LOOMIS, J.B. 1993. Integrated Public Lands Management: Principles and Applications to National Forests, Parks, Wildlife Refuges, and BLM Lands. Columbia University Press, New York. 544 p.

MUKHERJEE, C., H. WHITE, AND M. WUYTS. 1998. Econometrics and Data Analysis for Developing Countries. Routledge, New York. 496 p.

NAUTIYAL, J.C. 1988. Forest Economics: Principles and Applications. Canadian Scholars' Press Inc., Toronto. 581 p.

NORRIS FOUNDATION. Various Years, Timber Mart-South. The Daniel B. Warnell School of Forest Resources, University of Georgia, Athens.

PEARSE, P.H. 1990. Introduction to Forestry Economics. University of British Columbia Press, Vancouver. 226 p.

SADOULET, E., AND A. DE JANVRY. 1995. Quantitative Development Policy Analysis. Johns Hopkins University Press, Baltimore. 397 p.

SINGH, I., L. SQUIRE, AND J. STRAUSS (EDS). 1986. Agricultural Household Models. Johns Hopkins University Press, Baltimore. 335 p.

VARIAN, H.R. 1999. Intermediate Microeconomics: A Modern Approach. W.W. Norton & Company, New York, NY. 600 p.

WEAR, D.N. AND J.G. GREIS. 2002. The Southern Forest Resource Assessment – Summary Report. Gen Tech. Report SRS-53. Southern Research Station, USDA Forest Service, Asheville, NC. 103 p.

Chapter 2

GLOBAL FORESTS
Area, Management, and Ownership

Jacek P. Siry and Frederick W. Cubbage
University of Georgia and North Carolina State University

Data on the extent of different forest types provide the basis for monitoring the status of the world's forests, as well as analyzing the effects of markets and government on those forests. Estimates of the total area of the world's forests depend on how one defines forests, the year data were collected, the source of the data, and the organization compiling the information. This chapter summarizes data on world forest extent, plantation extent, and forest management and ownership. Forest types and ownership determine how forests grow, are altered by management, and are allocated in markets or by government.

1. FOREST TYPES AND AREAS

We collected data from many secondary sources and used our knowledge of forestry throughout the world to estimate world forest cover, plantation area, and forest ownership. These estimates are expected to vary considerably, due to the variety of definitions of forest area and differences in estimation methods. Forest cover may be estimated by continuous or periodic forest inventories; by aerial or remotely sensed data; by forestry or agricultural or environmental agencies or nongovernment organizations; or by scientific or bureaucratic guesses. The plethora of definitions and data collection methods cautions against uncritical use of the data that we summarize below.

Sills and Abt (eds.), Forests in a Market Economy, 9–21. ©*Kluwer Academic Publishers. Printed in The Netherlands.*

1.1 Forest Types

Forests can be considered as either closed (dense canopy forests) or open. Continuous grass cover can grow under open forest canopies; a closed forest is too shady for grass. Closed forests generally occur in moist or temperate climates, and open-canopied forests usually are found in drier climates. Frequent fires, such as in pre-Columbian North America, could encourage open canopies even in moist climates.

Closed forests cover 25% of the world's land area (FAO 2001a). Open forests cover another 3%, and shrubland, composed primarily of woody perennial plants, another 10% of the earth's land surface. Forest fallow occupies another 1% and includes areas where tree cover is returning on land that has been subjected to shifting agricultural practices and abandonment.

Of the world's total forest area, 47% is tropical forests, 9% is subtropical forests, 11% is temperate forests, and 33% is boreal forests (FAO 2001a). Forests in developed countries, mostly in the Northern Hemisphere, are comprised largely of coniferous species. Developed countries include North America, Europe, the Russian Federation, Australia, New Zealand, and Japan. Forests in the developing countries—Africa, Latin America, China, and the rest of Asia—are comprised mostly of broadleaved species.

1.2 Total Forest Area

The Forest Resource Assessment (FRA) 2000 provides the current standard world land and forest area (FAO 2001a). Table 2.1 summarizes these and other data as drawn from FRA data and modified by other sources described subsequently. The total world's forest cover amounts to nearly 3.9 billion ha. South America and the Russian Federation have the largest total forest cover in the world, at 886 million ha and 851 million ha, respectively. Brazil alone has 544 million ha. The Democratic Republic of Congo with 135 million ha has the most extensive forest cover in Africa, which has 650 million ha in total. Asia has 548 million ha of forest land, including China (163 million ha), and Indonesia (105 million ha). North America has 471 million ha of forest land, with 245 million ha in Canada and 226 million ha in the United States. Natural forest cover is calculated as the difference between FRA 2000 forest cover and estimated planted forest area. About 3.7 billion ha of the forests are of natural origin. These areas comprise about 94% of all forests. These statistics are only moderately accurate and reliable for many countries, but they are the best available data. Even the current Food and Agriculture Organization of the United Nations (FAO) estimates of the world's forest cover differ from previous assessments. The FRA 2000 forest cover estimate of 3.9 billion ha (FAO 2001a) is 15% higher than the

1995 forest cover estimate of 3.4 billion ha (FAO 1999). This difference does not indicate that the extent of forests has increased; rather, it results from new information and changes in forest definitions and measurement methods employed. For more information regarding variations in FAO forest cover estimates see, for example, FAO (2001b) and Matthews (2001). Throughout this chapter, we will employ the FRA 2000 estimate of 3.9 billion ha of forests, which includes all natural and planted forests.

1.3 Planted Forest Area

Data on world plantation coverage are sparse and incomplete. The FAO (1999) data enumerate total forest area and natural forest area. The difference between these data, when available, was computed to estimate the planted forest area reported in table 2.1. However, natural forest area was not reported for most major developed countries. Data for plantation area in these countries were taken from Brooks (1993), Flynn (1996), Edwards (1996), Brown (1998), Pisarenko et al. (2001), and our own current timber supply estimates. We estimate that there are 216 million ha of plantations in the world, or about 6% of the total forest area.

In several countries in temperate and boreal regions, especially in Europe, it is difficult to distinguish between plantations composed of native species and natural forests. Many European forests have been clearcut and then planted to provide wood for industrial purposes for decades, if not centuries. What distinguishes them from a typical plantation is a rather long rotation period. In order to assess the area of those forests, we assume that clearcutting, planting, and industrial wood production confer a plantation status on forests composed of native species. We estimated that there are about 108 million ha of plantations in Europe. The Russian Federation alone accounts for 17 million ha of plantations.

The amount of fast-grown industrial wood plantations also is difficult to quantify exactly, due to the general shortage of plantation data. For this analysis we assumed that fast-grown industrial wood plantations included all those grown on industrial, nonindustrial private, or on some public lands that could be used to grow and harvest industrial roundwood for manufacture of forest products. We further assumed that fast-grown industrial wood plantations were exotic or native species with growth rates of 5 m^3/ha/yr or more, and rotations typically shorter than 30 years.

Table 2.1. Land, forest, and population statistics by region and selected country, 2000

Region/Country	Land area	Total forest area	Natural forest area	Planted forest area	Population
	(000 ha)	(000 ha)	(000 ha)	(000 ha)	(million)
Africa	2,978,394	649,866	645,084	4,782	767
Congo, D. R.	226,705	135,207	135,165	42	50
Asia	3,084,746	547,793	479,070	68,723	3,634
China	932,743	163,480	129,680	33,800	1,274
Indonesia	181,157	104,986	98,861	6,125	209
Oceania	849,096	197,623	195,089	2,534	30
Australia	768,230	154,539	153,459	1,080	19
Europe	2,259,957	1,039,251	931,353	107,898	729
Russian Federation	1,668,851	851,392	834,112	17,280	147
North America	1,837,992	470,564	445,812	24,752	307
Canada	922,097	244,571	241,571	3,000	31
United States	915,895	225,993	204,241	21,752	276
Central America	298,974	78,740	78,255	485	171
Mexico	190,869	55,205	55,096	109	97
South America	1,754,741	885,618	878,539	7,079	341
Brazil	845,651	543,905	539,005	4,900	168
World	13,063,900	3,869,455	3,653,202	216,253	5,978

Flynn (1996) and Edwards (1996) provide relatively accurate estimates of industrial plantation areas for the Southern Hemisphere; Smith et al. (2001) do the same for North America. Brooks (1993) estimates broad classes of industrial plantation area for Japan, as does Yin (1998) for China, although probably only a fraction of these are for industrial production. Most of the Japanese, Chinese, Indian, and African forest plantations probably cannot be considered industrial wood plantations because they were established for watershed protection, fuelwood, or desertification protection. In addition, the plantations often were not very successful; industrial timber harvests may be excluded by higher recreational or development uses; and the accuracy of the data is unknown. For selected countries in Africa, Asia Major, and Central America, we assumed that about half of the total reported forest plantation area is devoted to fast-grown industrial wood production. Most planted stands in Europe have such long rotations that we excluded them from the fast-grown category.

In total, we estimate that there are about 68 million ha of fast-grown industrial wood plantations, or 1.8% of the world's forests. While the United States contains only about 10% of the world's planted stands, the U.S. South alone contains about 20% of the fast-grown industrial wood plantations per our definitions, with 13.5 million ha. The Southern Hemisphere of the world contains the most fast-growing exotic plantations. The faster growth rates for industrial plantations in the Southern Hemisphere may average 30 to 40 m^3/ha/yr, versus 10 to 15 m^3/ha/yr in the Northern Hemisphere.

Our estimate of 216 million ha of forest plantations in the world is greater than estimates made by prior approaches; Solberg (1996) estimated global plantation coverage at 130 million ha; Brown (1998) 119 million ha; Mattoon (1998) 180 million ha; Pandey and Ball (1998) 138 million ha; Jaakko Pöyry (1999) 116 million ha; and FAO (2001a) 187 million ha. This is primarily because we use a slightly broader definition of plantations that includes any planted stands, especially extensive areas in Europe and the Russian Federation.

1.4 Forest Management and Protection

Timber is the major market-based forest product and a primary reason for actively managing forests. The total standing wood volume of the world's forests is estimated at 386 billion m^3 (FAO 2001a). Europe and South America lead the world with largest forest inventories, 116 billion m^3 and 111 billion m^3, respectively. The world's forests produce approximately 1.6 billion m^3 of industrial roundwood and 1.8 billion m^3 of fuelwood and charcoal annually (FAO 2002). Plantation forests are estimated to supply at least 25% of global industrial roundwood (Brown 1998). The contribution of industrial roundwood production and manufacturing to the global economy is estimated at $400 billion in the early 1990s, or 2% of global GDP (WRI 2000). Timber production in developed regions is dominated by industrial roundwood, which in Europe accounts for 80% of wood harvest (FAO 2002). In developing regions, wood is primarily consumed for energy; in Asia 77% of harvest is for fuelwood and charcoal. A comparison of harvest and forest inventory data indicates that nearly 1% of the world's total standing volume is harvested annually.

Industrial wood plantations represent the minimum area that is subject to traditional or modern forest harvesting regulation techniques. Other planted stands, many of which are in Europe, are also managed actively and subject to sustained yield timber regulation and management. The 216 million ha of planted stands (6%) are probably the most actively managed forests in the world. The remaining 94% of the world's forests are managed in natural stands, left to grow naturally, or are reserved from wood production entirely.

While timber production is one of the most important forest uses, the role of the world's forests in supplying nontimber products and services is increasingly recognized. Forests provide a wide range of plant and animal products, including food, fodder, medicines, cosmetics, and utensils, but information about the quantity and value of nontimber products is limited. In some regions, nontimber uses such as tourism and recreation may generate higher values than timber production. Forests are also increasingly recognized as a source of important environmental benefits, including biodiversity, carbon storage, and watershed protection. Also in this case, data are limited throughout much of the world.

The existence of a forest management plan does not necessarily indicate intensive management. Even if forests are not managed intensively, however, management plans do indicate that at least forest resources were inventoried and that plans were made about their allocation to productive and protective uses. The FRA 2000 defines the area under forest management plans as the area managed for various purposes, such as productive or protective uses, in line with approved national plans covering 5-year periods or more. For developed countries, this category also includes informal management plans. Little is known, however, about more specific management objectives and their achievement.

Overall, FRA 2000 data suggest that about 43% of all forests have some type of management plan (table 2.2). This seems to be a pretty substantial share of forests. However, it includes large areas of public forests that are covered by management plans but see little if any active management or that are entirely set aside for environmental purposes. For example, this would be the case in the Russian Federation, with 851 million ha of forests under management plans, Canada (173 million ha), and Australia (155 million ha). In total, Europe and North America have the largest forest areas covered by management plans, 1,017 million ha and 299 million ha, respectively. On the other hand, in Africa management plans cover only 6 million ha, or about 1% of its forest area. A similar situation exists in South America.

About 12% of the world's forests are legally protected. To estimate protected area, FAO overlaid global forest cover maps and maps of protected areas with a legal protection status. Protected areas include areas classified by the World Conservation Union (IUCN) as categories I-VI. These categories include nature preserves, wilderness areas, national parks, natural monuments, protected landscape/seascape, and managed resource protection areas. The objectives for managing protected lands focus on conservation and protection of natural functions, values, and biodiversity. The statutory levels of protection surely have different levels of enforcement. Timber harvests and land clearing for settlement remain threats in protected areas of many countries. In total, FAO estimates that about 477 million ha of forests receive some form of protection. The largest protected forest areas are in South America (166 million ha), North America (103 million ha), and Europe (51 million ha). Relative to the total forest area in the region, North America protects the largest share of its forests—22%.

Forest certification may also indicate a drive towards more sustainable forest management and better forest protection. As part of the FRA 2000, FAO collected data from major forest certification organizations in the world. These organizations include American Tree Farm Program (ATFP),

Canadian Standard Association's (CSA) National Standard for Sustainable Forest Management, Forest Stewardship Council (FSC), Pan-European Forest Certification Council (PEFCC), Sustainable Forest Initiative (SFI) Program, and Green Tag (GT). FRA 2000 data indicate that about 79 million ha of forests were certified worldwide. This area amounts to only 2% of global forest area. Most of those forests are located in Europe, 47 million ha, and North America, 29 million ha. Interestingly enough, those forests are the best managed forests in the world, and benefits from certification, if they actually improve forest management, are likely to be smaller than in regions where forest management standards are lower.

Table 2.2. Forest management and protection statistics by region and selected country, 2000

Region	Forest area with management plans		Forest area protected		Forest area certified	
	(000 ha)	(%)	(000 ha)	(%)	(000 ha)	(%)
Africa	5,509	1	75,885	12	974	0
Congo, D.R.	-	-	11,527	9	-	-
Asia	133,708	24	49,831	9	158	0
China	-	-	5,011	3	-	-
Indonesia	72	0	17,246	16	-	-
Oceania	166,835	83	23,106	11	410	0
Australia	154,539	100	20,130	13	-	-
Europe	1,017,150	98	51,457	5	46,708	4
Russian Federation	851,392	100	29,059	3	33	0
North America	299,107	64	103,366	22	29,238	6
Canada	173,400	71	12,932	5	4,360	2
United States	125,707	56	90,434	40	24,878	11
Central America	10,468	13	7,496	10	427	1
Mexico	7,100	13	2,403	4	169	0
South America	25,809	3	166,232	19	1,551	0
Brazil	4,000	1	89,448	17	666	0
World	1,658,586	43	477,373	12	79,466	2

Note: - = not available

Despite growing environmental awareness and conservation efforts, forest decline continues. It is estimated that between 1980 and 1995, about 180 million ha of forests were lost, which represents an area larger than Mexico or Indonesia (FAO 1997). Developing countries lost 200 million ha of natural forests, while forests in developed countries expanded by 20 million ha. On average, about 12 million ha of forests were lost annually. Major causes include growing populations and income resulting in ever-increasing demand for wood and land for agriculture and for development. Forest decline has a broader meaning that goes beyond forest land loss and encompasses the decline in quality of existing forests resulting from overexploitation, fragmentation, and human-set fires. The extent of this process is largely unknown.

A comparison of FRA 1990 and FRA 2000 data indicates that the pace of deforestation has slowed somewhat (FAO 2001b). The current net annual deforestation rate (natural forest loss offset by planted forest gain) is estimated at 9 million ha annually. The total loss of the world's natural forests, which comprises deforestation and conversion to planted forests, is larger, estimated at about 16 million ha. Most natural forests, 94%, are lost in the tropics. Whether the pace of global deforestation has indeed slowed down is debated. Matthews (2001) attributes this result to changes in definitions and methods and indicates that natural forest loss in the tropics has likely accelerated. Net annual forest losses are largest in Brazil, with 2.3 million ha, Indonesia (1.3), Sudan (1.0), Zambia (0.9), Mexico (0.6), and the Democratic Republic of Congo (0.5). On a regional basis, forest cover decreases most rapidly in Africa with 5 million ha annually, followed by South America with nearly 4 million ha. The total forest cover is increasing only in Europe (0.9 million ha) and North America (0.4).

2. OWNERSHIP

Forests may be owned by firms or individuals in the private sector or by the public sector. Ownership implies that an entity claims land tenure rights to a forest. Tenure rights are the ability to acquire, use, control, and dispose of a piece of property—either the land itself or the produce derived therefrom. Tenure rights are often, but not always, exclusive. They are seldom absolute. Tenure greatly affects the ability of markets to allocate resources and to protect forests from destructive exploitation. It is possible for the government or the private sector to exercise strong tenure rights and control over forest land—each sector has advantages and disadvantages. One must look at the values of market and nonmarket goods and services, and the success of the government or private sector in providing them, in order to assess the merits of different ownerships.

Public ownership describes the case where a government body exercises ownership jurisdiction over land. Private ownership describes the situation where individuals, firms, businesses, corporations, or even nongovernment organizations control the tenure rights to forests. Corporations act as collections of individuals that usually try to maximize profits, thus achieving economic efficiency for the production of market goods. Cooperative associations of forest landowners, such as is common in Scandinavia, manage and bargain with buyers collectively in private markets. Cooperative forest ownership groups also exist in some developing countries. Depending on their structure or their country, indigenous people may be organized as cooperatives or firms, or they may be considered a dependent entity under some other government control.

Forest land (timberland) ownership in the United States can be classified into four broad classes: National Forest, Other Public, Forest Industry, and Nonindustrial Private. National Forests contain 19% of all forests, primarily in the West where the majority of National Forests were established (Smith et al. 2001). Other Public ownership includes all forests managed by public agencies other than the USDA Forest Service. These include the Bureau of Land Management, states, counties, and municipalities, which together control about 10% of forests. Private owners control 71% of forests in the country. Primary wood products manufacturers, classified as Forest Industry, own about 13% of forests. Nonindustrial Private ownership includes individuals, trusts, and corporations, which control about 58% of forests. This group, for example, includes timber management organizations, which buy forest land and manage timber or land conversions to generate financial returns for pension funds or large institutional or individual investors in forest land. They own at least 3 million ha in the United States.

Various other intermediate forms of forest land, timber, or other product ownerships may exist. Nongovernment organizations often own forest land in order to maximize environmental benefits. Costa Rica has entered into agreements with pharmaceutical companies for prospecting and patent rights to medicinal plants. The USDA Forest Service owns public land, but leases it for ski resorts. Large individual investors and timber management organizations lease land to forest products firms. Hunting leases are common throughout Europe and North America. These intermediate mechanisms provide various means to allocate the rights to the use and disposition of forest goods and services in manners that compensate the various owners adequately for their capital, labor, and entrepreneurial efforts.

The tenure rights granted by the government are fundamental in determining how forests will be managed, protected, or neglected. They may assign all control above and below the surface to the owner of a piece of land, or they may only allow partial uses, such as rights to the timber or nontimber forest products but not to the mineral or oil products. They may allow personal use only or selected commercial uses, and they may be restricted by zoning or tax controls. Landowners may be allowed to sell some development rights, so that the land remains in a natural state. Government has the power to determine the tenure rights to a piece of property. These rights change periodically as government evolves, as demonstrated by privatization of forests in New Zealand and South Africa and increasing legal restrictions on forest practices in the United States.

In addition to various ownership strategies, forest land owners may employ a variety of management approaches. Governments may employ their own forest managers to plan, monitor, manage, and protect their land. So might individuals, at least on a part-time basis as needed for specific projects. Companies may employ their own personnel; hire consultants on a contract basis; or lease land to other management firms. Again, these various arrangements provide means for forest land owners to exercise adequate control over their land at reasonable costs. In some cases, control and management may be extensive, or even ineffective, if firms or governments do not have adequate capital or budgets to maximize profits. Forests have usually been less valuable than more intensive land uses, and harder to exercise control over at reasonable costs.

Data on public and private forest land ownership are hard to find, let alone on the smaller or more creative types of ownership and separated tenure rights for land and trees. FAO (1963) examined ownership structure for three quarters of the world's forest land, finding that nearly 80% of forests were in public ownership. The share of public forests was probably higher as this estimate did not account for some countries with large public forests, such as China. Laarman and Sedjo (1992) estimated that up to 90% of the world's forests were publicly owned as of 1990. This includes the vast forests of the Russian Federation and the Amazon, as well as most of Continental and Insular Asia. White and Martin (2002) estimated that in the 24 most forested countries, 81% of forests are in public ownership.

In contribution to the FRA 2000, the United Nations Economic Commission for Europe, Timber Section, carried out the Temperate and Boreal Forest Resources Assessment (TBFRA) 2000 (UNECE 2000). TBFRA estimated forest ownership structure in industrialized temperate and boreal countries, including Europe with the former Soviet Union, North America, Australia, New Zealand, and Japan. Forests in these countries grow on 1,720 million ha and account for about 44% of the world's forest area (FAO 2001a). Nearly 81% of forests in the region are in public ownership (UNECE 2000). Private owners control 18% of forests, and indigenous people own another 1%. The large share of public ownership in the region is due largely to vast government forest holdings in the Russian Federation and Canada.

Forest ownership structure, however, varies substantially throughout the region. All forests are publicly owned in the Russian Federation and other countries of the Commonwealth of Independent States. Public forests also dominate in Canada (90%) and in many former communist countries in Central and Eastern Europe, including the Czech Republic (84%), Poland (83%), and Romania (95%). Some countries in Western Europe also have large public forests, including Germany (54%), Greece (77%), Ireland

(66%), and Switzerland (68%). Private forest ownership dominates in Austria, Denmark, Finland, France, Norway, Portugal, Slovenia, Spain, and Sweden, which all have about 70% or more of their forests privately owned. Portugal has the highest private forest share in the region—93%.

Given the debate regarding the virtue of government forest ownership over private control, it is surprising how little information about forest tenure rights is available throughout much of the world. Our limited data indicate that the vast majority of forests are government owned and that many of those forests are being destroyed and degraded. This outcome, which links government ownership and regulation with forest decline, may result from a variety of reasons. Governments may convert forests to other uses to promote the achievement of social and development goals, which in some cases are justified and increase social welfare. In many situations, governments simply lack resources and expertise to adequately manage their forest resources. Yet in other cases, governments' ill-conceived policies and corruption result in forest destruction.

In situations where governments lack resources or are unable to combat corruption, a greater reliance on private or communal property and free markets may be considered. Market failures are often used to justify government ownership, but widespread losses in government owned forests suggest that government policy failures are equally serious factors behind forest decline. While it is impossible to define universal tenure rights approaches which would work best towards good stewardship in forest management, in countries that experienced success in developing and protecting their forest resources, a combination of private and public ownership, secure and effective tenure rights, and efficient markets appeared to be the key elements ensuring an adequate supply of forest products and environmental services.

3. CONCLUSION

World forest cover is estimated at 3.9 billion ha as of 2000. Natural forests cover about 94% of forest land area; the remainder is planted forests. The world's total forest area continues to decline, although deforestation has slowed to about 9 million ha annually, according to FRA 2000 data. About 68 million ha of planted forests are devoted to fast-grown industrial wood production. They become increasingly important as they supply about a quarter of global industrial roundwood production. These developments may indicate that appropriately designed and managed planted forests, which occupy only a small fraction of the world's forest area, may be able to fulfil much of global roundwood demand, reducing harvest pressures on remaining forests (Sedjo 1995, Sedjo and Botkin 1997, Mattoon 1998, World Wildlife Fund 2003). Other forests are managed less intensively.

Overall, information about the state of the world's forests is scarce. Even the most extensive information about forest cover is only moderately accurate; forest statistics from several countries that were used for the FRA 2000 are as much as 20 years old (FAO 2001a). We even have difficulties determining the area of forest plantations, which are the most intensively managed forests worldwide. And very limited information is available about nontimber forest products, management, and protection. These information shortages make monitoring the state of the world's forests difficult. Consequently, the development of effective global and regional polices addressing forest decline is hindered as well.

Developed regions of the world, where the state of forests is generally satisfactory, usually have sufficient forestry data to monitor resource developments. Developing regions, where the vast majority of most endangered forests are located, have little or no information about their resources. This leads us to believe that in order to develop effective polices addressing forest management and protection, we need to strengthen our efforts to provide the basic, standardized, and reasonably accurate information needed to effectively monitor and analyze global forest trends. Such information will help to reach an international consensus on the use, misuse, and protection of forests and actions that need to be taken to improve forest management and resource conditions worldwide.

4. LITERATURE CITED

BROOKS, D. 1993. U.S. Forests in a Global Context. General Technical Report RM-228. Fort Collins, CO: USDA Forest Service, Rocky Mountain Range and Experiment Station. 24 p.

BROWN, C. 1998. Global Forest Products Outlook Study: Thematic Study on Plantations. Consulting Report. Food and Agriculture Organization of the United Nations. 80 p.

EDWARDS, M. 1996. The South African forestry and forest products industry: A synopsis, *in* Proceedings of the International Woodfiber Conference. Omni Hotel at CNN Center. Atlanta, GA. May 13-14, 1996. Published by the Pulp & Paper Report, Miller Freeman. Looseleaf.

FLYNN, B. 1996. Latin America: The future of fiber exports, *in* Proceedings of the International Woodfiber Conference. Omni Hotel at CNN Center. Atlanta, GA. May 13-14, 1996. Published by the Pulp & Paper Report, Miller Freeman. Looseleaf.

FAO (Food and Agriculture Organization of the United Nations). 1963. World Forest Inventory, 1963. Food and Agriculture Organization of the United Nations. 113 p.

FAO. 1997. State of the World's Forests, 1997. Food and Agriculture Organization of the United Nations. Available at: http://www.fao.org/fo/sofo/sofo97/97toc-e.stm.

FAO. 1999. State of the World's Forests, 1999. Food and Agriculture Organization of the United Nations. Available at: http://www.fao.org/fo/sofo/sofo99/pdf/sofe_e/coper_en.pdf.

FAO. 2001a. Global Forest Resources Assessment 2000. Food and Agriculture Organization of the United Nations. Available at: http://www.fao.org/forestry/fo/fra/index.jsp.
FAO. 2001b. Comparison of Forest Area and Forest Area Change Estimates Derived from FRA 1990 and FRA 2000. Working Paper 59. Food and Agriculture Organization of the United Nations. Available at: http://www.fao.org/forestry/fo/fra/index.jsp.
FAO. 2002. Yearbook of Forest Products. FAO Forestry Series No. 35. FAO Statistics Series No. 158. Food and Agriculture Organization of the United Nations. 243 p.
JAAKKO PÖYRY. 1999. Global Outlook for Plantations. Research Report 99.9. Canberra, ABARE. 99 p.
LAARMAN, J., AND R. SEDJO. 1992. Global Forests. New York, McGraw-Hill. 337 p.
MATTHEWS, E. 2001. Understanding the FRA 2000. Forest Briefing No. 1. Washington, D.C., World Resources Institute. 11 p.
MATTOON, A. 1998. Tree plantations taking root. P.126-127 in Vital Signs. L. Starke (ed.). WorldWatch Institute. New York, W.W. Norton & Company.
PANDEY, D. AND J. BALL. 1998. The role of industrial plantations in future global fibre supply. Unasylva 193(49): 37-43.
PISARENKO, A., V. STRAKHOV, R. PÄIVINEN, K. KUUSELA, F. DYAKUN, AND V. SDOBNOVA. 2001. Development of Forestry Resources in the European Part of the Russian Federation. Research Report 11. European Forest Research Institute. Boston, Brill. 102 p.
SEDJO, R. 1995. The Potential of High-Yield Plantation Forestry for Meeting Timber Needs: Recent Performance and Future Performance. Discussion Paper 95-08. Washington, DC, Resources for the Future. 30 p.
SEDJO, R., AND D. BOTKIN. 1997. Using forest plantations to spare natural forests. Environment 39(10): 14-20.
SMITH, B., J. VISSAGE, R. SHEFFIELD, AND D. DARR. 2001. Forest Resources of the United States, 1997. General Technical Report NC-219. St. Paul, MN: USDA Forest Service North Central For. Exper. Station. 190 p.
SOLBERG, B. 1996. Long Term Trends and Prospects in World Supply and Demand for Wood and Implications for Sustainable Forest Management. Research Report 1996-6. European Forest Institute. 150 p.
UNECE (United Nations Economic Commission for Europe). 2000. Forest Resources of Europe, CIS, North America, Australia, Japan and New Zealand. Geneva Timber and Forest Study Papers, No. 17. United Nations Economic Commission for Europe, Timber Section. Available at: http://www.unece.org/trade/timber/fra.
WHITE, A., AND A. MARTIN. 2002. Who Owns the World's Forests? Washington, D.C., Forest Trends. Available at: http://www.forest-trends.org.
WRI (World Resources Institute). 2000. World Resources 2000-2001. Joint publication by The United Nations Development Programme, The United Nations Environment Programme, The World Bank, and The World Resources Institute. Washington, D.C., World Resources Institute. 389 p.
WORLD WILDLIFE FUND. 2003. The Forest Industry in the 21st Century. WWF International, Global Forest and Trade Network. Available at: http://www.panda.org/forestandtrade/latest_news/publications.
YIN, R. 1998. Forestry and the environment in China: The current situation and strategic choices. World Dev. 26(12): 2152-2167.

Chapter 3

Private Forests
Management and Policy in a Market Economy

Frederick W. Cubbage, Anthony G. Snider, Karen Lee Abt, and Robert J. Moulton
North Carolina State University, University of Minnesota, USDA Forest Service, and USDA Forest Service (Retired)

This chapter discusses privately owned forests and timber management in a market economy, including private property rights and tenure, landowner objectives and characteristics, markets, and government policies. Private forest land ownership and management—whether it be industrial or nonindustrial—is often assumed to represent the classic model of atomistic competition in a free market, private enterprise system. Private stumpage markets for timber are perhaps the best example of how this kind of market competition allocates scarce inputs such as land, capital, and labor for efficient production of wood fiber outputs, for example, pulpwood and sawtimber. Where strong private markets for timber exist, there are usually many private forest landowners (producers) and a moderate number of timber buyers (consumers).

Property rights, landowner characteristics and objectives, commodity and land markets, and political processes determine how forests are managed, protected, or reserved. These factors and their effects on forest management and allocation are discussed in this chapter. Most studies have employed various types of econometric techniques to assess the effect of various independent variables (e.g., landowner objectives, income, timber prices, subsidies) on dependent variables (e.g., tree planting, timber harvesting).

Landowner characteristics and objectives influence all types of forest management, including timber harvesting, reforestation, and participation in public programs. Specific factors tested for influence on timber harvesting include timber prices, reforestation costs, and nontimber values.

Sills and Abt (eds.), Forests in a Market Economy, 23–38. ©Kluwer Academic Publishers. Printed in The Netherlands.

Reforestation is hypothesized to be affected by timber prices, reforestation costs, interest rates, nontimber values, and public programs.

1. PROPERTY RIGHTS AND TENURE

Various individuals, firms, communities, and units of government may own forests and timber. Different types of forest land ownership and forest landowner objectives affect how forest land will be managed. The spectrum of forest land and timber ownership could range from absolute fee simple ownership, where the land and timber belong entirely to a private individual or firm, to complete government control and production. Within each of these different types of ownership, landowners may have different degrees of rights of control and exclusion in different countries or in different regions of the same country.

Forest land may be bought and sold in most countries where private ownership dominates. However, transactions costs may restrict entry and exit into forest land markets. Fee simple ownership implies that individual forest landowners have exclusive rights to use and dispose of their property and its produce. These rights are still not absolute, but rather are conditioned by the overall interests of the state in protecting health, safety, public welfare, and, more recently, the environment. Nonindustrial private forest (NIPF) landowners (farmers and, increasingly, urban residents) and industrial private forest landowners in most countries hold some type of fee simple ownership.

Government ownership and management of forests, usually with private market production of goods, is at the other end of the spectrum. Governments also may own or co-own processing facilities, such as sawmills or plywood mills, but this is becoming less common. More often, governments that do own forests will provide services for recreation, wilderness, or amenities, or will contract with private vendors to provide those services.

NIPF landowners may also hold title to land, but enter into conservation easements to protect the land from development. This transfers some of the rights of disposition from landowners to the state or other public or quasi-public entity. For example, a transfer of development rights allows private owners to maintain use of their land, but may require public access to that land in exchange for direct payments or indirect tax benefits.

Industrial forest landowners are creating new forest and timber ownership vehicles, separating the various components of land and timber into bundles of rights. Traditionally, the forest industry purchased forests or bare land outright, and then managed the existing forests or planted new forests for timber production. Even in the most traditional forest products firms, this single-purpose landownership model is waning. Firms that have land near expanding urban areas often have real estate development divisions. Several large firms have separated forest land ownership from wood processing, purchasing all their wood on the open market or from long-term lease arrangements. Even companies that have large land bases—often amassed during corporate mergers and buyouts—conduct active land sales programs to consolidate their holding near strategic centers or to generate revenue to pay off debt. Three large forestry firms in the United States have separated stock for their forest land holdings and manufacturing facilities since the 1980s, and several now have no company forest lands.

As discussed in greater detail in chapter 6, nontraditional corporate or other private entities also are becoming owners and managers of forest land for timber production. Timber investment management organizations (TIMOs) have become popular in the United States and now own in excess of three million hectares of forest land, mostly in the South. These TIMOs obtain funds from pension funds or from other large investors and then purchase and manage forest land. Thus, there is a trend toward more separation of timber rights from forest land rights, although most private forest land ownership still is in traditional NIPF or industry ownership and management.

2. LANDOWNER OBJECTIVES

Two basic economic theories, utility maximization and profit maximization, have been used to test hypotheses about private forest management. Utility is assumed to include both monetary and nonmonetary benefits associated with owning forest land. NIPF owners are often assumed to choose among the various benefits that forests produce in order to achieve the greatest utility for themselves. The profit-maximizing approach represents the landowner as a commercial entity that uses forest as a means of production, usually of timber products (see chapters 4 through 9). Utility maximization may be more relevant for owners of small forest tracts (see chapters 14 through 16) and profit maximization more germane for large tracts, forest industry, and TIMOs. Application of utility-maximization theory is challenging because of difficulties in assessing nonmarket values. The profit-maximization approach is narrower in its perspective but offers considerable power in examining nonmarket values through shadow prices from profit functions (Newman and Wear 1993, Prestemon and Wear 2000, Wear and Newman 1991).

Private industrial firms are generally presumed to be economically rational decision makers that seek to maximize the timber profits from their forest lands, within the constraints of society and politics. This may not always be the case, especially in the short run, since many large firms own timberlands as a hedge against large price swings for wood fiber in the open market or to prevent an expensive paper mill (establishment costs exceeding $1 billion) from running out of wood. Large firms also may have capital constraints that limit their timber investments, even if they can meet their nominal cost of capital or hurdle rates.

NIPF owners may have multiple reasons for owning forest land, including timber production, carbon storage (chapter 13), amenities (chapters 4 and 14), and personal identity (Bliss and Martin 1989). Young and Reichenbach (1987) found that the intent to harvest timber is a function of attitudes and beliefs, and traditional assistance programs will not influence landowner behavior. Timber production is often one objective, but not the dominant one in many cases. Thus these landowners may manage forests less intensively, often for longer rotations and for more diverse benefits than industry owners.

The new land-owning organizations, such as TIMOs and pension funds, as well as wealthy private investors, generally own forest land to earn a reasonable rate of return on their capital. They seek profits as one of their key objectives, and their timber investments compete with other potential investment vehicles. Timber offers some security and less volatility than other investments such as stocks or commercial real estate, but it must earn good investment returns to be viable (chapter 6).

These landowners manage their forest lands within the context of implicit social norms and explicit laws, regulations, and incentives in their countries, states, or localities. Most countries now have explicit regulations to ensure continuous forest production and adequate environmental protection and biodiversity. For example, Indonesia requires that the annual harvest from public lands by timber concessionaires cannot exceed the net annual growth of their forests (about 1 cubic meter per hectare per year).

Social norms, as reflected by pressure from environmental groups and perhaps from individual consumers, are also being reflected in worldwide retail market demands for sustainable forest management. This adds another dimension to successful profit maximization, even for narrowly focused firms. For example, some firms are seeking certification for their forests in order to guarantee that they can sell their products in the green European markets and in U.S. retail home improvement stores.

The objectives and motivations of NIPF landowners throughout the world have been examined in a plethora of studies. Table 3.1 presents the results of a national survey of NIPF landowners in the United States.

Conclusions from this survey depend on whether statistics are summarized by the number of owners or by forest area. Private landowners holding land with timber production as the main objective comprised only 3% of the total private forest landowners but controlled 29% of the private forest area in the United States (Birch 1996a and b).

Table 3.1. Nonindustrial private forest landowner objectives, United States and the U.S. South (Birch 1996a and b)

Objective	United States		U.S. South	
	% of acres	% of owners	% of acres	% of owners
Timber	29	3	35	4
Other economic	19	17	20	20
Recreation and aesthetic	27	23	13	15
Part of farm or residence	17	39	17	37
Other	16	15	14	18
No answer	2	4	1	5

This table also summarizes survey results from the U.S. South, which is 90% privately owned and is a productive timber-growing region. More than 60% of the acres in the South have timber production as their main or secondary objective, and various commodity interests are important for over 75% of the southern land base. Bliss and McNabb (1992) concluded that the most prevalent reason for owning property was to keep it in the family, while NIPF landowners in Illinois wanted to (1) provide wildlife habitat, (2) preserve natural beauty, and (3) provide a heritage to pass on to future generations (Young et al. 1984).

3. LANDOWNER CHARACTERISTICS

Examination of the influence of landowner characteristics has focused exclusively on NIPF landowners. In addition, because this type of information is available only with micro data (for individual landowners), characteristics are tested only when landowner survey data are used. One exception to this can be found in chapter 14, where aggregate income and education data were incorporated into a model of amenity demand and harvest probability.

Several of the studies considered in this chapter used a survey conducted by Fecso et al. (1982) of reforestation by NIPF landowners in the South (Hardie and Parks 1991, Royer 1987). A more recent survey (Birch 1996a and b) provided information on NIPF owners by region, but has not been used in empirical hypothesis tests. In addition, many of the studies have relied on smaller, focused surveys, often addressing only one region or state.

Following the application of utility theory to forest management (Binkley 1981), empirical tests of NIPF management began to include such characteristics as landowner income, occupation (primarily focusing on farmer/nonfarmer), residency status (absentee or resident), education, and age. Perhaps the most important characteristic is income, because increasing income is hypothesized to reduce landowner incentives to harvest timber. Alig et al. (1990) found that owner characteristics had a greater effect on the probability of timber harvesting by NIPF owners than on the probability of tree planting. This probably means that immediate market feedback (timber income) induces a more direct response than the delayed response required for tree planting. Below we discuss the general conclusions from the literature, with key citations.

Table 3.2. Influence of landowner characteristics on NIPF forest management

Characteristic	General conclusion		Selected citations
	Planting	Harvesting	
Income	Usually positive, not unanimous	Negative, limited studies	Hyberg and Holthausen (1989) Dennis (1989) Alig (1986) Chapter 14 Hardie and Parks (1991) Zhang and Flick (2001) Romm et al. (1987) Greene and Blatner (1986) Royer (1987) deSteiguer (1984)
Occupation as farmer	No effect	Weak positive	Hyberg and Holthausen (1989) Binkley (1981) Boyd (1984) Dennis (1989) Romm et al. (1987)
Age	No effect	No studies	Romm et al. (1987) Zhang and Flick (2001)
Education	No studies	Mixed	Greene and Blatner (1986) Chapter 14 Dennis (1989) Binkley (1981) Boyd (1984)

As noted in table 3.2, income is the landowner characteristic that has been addressed most frequently and provides the most consistent results. In the utility model of NIPF behavior (Binkley 1981, Dennis 1989), income is hypothesized to have an ambiguous effect on harvesting and planting. The empirical studies, while not unanimous, indicate that higher income landowners are more likely to reforest, but that the substitution effect of harvesting dominates the income effect, resulting in higher incomes leading to lower harvest probabilities.

Occupation as a farmer is not a distinct variable in the utility model, but farmers are often hypothesized to behave more as profit than utility maximizers, and thus to have a higher probability of harvesting. This was weakly confirmed by the studies we examined. The hypothesized effect on planting is ambiguous, and our survey found no empirical support for any differences between nonfarmers and farmers. Hardie and Parks (1991) used a theoretically elegant study to examine the effect of owner, market, and policy variables on the probability of reforestation. They found that household income, income as a farmer, and income as a forest resident were significant determinants of the probability of reforestation, lending very limited support the influence of farm occupation on management. Full- or part-time residence on the forest versus absentee landownership is often hypothesized to influence forest management. Romm et al. (1987) found that full-time residents had a higher probability of planting, and part-time residents had a lower probability.

Age has been included in only a few studies. Romm et al. (1987) found that older age reduced the probability of investment, but Zhang and Flick (2001) found that age had no influence on planting. Education has been shown to have a significant negative influence on harvesting. Hardie and Parks (1991) also found that having inherited a forest had a significant negative impact on reforestation probability.

One significant complication with the empirical results of the characteristics examined above is that there should be multicollinearity between education, age, occupation, and income. All of these characteristics, as well as residency status and estate circumstances, contribute to landowner wealth, and wealth may be a better indicator of forest management plans, but data are generally unobtainable. More complex treatments of landowner wealth may be necessary to address the influences on management behavior.

4. MARKETS

Markets for timber and other forest products are complex and often assumed to be imperfect. Timber sales generally entail high transactions costs for professional consulting fees, landowners' time, government agency administration, and timber buyer cruising and bidding. These costs range from as low as a few percentage points of the total value of the product sold to more than 20% of the total sale value. For local timber products such as fuelwood and poles, the transactions costs may largely involve time collecting wood from state or community lands and perhaps paying a small fee for doing so. These markets for household use of forest products are less well defined than for industrial forest products (see chapter 15 for additional discussion of nontimber forest products).

The United States has fairly active markets for sawtimber products, but forest landowners may have few markets for their small-diameter pulpwood trees. With low timber values, transport costs may restrict sales to a local area. In many areas, only one or two timber buyers for any product operate within this range, while there may be many forest landowners. This oligopsonistic market structure may lead to some loss of efficiency in timber markets and thus lower prices and less production than would be the case if perfect competition existed.

Statistical tests of the influence of market factors on private forest management are numerous, although the findings are weak. Landowners, both industrial and NIPF, are hypothesized to increase both planting and harvesting in response to higher product prices. The joint nature of forest production, both the production of timber and amenities from the forest and the production of sawtimber and pulpwood from the timber, complicates this response. Investigators have devised numerous techniques to isolate the effects of prices on management, including application of principal components to develop a price index (chapter 14) and using a site-specific growing stock weighted price (chapter 4; Lee 1997, Prestemon and Wear 2000). Table 3.3 provides our general conclusions on the influence of market factors on private forest management and lists some key citations.

Landowners are hypothesized to increase both harvesting and planting when output prices (timber prices) increase. Other market factors addressed by various studies include reforestation cost, tract size, and the discount rate. Reforestation costs and the discount rate are considered the cost of inputs to the production of timber, and thus the higher these costs are, the lower the probability of planting and harvesting. Fixed costs associated with initiating harvests (see chapter 5) make costs per unit of output lower on larger tracts, and thus tract size is hypothesized to proxy for harvesting costs or possibly for amenities (Amacher et al. 1998, Newman and Wear 1993).

One potential concern is that most studies isolate the harvesting and planting decision, addressing only one of these clearly related decisions. An analysis by Newman and Wear (1993) jointly addressed these two forest management decisions and responses to output prices and input costs.

Overall, there is some support for increasing sawtimber prices leading to increased harvest. Pulpwood prices and the effect of sawtimber prices on planting are generally not significant (no effect). Some studies have found that industrial landowners are more price responsive, and one study (Newman and Wear 1993) found more responsiveness in the long run, although statistical significance tests could not be conducted on these estimates. Newman and Wear also found that pulpwood and sawtimber appear to be substitutes in short-run production but complements in long-run production.

Table 3.3. Influence of market factors on private forest management

Market factor	General conclusion		Selected citations
	Planting	Harvesting	
Timber price	Limited price responsiveness for NIPF and industrial owners, not unanimous	More price responsiveness than planting	Alig (1986) Brooks (1985) Lee et al. (1992) Hyberg and Holthausen (1989) Dennis (1989) Newman and Wear (1993) Royer (1987) Lee (1997) Chapter 4 Chapter 14 Boyd (1984) Binkley (1981)
Reforestation cost	Negative, less conclusive for industrial owners	Mixed results	Hyberg and Holthausen (1989) Newman and Wear (1993) Royer (1987)
Discount rate	No effect, but not unanimous	Inadequate number of studies	Lee et al. (1992) deSteiguer (1984) Hyberg and Holthausen (1989)
Tract size	No effect	Generally no effect, not unanimous	Royer (1987) Hyberg and Holthausen (1989) Greene and Blatner (1986) Dennis (1989) Romm et al. (1987)
Amenity values	No studies	Negative	Chapter 14, Lee (1997), Chapter 4

Evidence suggests that increased reforestation costs lead to reduced reforestation. Results of the influence of input costs on harvesting are mixed. The discount rate is hypothesized to negatively influence planting. Empirical evidence is weak and contradictory. Prestemon and Wear (2000) estimated the probability of harvesting as a function of price change and calculated an implied discount rate for both landowner groups, concluding that NIPF landowners have higher rates than industrial landowners.

Tract size was not found significant in empirical tests on the influence on planting or harvesting, although two studies provide limited support for the hypothesis that larger tracts are more likely to be harvested. Moulton and Birch (1995) found that an emphasis on commodity production was directly related to tract size.

5. POLICY IMPACTS

While markets are imperfect, they are the principal mechanism that allocates forest management and timber harvests throughout the world. Markets for other private goods produced from forests have more limitations than timber markets but remain the principal means of resource allocation. Markets will allocate forest land and timber and are usually considered reasonably efficient. However, perceived market imperfections such as imperfect competition (oligopsony), long waits for investment returns, and external costs (e.g., water pollution) or benefits (e.g., carbon storage) have led to government interventions in private timber production. Almost every country has some form of government education, research, subsidy, protection, or regulation for public and private forest lands (Cubbage and Haynes 1988).

Public policies can augment market incentives, but they do not supplant or negate market prices and mechanisms except in the rare instances of total government control and production of forest goods and services. Cooperation among landowners or forest producers, government subsidies or regulation, redefinition of property rights, controls on monopoly power or other market or government methods may be able to promote efficient and equitable outcomes for both buyers and sellers, as well as prevent negative externalities from adversely affecting other owners or the public.

Economists generally believe that forest regulations have an adverse effect on forestry investments, although landowner opposition to regulation is not universal, and little empirical work has substantiated investment effects (Johnson et al. 1997). Boyd and Hyde (1989) found no significant differences between reforestation levels in Virginia (with a regulatory seed tree law) and North Carolina (with no regulations but the largest state-funded cost-share program in the United States). This led them to conclude that regulation was not effective and has high administrative costs as well.

It has generally been presumed that forest industry manages its forest land for timber production in an economically rational manner and achieves reasonably good levels of productivity based on economic efficiency criteria.

Nonindustrial private forests, however, have been widely viewed as underproductive and subject to some criticism for not achieving their potential timber production capacity, thus prompting calls for public policy interventions. The means of influencing private landowners depends on the severity of the perceived difference between private market outcomes and public goals, the influence of interest groups seeking public policies to alter market outcomes, and the ability of governments to pay for public programs. Most studies have examined NIPF timber production policies, but more recently, forest stewardship and multiple use policies have been examined. Program participation has also been studied to attempt to determine if the programs are reaching the intended audience (table 3.4).

Table 3.4. Influence of forestry programs on private forest management

Forestry program	General conclusion		Selected citations
	Planting	Harvesting	
Knowledge of cost-share programs	Positive	Generally positive, not conclusive	Hyberg and Holthausen (1989) Royer (1987) Zhang and Flick (2001) Megalos (2000) Hardie and Parks (1991)
Knowledge of public technical assistance	Positive	Positive	Hyberg and Holthausen (1989) Royer (1987) Zhang and Flick (2001) Hardie and Parks (1991)
Regulation	Negative	No studies	Zhang and Flick (2001), Boyd and Hyde (1989)

Cost-share assistance directly influences planting but not harvesting. In contrast, most technical assistance is aimed at harvesting practices. Because of the simultaneity problem with receiving cost-share assistance and, to a smaller degree, technical assistance, many surveys ask if landowners had prior knowledge of these programs. The empirical results indicate that landowners with knowledge of these programs were more likely to plant and harvest than landowners who were unaware of these programs. One study (Lee et al. 1992) addressed only the non-cost-shared planting done by both landowner groups. Of the four programs evaluated (Soil Bank, Forestry Incentives, Conservation Reserve [CRP], and Agricultural Conservation), three had no influence on non-cost-shared acres, implying that (1) the influence of the programs did not extend beyond the participants, and (2) the use of federal funding did not substitute for private funding for reforestation. The fourth program, CRP, had a positive influence on both NIPF and industrial planting, implying that there may be a positive externality beyond program participants.

The econometric models by Hardie and Parks (1991) indicated that the parameter for technical assistance by public foresters had the greatest positive magnitude by far and the highest level of statistical significance in predicting area reforested by private landowners. They also examined the interaction of technical assistance programs and state and federal cost-share programs. They concluded that cost-share programs and rates had a strong impact on reforestation performed in consultation with public forestry assistance, with 85% of the predicted acres regenerated being due to these programs. Changing the amount of publicly owned technical assistance was effective but did not produce the same magnitude of increase as the cost-share instruments.

Cubbage (2003) reviewed field surveys of harvested tracts to evaluate the effects of forestry assistance conducted in the 1980s. Cubbage reports that the studies consistently found that public forestry or consulting assistance increased net revenues from timber sales to forest landowners. In addition, prospective returns from future management were higher when technical assistance was used in the current harvest. Forestry assistance resulted in positive impacts on both revenues and residual stand quality (Cubbage 2003).

Several studies have examined program participation, primarily for the cost-share programs. English et al. (1997) found that higher income and lower costs led to increased participation. Nagubadi et al. (1996) found that higher age, larger size, and forest association membership led to increased participation. In contrast, Stevens et al. (1999) found that age reduced participation, but that income increased participation. Megalos (2000) and Lorenzo and Beard (1996) also found that owners of larger tracts and nonfarm owners were more likely to participate.

Esseks and Moulton (2000) performed a survey of NIPF owners, assessing their use of the Forest Stewardship Program (FSP) and Stewardship Incentives Program (SIP). These programs were initiated in the late 1990s and continue as part of the Forest Land Enhancement Program enacted as part of the 2002 Farm Bill. They provided cost-share assistance for a broad range of forest practices, including reforestation/afforestation, forest improvement, wildlife, forest stewardship plans, agroforestry, recreation, soil and water quality, riparian areas and wetlands, and fisheries. This broader subsidy program attracted many new persons to nontimber forest practices. More than half of the active FSP landowners in the South (58%) had not previously received professional forestry advice.

More than 80% of all landowners had begun to carry out at least one management activity; 69% of the southern forest landowners had begun at least two management activities; and 44% had begun three or more activities. Nationwide, 25% of SIP participants performed reforestation; 44% performed forest improvement; 11% established wildlife practices; 3% had agroforestry practices; 6% had recreation practices; and 10% had soil and water practices. Less than 1% of the area involved riparian or fisheries practices. These statistics indicate that subsidies can be effective at inducing a wide range of multiple-use forest practices. Landowners in the South remain more utilitarian than the rest of the United States but are still quite amenable to enrolling in nontimber forest management practices.

6. CONCLUSIONS

Timber production by private forest landowners is the classic forestry case of supply and demand in private markets. Owners of private forests in most developed countries own both the land and the timber that grows on that land. Privately owned timber is generally sold in markets, either as stumpage for individual stands, prices for individual logs, concessions for large forest areas, or as transfer prices between divisions of vertically integrated firms.

Economic studies of NIPF forest management (reforestation) and timber harvesting provide excellent examples of the power of economics and econometric analyses. Timber harvesting decisions are not always influenced by the same factors as reforestation. Owner income had a positive influence on reforestation probability but a negative effect on harvesting probability. Farming as an occupation, level of education, and tract size usually were found not to influence tree planting but positively influenced timber harvesting. Timber prices were only moderately effective at influencing tree planting but almost always influenced timber harvesting. Public policy interventions—subsidies and technical assistance—were usually the largest factors influencing reforestation, and technical assistance was important in timber harvesting decisions.

In two explicit studies that have been performed in the United States, regulations have been found to discourage timber investments by NIPFs. Those studies were narrowly construed, but the results are not surprising. Regulation is almost by definition meant to alter free market outcomes in order to protect nonmarket values, so some efficiency loss should be expected. Economic analyses can assess and compare the efficiency losses to the environmental benefits. Nontimber forest values are becoming more important to NIPFs and demonstrably reduce timber production. Subsidies encourage NIPF landowners to enroll in a broader set of multiple use forestry practices.

These research findings can help analysts, planners, and policy makers decide when markets will achieve desired forest management outcomes, when policy interventions will be required to achieve outcomes different than those prompted by markets, and the magnitude of the public investment required to achieve new program goals.

In a market system, private forest landowners will produce timber to maximize profits or will maximize the utility of the bundle of all goods and services on their forests. Subsequent chapters in this book expand the set of economics tools for analyzing both timber and nontimber forest products and nonmarket forest goods and services. Theoretically sound and technically accurate analyses of the values of all these goods and services from forests can help us better allocate scarce resources.

7. LITERATURE CITED

ALIG, R. 1986. Econometric analysis of the factors influencing forest acreage trends in the southeast. For. Sci. 32(1):119-134.

ALIG, R.J., K.J. LEE, AND R.J. MOULTON. 1990. Likelihood of timber management on nonindustrial private forests: Evidence from research studies. Gen. Tech. Rpt. SE-60. USDA Forest Service, Southeastern For. Exper. Station, Asheville, NC 17 p.

AMACHER, G., C. CONWAY, J. SULLIVAN, AND C. HENSLEY. 1998. Effects of shifting populations and preferences on behavior of nonindustrial private forestland owners and forest industry: Empirical evidence from Virginia. SOFAC Report No. 12. Southern Forest Resource Assessment Consortium, Research Triangle Park, NC.

BINKLEY, C.S. 1981. Timber supply from nonindustrial forests. Bull. 92. School of Forestry and Environmental Studies, Yale University, New Haven, CT. 97 p.

BIRCH, T.W. 1996a. Private forestland owners of the United States, 1994. Res. Bull. NE-134. USDA Forest Service, Northeastern For. Exper. Station, Radnor, PA. 183 p.

BIRCH, T.W. 1996b. Private forest-land owners of the Southern United States, 1994. Res. Bull. NE-138. USDA Forest Service, Northeastern For. Exper. Station, Radnor, PA. 195 p.

BLISS, J.C. AND MARTIN. 1989. Identifying NIPF management motivations with qualitative methods. For Sci. 35(2):601-622.

BLISS, J.C. AND K. MCNABB. 1992. Landowners reveal some surprising attitudes about regulation. For. Farmer 52(1):14-15.

BOYD, R. 1984. Government support of nonindustrial production: The case of private forests. So. J. Econ. 51:89-107.

BOYD, R. AND W. HYDE. 1989. Forestry sector intervention: The impacts of public regulation on social welfare. Iowa State University Press, Ames, IA. 295 p.

BROOKS, D.J. 1985. Public policy and long-term timber supply in the South. For. Sci. 31(2):342-357.

CUBBAGE, F.W. 2003. The value of foresters. Forest Landowner 62(1):16-19.

CUBBAGE, F.W. AND R. HAYNES. 1988. Evaluation of the effectiveness of market responses to timber scarcity problems. Mktg. Res. Rpt 1149. USDA Forest Service, Washington DC. 87 p.

DENNIS, D.F. 1989. An economic analysis of harvest behavior: Integrating forest and ownership characteristics. For. Sci. 35(4):1088-1104.

DE STEIGUER, J.E. 1984. Impact of cost-share programs on private reforestation behavior. For. Sci. 30:697-704.

ENGLISH, B.C., C.D. BELL, G.R. WELLS, AND R.K. ROBERTS. 1997. Stewardship incentives in forestry: Participation factors in Tennessee. S. J. Appl. For. 21(1): 5-10.

ESSEKS, J.D. AND R.J. MOULTON. 2000. Evaluating the forest stewardship program through a national survey of participating forestland owners. The Center for Governmental Studies, Social Science Research Institute, Northern Illinois University, DeKalb, IL. 113 p.

FECSO, R.S, H.F. KAISER, J.P. ROYER, AND M. WEIDENHAMER. 1982. Management practices and reforestation decisions for harvested southern pinelands. Staff Report No. AGES921230, USDA Statistical Reporting Service, Washington DC.

GREENE, J AND K. BLATNER. 1986. Identifying woodland owner characteristics associated with timber management For. Sci. 32(1):135-146.

HARDIE, I.W. AND P.J. PARKS. 1991. Individual choice and regional acreage response to cost-sharing in the South, 1971-1981. For. Sci. 37(1):175-190.

HYBERG, B. AND D. HOLTHAUSEN. 1989. The behavior of nonindustrial private forestlandowners. Can. J. For. Res. 19:1014-1023.

JOHNSON, R.L., R.J. ALIG, E. MOORE, AND R.J. MOULTON. 1997. NIPF: landowners' view of regulation. J. For. 95(1):23-28.

LEE, K., F. KAISER, AND R. ALIG. 1992. Substitution of public for private funding in planting southern pine. So. J. Appl. For. 16(4):204-208.

LEE, K.J. 1997. Hedonic estimation of nonindustrial private forest landowner amenity value. Unpublished Ph.D. Dissertation. North Carolina State University, Raleigh, NC. 71 p.

LORENZO, A.B., AND P. BEARD. 1996. Factors affecting the decisions of NIPF owners to use assistance programs. P. 264-275 *in* Proceedings: Symposium on Nonindustrial Private Forests: Learning from the Past, Prospects for the Future. Baughman, M.J. (ed.). Minnesota Extension Service, University of Minnesota, St. Paul, MN.

MEGALOS, M.A. 2000. North Carolina landowner responsiveness to forestry incentives. Unpublished Ph.D. dissertation, North Carolina State University, Raleigh, NC. 119 p.

MOULTON, R.J., AND T.W. BIRCH. 1995. Southern private forestland owners: A profile. For. Farmer 54(5):44-46.

NAGUBADI, V., K.T. MCNAMARA, W.L. HOOVER, AND W.L. MILLS, JR. 1996. Program participation behavior of nonindustrial forestland owners: A probit analysis. J. Agric. Appl. Econ. 28(2):323-336.

NEWMAN, D.H., AND D.N. WEAR. 1993. Production economics of private forestry: A comparison of industrial and nonindustrial forest owners. Am. J. Agric. Econ. 75:674-684.

PRESTEMON, J.P., AND D.N. WEAR. 2000. Linking harvest choices to timber supply. For. Sci. 46(3):377-389.

ROMM, J., R. TUAZON, AND C. WASHBURN.1987. Relating forestry investment to the characteristics of nonindustrial private forestland owners in Northern California. For. Sci. 33(1):197-209.

ROYER, J. 1987. Determinants of reforestation behavior among southern landowners. For. Sci. 33(3):654-667.

STEVENS, T.H., D. DENNIS, D. KITTREDGE, AND M. RICKENBACH. 1999. Attitudes and preferences toward co-operative agreements for management of private forestlands in the Northeastern United States. J. Environ. Manage. 55(2):81-90.

WEAR, D.N., AND D.H. NEWMAN. 1991. The structure of forestry production: Short-run and long-run results. For. Sci. 37(2):540-551.

YOUNG, R.A., M.A. REICHENBACH, AND F. PERKUN. 1984. A survey of non-industrial private forest owners in Illinois: A preliminary report. For. Res. Rpt. 84-2. University of Illinois, Agricultural Experiment Station, Champaign-Urbana, IL.

YOUNG, R.A., AND M.R. REICHENBACH. 1987. Factors influencing the timber harvest intentions of NIPF owners. For. Sci. 33:381-393.

ZHANG D., AND W. FLICK. 2001. Sticks, carrots, and reforestation investment. Land Econ. 77(3):443-456.

Section One

TIMBER PRODUCTION AND MARKETS

Jeffrey P. Prestemon and Brian C. Murray
USDA Forest Service and Research Triangle Institute

Timber production has been the foundation of active forest management for over a century. The science and economics of forest management were developed 150 years ago, but for years, the focus was on activity at the stand level, with very little attention to market phenomena such as price behavior, demand factors, substitution, and market structure. That has changed as advances in economic theory and methods have enhanced researchers' interest and understanding of the interaction between individual decisions (e.g., land allocation, management, harvest timing) and market outcomes (e.g., price, volumes, and trade patterns). The following chapters take the reader from these individual stand and forest-level decisions to the collective result of these decisions in the market place. The section begins with a treatment of the classical production problems in forest economics, including optimal stand management (chapter 4), and the application of modern production theory (chapters 5 and 7) and finance theory (chapter 6) to timber production.

The section then turns to market-level concepts. The market is, figuratively, where and when buyers and sellers come together to make an exchange. The market is not one physical location, but rather a collection of transactions ruled essentially by the same terms of exchange. Chapters 8 and 9 address market-level analysis of the supply and demand for timber. Space plays a particularly important role in timber markets because timber is costly to transport. This raises unique aspects of market structure and spatial market integration that are addressed in chapter 10. Finally, despite the costliness of transport, there is nonetheless a vibrant international trade in forest products (chapter 11), as there is a geographic asymmetry between where the products are demanded and where they are most abundantly supplied. While this section covers the territory from reforestation to

Sills and Abt (eds.), Forests in a Market Economy, 39–40. ©Kluwer Academic Publishers. Printed in The Netherlands.

exchange of processed wood products in world markets, these products are only part of the reason that forests are valued and managed. Other forest outputs are addressed in subsequent sections.

Chapter 4

Optimal Stand Management
Traditional and Neotraditional Solutions

Karen Lee Abt and Jeffrey P. Prestemon
USDA Forest Service

The traditional Faustmann (1849) model has served as the foundation of economic theory of the firm for the forestry production process. Since its introduction over 150 years ago, many variations of the Faustmann have been developed which relax certain assumptions of the traditional model, including constant prices, risk neutrality, zero production and management costs, and the single management objective. We describe the traditional Faustmann and provide an overview of the neotraditional Faustmann and Hartman (1976) models. We then use the neotraditional Hartman model to develop testable hypotheses regarding harvest response to timber, land, and amenity values from forests. Using data from the North Carolina coastal plain, we test for inclusion of several often omitted variables in models of industrial and nonindustrial harvest behavior.

1. TRADITIONAL AND NEOTRADITIONAL OPTIMAL HARVEST MODELS

Efforts to model maximization of economic returns from forest land began with a treatise by Faustmann (1849). Faustmann derived a formula for determining the economically optimal rotation length for timber production alone. This model assumed a constant price for a single wood output and had no input costs other than a constant and known price of capital. In 1999 a symposium was held to mark 150 years since the publication of Faustmann's seminal paper (Brazee 2001). Several of the papers from this symposium have been published and are included in this summary.

Sills and Abt (eds.), Forests in a Market Economy, 41–57. ©*Kluwer Academic Publishers. Printed in The Netherlands.*

Faustmann's model, also referred to as land expectation value (LEV) maximization, languished in obscurity in forestry circles in America until the middle of the 1900s (Gaffney 1957). Practical applications continue to be limited because owners of forest land have other objectives in addition to maximizing income from timber (Gregory 1972). Early data for private landowners showed a consistently longer rotation for nonindustrial private forest (NIPF) landowners than for industrial landowners. The other objectives are one explanation for the variation in rotation ages.

In addition, the real world is more complicated than this model. Based solely on forest outputs for which there are functioning markets in the economy, LEV may not be the appropriate maximization criterion for a vast share of the land-owning population. Nonetheless, models simplify the world in ways that can reveal the relationships among many or all of the important factors affecting observed phenomena. This section describes the latest stand-level models of economic optimization for private landowners.

The traditional Faustmann model assumes bare land and determines the rotation length of an even-aged stand that maximizes the discounted timber revenues minus timber production costs, for the first and all subsequent rotations, assuming constant prices and a known timber yield production function. This approach can be adjusted to include factors that bring the model closer to the reality of forest product and land markets (table 4.1). These neotraditional Faustmann models (table 4.2) may include input costs (e.g., Hyde 1980, Nautiyal and Williams 1990), values other than timber (e.g., Hartman 1976), stochastic prices (e.g., Norstrøm 1975), or production risk (e.g., Martell 1980). Another type of neotraditional Faustmann model addresses uneven-aged management, which is in the spirit of Faustmann but must be derived and tested differently (e.g., Adams and Ek 1974).

Table 4.1. Development of the traditional Faustmann model

Author	Innovation
Faustmann (1849)	First development
Gaffney (1957)	Rediscovery in United States
Samuelson (1976)	Second rediscovery in United States
Binkley (1981)	Used Faustmann to derive supply
Heaps (1984)	Steady-state optimal is Faustmann

Efforts to understand optimal harvest decisions have been along two central tracks: normative and positive analyses (table 4.2). Normative analyses are theoretical developments of the Faustmann and variants. One objective of these theoretical analyses is to provide a tool for landowners that incorporates more real-world factors into decision models. Positive analyses of the timber harvest decision model have been reflective attempts to understand whether behavior of timberland owners is consistent with

Optimal Stand Management

Table 4.2. Normative and positive analyses of neotraditional Faustmann model

Innovation	Normative	Positive
Variable inputs	Hyde (1980) Jackson (1980) Chang (1983, 1998) Nautiyal and Williams (1990)	
Nontimber outputs	Hartman (1976) Binkley (1981) Strang (1983) Calish et al (1978) Swallow and Wear (1993) Plantinga and Birdsey (1994) Plantinga (1998) Dole (1999) Tahvonen and Salo (1999) Englin et al. (2000) Koskela and Ollikainen (2001)	Max and Lehman (1988) Dennis (1989, 1990) Kuuluvainen and Salo (1991) Provencher (1995, 1997) Lee (1997) Newman and Wear (1993) Kuuluvainen and Tahvonen (1999) Prestemon and Wear (2000) Pattanayak et al. (2002) Pattanayak et al. (chapter 14)
Uncertainty	Norstrøm (1975) Martell (1980) Routledge (1980) Reed (1984) Lohmander (1988) Brazee and Mendelsohn (1988) Clarke and Reed (1989) Thomson (1992) Gong (1994, 1999) Yin and Newman (1995a, 1995b, 1997) Yin and Newman (1996) Forboseh et al. (1996) Abildtrup et al. (1997) Willassen (1998) Brazee and Bulte (2000) Lohmander (2000) Fina et al. (2001) Zhang (2001) Buongiorno (2001)	
Risk aversion	Caulfield (1988) Pukkala and Miina (1997) Gong (1998) Peltola and Knapp (2001) Uusivori (2002)	
Uneven-aged	Adams and Ek (1974) Buongiorno and Michie (1980) Bare and Opalach (1987) Buongiorno and Lu (1990) Lu and Buongiorno (1993) Buongiorno et al. (1994)	Raunikar et al. (2000) Scarpa et al. (2000)

theories developed by normative models. Both kinds of analyses are useful because they serve as complements: Normative analyses provide tools, positive analyses evaluate their usefulness and test the importance of other potential factors in decisions, and then normative models are updated to reflect the new knowledge.

Many of the studies in table 4.2 do not directly analyze or test the Faustmann but address various assumptions of the traditional model. Seminal works in this area include Faustmann (1849), Samuelson (1976), Hartman (1976), and Binkley (1981). Samuelson is basically a rediscovery of the Faustmann model, Hartman was the first to address nontimber outputs, while Binkley, using a household production framework, incorporated landowner characteristics into the decision framework. Other papers using the household production model include Max and Lehman (1988) and Dennis (1989, 1990). Several papers use optimal control methods, which have been shown by Heaps (1984) to converge to the Faustmann model in the steady state. Some of the papers directly address the incorporation of Faustmann models into market analyses (e.g., Binkley 1993, Brazee and Mendelsohn 1988). Swallow and Wear (1993) and Tahvonen and Salo (1999) address multiple stands, including adjacency issues.

As noted by many studies, both normative and positive, the traditional Faustmann has many limitations. We recognize and discuss several of these limitations, including (1) variable inputs, (2) nontimber outputs, (3) uncertainty, (4) risk aversion, and (5) uneven-aged management.

The first addition to the traditional models was the inclusion of variable inputs and input costs. Because the algebra for optimal solutions is complicated, in most models these inputs are ignored. Yet, it is commonly recognized that these inputs matter to the optimal harvest decision.

The second neotraditional innovation was the addition of nontimber outputs to the decision framework. Hartman (1976) demonstrated that a stand with age-based amenity outputs would be harvested later or not harvested at all. Because NIPF landowners are assumed to value amenity outputs, several studies have tested for the influence of amenities on harvest decisions. These tests all suffer from the use of proxies to measure amenity outputs.

A third limitation of the traditional Faustmann model stems from its reliance on static and known prices and production functions. Analyses attempting to correct this shortcoming are primarily normative (see table 4.2), focusing on stochasticity of output and input prices. Efforts to evaluate optimal strategies in the context of multiple investment vehicles include Redmond and Cubbage (1988) and Zinkhan et al. (1992). While these efforts lie outside the Faustmann and variants, they represent a potential approach that would broaden the stochastic harvest timing models cited above. Other refinements of the traditional Faustmann address variations in timber yield resulting from uncertainty regarding stand growth functions and from random natural events.

Fourth, the traditional Faustmann does not address risk aversion, and the majority of neotraditional analyses assume risk neutrality. Because there is positive value to managing risk, the risk-averse case is relevant to many producers. Dispersion of returns, in fact, plays a central role in portfolio theory (Lintner 1965, Sharpe 1964). Hence, it seems appropriate to understand the effect of risk aversion on optimal decisions.

A normative study in forestry by Caulfield (1988) used stochastic dominance analysis to improve timber harvest decisions of risk-averse timberland owners. Putting the forest investment in the context of other possible investments is another innovation (Wagner et al. 1995, Washburn and Binkley 1990, Zinkhan et al. 1992). Positive analyses searching for evidence that risk-averse behavior is accounted for in the production decision are lacking. However, evidence of higher discount rates for private timberland owners than for industrial landowners (Prestemon and Wear 2000) is consistent with risk aversion (Lintner 1965, Sharpe 1964). Positive analyses of the effects of production and other kinds of risk would be useful in revealing the qualitative implications for management, would suggest the value of mitigating risk, and would be beneficial in attempts to understand quantitatively the aggregate effects of risk on producers with different ownership characteristics. Those kinds of studies could lead to improved understanding of aggregate supply behavior.

Faustmann and its neotraditional variants do not apply to uneven-aged forests. Uneven-aged management is used to address biological diversity and minimize visual and ecological impacts of forest management and is becoming increasingly important as a management tool in the United States. The work on the economics of uneven-aged management has been both normative and positive. Normative models are based on growth models, and positive analyses measure how closely the harvest behavior of owners of multi-aged stands conforms to a criterion analogous to LEV. No work has been done to develop a model to empirically evaluate the effects of uncertainty or risk aversion on optimal decision making under uneven-aged management.

Although many advances have been made to include real-world phenomena in the land value maximization decision in both even-aged and uneven-aged timber production, a fully stochastic model for both is lacking. Ultimately, if the correct stand-level optimization model for timber growing includes nontimber values, risk aversion, production risk, and price stochasticity, then these inclusions should have implications for what appears in specifications of aggregate market models. Aggregate market models would then need to be updated from the traditional specification. The result of better aggregate models would be more accurate and spatially refined timber market and land use projection models (e.g., Abt et al. 2000, Adams and Haynes 1980), enabling new kinds of research into the market and land use effects of demographic, macroeconomic, and public policy changes.

2. THE HARTMAN MODEL: A NEOTRADITIONAL FAUSTMANN

The Hartman model (1976) modified the traditional Faustmann model by including amenity outputs. According to Hartman's calculus, a stand with amenity values in standing timber may be harvested later or may never be harvested. This could have important implications for timber supply modeling and in understanding the market and economic welfare implications of policies and catastrophic shocks. In econometric analyses of timber harvesting behavior, if such values affect the production decisions of landowners but are not included in empirical specifications, then incorrect inferences on the effects of included variables might result. If the Hartman model is an accurate reflection of timber production decisions of a large proportion of NIPF landowners, then this could help explain differences in harvest timing between these two broad classes of owners.

In this section we develop a model to test whether amenity values influence private harvest decisions. We model southern pine harvesting in the North Carolina coastal plain for the period 1983-1989. Harvest choice is modelled as a function of timber values, land values, and amenity proxies. The obvious risk in choosing a proxy for amenity values it that the hypotheses are joint: if we find that our measures of amenity values are not significantly related to the harvest decision, then we cannot be sure whether this is because the measure is incorrect or because amenities truly play no role in the decision. On the other hand, if we find that measures of amenity values are significantly related to the harvest decision, and in the manner hypothesized (i.e., high amenity values are negatively associated with the harvest decision), then we can conclude that these values are linked to the harvest decision and should be accounted for when evaluating the response of landowners to market shocks and government policies.

The bare land Hartman land value (HLV) can be characterized as:

$$HLV = \frac{\sum_{j=1}^{J} V_j e^{-rt} - k + \int_{t=1}^{T} A(x)e^{-rx}dx}{1-e^{-rt}} \qquad 4.1$$

where V_j is the timber value for product j ($j = 1,...,J$), k are establishment costs, t ($t = 1,...T$) is the age of the stand, r is the discount rate, and A are the age-dependent stand-level amenities. For an existing stand, this model is modified through the inclusion of discounted current stand values and by discounting the HLV from harvest of the current stand back to today. We posit that in every period, landowners compare the benefits of harvesting (π_t^1) with the benefits of delaying harvest (π_t^0). The benefits include the revenues from the current stand of timber as well as the benefits and costs of the delay of future rotations and the benefits and costs of amenities from both current and future rotations.

We define the binary variable, y_t, as:

$$y_t = \begin{cases} 1 \text{ if } y_t^* = \pi_t^1 - \pi_t^0 > 0 \\ 0 \text{ otherwise} \end{cases} \qquad 4.2$$

Thus, if the net benefits of harvesting today are greater than the net benefits of delaying harvest, the landowner will harvest, and the observed variable, y_t, will equal 1. The latent variable y_t^* is equivalent to an intertemporal value comparison. We can specify the harvest decision as a function of the latent variable:

$$y_t^* = [V_t(a) - \rho V_t(a+m)] + [A(a) - \rho A(a+m)] + [HLV(a) - \rho HLV(a+m)] + \varepsilon^1 - \varepsilon^0 \qquad 4.3$$

where $V_t(a)$ is a vector of timber product revenues at time t for age a, m is the number of years between decisions, $A(a)$ is the amenity utility derived from a stand of age a, $\rho = (1 + r)^{-1}$, and ε_i is the error associated with inaccurately calculating the benefits associated with current harvest (1) or future harvest (0). The errors in equation 4.3 may be associated with inaccurate calculations by the landowner or with factors unobserved by the analyst but observed by the landowner.

This model incorporates (1) different values for pulpwood and sawtimber; (2) different estimates for nonindustrial and industrial landowners; (3) the value of future infinite series of rotations, which is often dismissed in empirical tests as being too small to have an influence on the harvest decision although this is precisely what makes the LEV model different from the Fisherian model of net present value; and (4) proxies for the Hartman old growth, a parklike vegetation profile that has a dense ground cover and overstory.

We hypothesize that current timber and land values will positively correlate with harvesting, while future timber and land values will negatively correlate with harvesting. Opposite results will hold for current and future rotation values of amenities, where higher levels of current amenities will correlate with lower harvesting and higher levels of future amenities will correlate with decreased harvesting. However, because North Carolina's coastal plain forests are managed fairly intensively for timber, at least compared with forests of other regions of the United States, we hypothesize that industrial harvest behavior would not be influenced by amenity values.

3. DATA AND ESTIMATION

In decision-making, landowners are assumed to compare the present with the expected future. We explicitly recognized that harvest decisions are based on changes in all stand attributes: current stumpage values, LEVs, and amenity values. In addition, because we do not have adequate data regarding the expected amenity conditions, we use only the current values for the vegetative profile as a proxy for Hartman's parklike stand conditions. The determination of the amenity values is described below.

The empirical version of equation 4.3 is:

$$y_t^* = \beta_0 + \beta_1 V(a) + \beta_2 V(a+m) + \beta_3 LEV(a) + \beta_4 LEV(a+m) \\ + \beta_5 A(a) + \varepsilon_t \qquad 4.4$$

where t indexes time, a is the age at the initial survey and $a + m$ is the age of the stand at the final survey, V is timber revenues from both sawtimber and pulpwood, LEV includes only timber as an output, A is a vector of vegetative profile proxies for amenities, and ρ is the discount rate.

Data for all variables except timber prices and future product volumes were taken from the USDA Forest Service's Forest Inventory and Analysis (FIA) surveys of the forests of the coastal plain of North Carolina. Sampled stands were measured during the summers of 1983 and 1989 so that the time elapsed between periods was 6 years ($m = 6$).

While standing volumes of sawtimber and pulpwood in period t were observed for all sampled stands, expected volumes in period $t + m$ were not (actual volumes were not available for the harvested stands). The expected period $t + m$ volumes were estimated by fitting quadratic models of pulpwood and sawtimber volume to unharvested stands. These quadratic equations predicted 1989 volumes of pulpwood and sawtimber as a function of 1983 volumes, 1983 stand age in years, 1983 stand basal area, and 1983 site index (base age 50).

Harvest (y_t, in equation 4.4) was specified as a binary (1,0) variable indicating whether or not the stand was harvested between 1983 and 1989. Partially cut stands were dropped from the data set.

LEV was calculated by assuming an infinite series of rotations identical to the current rotation. Thus, the current and expected timber values were used, and a plantation establishment cost of $150/acre was assumed. The revenues were discounted using a 5% rate.

Based on the Hartman vision of parklike stands, we used stand structure variables from the FIA inventory data to construct overstory, shrub, and ground cover vegetation profiles. Using data for trees and all other vegetation, we calculated the percent of space occupied by ground covers (0 to 2 feet), shrubs (2 to 15 feet), and overstory (80+ feet).

Stumpage price data were obtained from Timber Mart-South (Norris Foundation 1977-1989). Real stumpage prices for both t (1983) and $t + m$ (1989) were $11.00 per cord for pulpwood and $158.60 per thousand board feet for sawtimber. Constant real prices reflect our assumption that timberland owners in North Carolina in the mid-1980s did not expect real increases in prices for southern yellow pine pulpwood and sawtimber. Variation by stand occurs because stumpage volumes differ by stand and by year, resulting in a different timber value for each oservation. The differences in stand volumes result from differences in growth and variation in the mix of sawtimber and pulpwood across stands.

4. RESULTS

The results of the estimation of equation 4.4 are shown in table 4.3 for industrial landowners and table 4.4 for NIPF landowners. A series of models was estimated to examine model response to severe multicollinearity expected in the estimations. The χ^2 statistics indicate that all the models are significant.

Table 4.3 shows the results of four estimations of the probability of harvest by industrial landowners. The models do indicate multicollinearity may be influencing the standard errors for LEV and timber values, with both current and future expected values significant in the timber only and LEV only models but insignificant in the timber plus LEV model. The addition of the amenity characteristics, which are not significant influences on industrial

Table 4.3. Estimates of harvest choices by industrial owners of southern pine stands in the coastal plain of North Carolina between 1983 and 1989 (n = 268)

Variable	Timber	LEV	Timber plus LEV	Timber plus LEV plus amenity
Intercept	-0.1217	-0.0902	-0.0631	-0.0509
	0.0356[a]	0.0292	0.0309	0.0492
Timber value (1983)	0.0004 *		0.0003	0.0003 *
	0.0001		0.0002	0.0002
Expected timber value (1989)	-0.0003 *		-0.0002	-0.0002 *
	0.0001		0.0001	0.0001
LEV (1983)		0.0003 *	0.0002	0.0002
		0.0001	0.0001	0.0001
Expected LEV (1989)		-0.0003 *	-0.0001	0.0000
		0.0001	0.0002	0.0002
Ground cover occupancy (0-2 feet)				0.0007
				0.0008
Shrub layer occupancy (2-15 feet)				-0.0010
				0.0009
Overstory occupancy (>80 feet)				-0.0018
				0.0016
Log-likelihood	-57.777	-55.306	-53.447	-51.638
χ^2	11.108	16.050	19.769	23.386
Significance level	0.004	0.000	0.006	0.001
Pseudo-R^2	0.274	0.396	0.359	0.364

* significant at the 5% level
[a] Standard errors in italics

harvest choices, alters the results slightly to return the timber values to significance.

Industrial landowners respond as expected to current timber values by increasing their harvests, while an increase in future timber values reduces harvests. Similarly, it appears that these landowners also respond as if they were aware of LEV, by increasing harvests with a higher LEV and reducing harvests with a higher future LEV. While it is likely that few landowners, industrial or NIPF, actually calculate the LEV for any of their timber stands, we are testing whether or not their behavior is consistent with wealth maximization as represented by LEV. We conclude that industrial harvest choices are consistent with maximizing LEV. As expected, industrial landowners do not respond to the proxies for amenity characteristics.

Table 4.4 shows similar results for NIPF landowners, with significant χ^2 statistics and pseudo-R^2 ranging from 0.34 to 0.39. Current and expected future timber values were not significant in any of the models. Current and expected future LEV were significant in the models. Again, as with industrial landowners, increases in LEV led to increased probability of harvest, and higher expected LEV led to decreased probability of harvest. Thus, landowners delayed harvest when future LEV increased, holding current LEV constant.

The hypothesized Hartman effect is given some support with these results. The Hartman model proposes that harvest will be delayed if standing timber has value. In this analysis, we developed ecological proxies for a parklike stand condition—dense ground cover and overstory and limited vegetation in the shrub layer (2 to 15 feet above ground). We hypothesized that a dense shrub layer, which corresponds to the area most likely to block human visibility while standing, would be undesirable, and thus a landowner's probability of harvest would increase with increasing shrub density. The estimated coefficients on ground cover and shrub layer are not significant at the 5% level but do have the correct sign. Multicollinearity is also possible with these ecological proxies, with higher ground cover and lower shrub density occurring naturally with higher overstory density.

Dense overstory significantly influences NIPF harvest choices, with more overstory corresponding to lower harvest probability. The overstory measure used, vegetative occupancy over 80 feet, is highly correlated with stand age. Older stands are expected to be taller and have denser overstory at this height (excluding true old growth stands, which rarely occur in coastal plain southern pine stands). Thus, in the absence of other variables, harvest and overstory should be positively correlated. However, including timber values and LEV to explicitly account for the impacts of age on increasing value, we found a negative correlation between overstory and harvest probability.

5. CONCLUSIONS

Stand-level economic optimization theory has advanced substantially in the last quarter century, an effort that has permitted the development and

Table 4.4. Estimates of harvest choices by NIPF owners of southern pine stands in the coastal plain of North Carolina between 1983 and 1989 (n = 451)

Variable	Timber	LEV	Timber plus LEV	Timber plus LEV plus amenity
Intercept	-0.3280	-0.2150	-0.3046	-0.3132
	0.0320 [a]	0.0202	0.0324	0.0587
Timber value (1983)	0.0001		-0.0003	-0.0002
	0.0002		0.0002	0.0002
Expected timber value (1989)	0.0001		0.0003	0.0003
	0.0002		0.0002	0.0002
LEV (1983)		0.0007 *	0.0007 *	0.0006 *
		0.0002	0.0003	0.0003
Expected LEV (1989)		-0.0007 *	-0.0010 *	-0.0009 *
		0.0003	0.0004	0.0004
Ground cover occupancy (0-2 feet)				-0.0007
				0.0011
Shrub layer occupancy (2-15 feet)				0.0005
				0.0010
Overstory occupancy (>80 feet)				-0.0019 *
				0.0009
Log-likelihood	-225.480	-231.406	-225.480	-219.466
χ^2	26.219	14.367	26.219	38.248
Significance level	0.000	0.001	0.001	0.000
Pseudo-R^2	0.342	0.342	0.373	0.387

*significant at the .05 level
[a] Standard errors in italics

refinement of economics-based timber supply projection models. From these advancements, we now have a better understanding of the effects of incorporating other inputs, other outputs, and price and production risks into the harvest decision model. Most work on the harvest decision model has been normative, describing optimal choices with additional complexity of the neotraditional Faustmann models. More recently, positive analyses have found statistical evidence that traditional and neotraditional models can be used to represent harvest choices made by land managers.

In this chapter, we developed a test of a neotraditional Hartman model. The model and data we used indicated that both the traditional and neotraditional models can be used to explain harvest decisions by NIPF landowners. For industrial landowners, the addition of the neotraditional Hartman elements (amenity proxies) did not improve the model, nor were any of the individual measures significant in predicting industrial harvest choices in the coastal plain of North Carolina.

One important result from our tests is that LEV may be an important predictor of harvest choices, especially choices made by NIPF landowners. Because the calculations for this value are complex and dynamic, and because the value of future rotations is small when compared to the value of the current rotation, most empirical research does not include a measure of LEV. A third reason for not including LEV is that NIPF landowners are assumed to be unaware or incapable of understanding and calculating LEV for a stand of trees. We found, however, that private landowners behave *as if* they had knowledge of LEV and, in particular, knowledge of how changes in LEV affect the decision to harvest.

Our tests also demonstrated the use of ecosystem measures as proxies for amenity values. The results support our use of vegetative occupancy at over 80 feet as a measure of a desirable amenity. We did not find statistical support for private landowner valuation of ground cover or shrub layer as proxies for Hartman's parklike stands. One possibility is collinearity between the overstory, shrub, and ground cover layers. Difficulties associated with these proxies include (1) the maintained assumption that amenity values increase with the age of the stand, (2) the use of vegetation measures that may represent the amenity values but may also represent other omitted variables in the model, (3) the inclusion of only current amenity conditions, and (4) the use of a 5% discount rate to calculate current and future LEV.

Overall, the models explained significant variation in harvest decisions of NIPF and industrial landowners of pine stands in the coastal plain of North Carolina. Considering the potential multicollinearity, both timber value and land value should be included in future tests of harvest choices. Ecosystem characteristics are influential in predicting harvest decisions only of NIPF landowners, and to the extent that these characteristics can proxy for amenity values, the Hartman neotraditional model can represent NIPF landowner harvest behavior.

6. LITERATURE CITED

ABILDTRUP, J., J. RIIS, AND B.J. THORSEN. 1997. The reservation price approach and informationally efficient markets. J. For. Econ. 3:229-245.

ABT, R.C., CUBBAGE, F.W., AND G. PACHECO. 2000. Southern forest resource assessment using the Subregional Timber Supply (SRTS) model. Forest Prod. J. 50 (4):25-33.

ADAMS, D.M., AND A.R. EK. 1974. Optimizing the management of uneven-aged forest stands. Can. J. For. Res. 4:274-287.

ADAMS, D.M., AND R.W. HAYNES. 1980. The 1980 Softwood Timber Assessment Market Model: structure, projections and policy simulations. For. Sci. Mon. No. 22.

BARE, B.B., AND D. OPALACH. 1987. Optimizing species composition in uneven-aged forest stands. For. Sci. 33(4):958-970.

BINKLEY, C.S. 1981. Timber supply from private forests. New Haven, CT: Yale Univ. School of Forestry & Environmental Studies Bull. No. 92.

BINKLEY, C.S. 1993. Long-run timber supply: price elasticity, inventory elasticity, and the use of capital in timber production. Nat. Res. Mod. 7(2):163-181.

BRAZEE, R.J. 2001 Introduction—the Fuastmann Formula: Fundamental to forests economics for 150 years. For. Sci. 47(4):441-442.

BRAZEE, R.J., AND R. MENDELSOHN. 1988. Timber harvesting with fluctuating prices. For. Sci. 34:359-372.

BRAZEE, R.J., AND E. BULTE. 2000. Optimal harvesting and thinning with stochastic prices. For. Sci. 46(1):23-31.

BUONGIORNO, J. 2001. Generalizationof Faustmann's formula for stochastic forest growth and prices with Markov decision process models. For. Sci. 47(4):466-474.

BUONGIORNO, J., AND B. MICHIE. 1980. A matrix model of uneven-aged forest management. For. Sci. 26(4):609-625.

BUONGIORNO, J., AND H.-C. LU. 1990. Economic stocking and cutting-cycle in a regulated selection forest. For. Ecol. Manage. 32:203-216.

BUONGIORNO, J., S. DAHIR, H.-C. LU, AND C.-R. LIN. 1994. Tree size diversity and economic returns in uneven-aged forest stands. For. Sci. 40(1):83-103.

CALISH, S., R.D. FIGHT, AND D.E. TEEGUARDEN. 1978. How do nontimber values affect Douglas Fir rotations? J. For. 76(4):217-222.

CAULFIELD, J.P. 1988. A stochastic efficiency approach for determining the economic rotation of a forest stand. For. Sci. 34(2):441-457.

CHANG, S.J. 1983. Rotation age, management intensity, and the economic factors of timber production: do changes in stumpage price, interest rate, regeneration cost, and forest taxation matter? For. Sci. 29(2):267-278.

CHANG, S.J. 1998. A generalized Faustmann model for the determination of optimal harvest age. Can. J. For. Res. 28(5):652-659.

CLARKE, H.R. AND W.J. REED. 1989. The tree-cutting problem in a stochastic environment: the case of age-dependent growth. J. Econ. Dyn. Contr. 13:565-595.

DENNIS, D.F. 1989. An economic analysis of harvest behavior: integrating forest and ownership characteristics. For. Sci. 35(4):1088-1104.

DENNIS, D.F. 1990. A probit analysis of the harvest decision using pooled time-series and cross-sectional data. J. Environ. Econ. Manage. 18:176-187.

DOLE, D. 1999.Implicit valuation of non-market benefits in even-aged forest management. Envtl. and Res. Econ. 13(1):95-105.

FAUSTMANN, M. 1849. On the determination of the value which forest land and immature stands possess for forestry. Institute Paper 42 (1968), M. Gane, ed. Oxford: Commonwealth Forestry Institute, Oxford University.

FINA, M., G.S. AMACHER, AND J. SULLIVAN. 2001.Uncertainty, debt, and forest harvesting: Faustmann revisited. For. Sci. 47(2):188-196.

FORBOSEH, P.F., R.J. BRAZEE, AND J.B. PICKENS. 1996. A strategy for multiproduct stand management with uncertain future prices. For. Sci. 42:58-66.

GAFFNEY, M.M. 1957. Concepts of financial maturity of timber and other assets. Agric. Inf. Ser. 62. Raleigh: North Carolina State College. 105 pp.

GONG, P. 1994. Adaptive optimization for forest-level timber harvest decision analysis. J. Environ. Econ. Manage. 40:65-90.

GONG, P. 1998. Risk preferences and adaptive ahrvest policies for even-aged stand management. For. Sci. 44(4):496-506.

GONG, P. 1999. Optimal harvest policy with first-order autoregressive price process. J. For. Econ. 5:413-439.

GREGORY, G.R. 1972. Forest Resource Economics. New York: John Wiley and Sons. 548p.

HAIGHT, R.G., AND T.P. HOLMES. 1991. Stochastic price models and optimal tree cutting: results for loblolly pine. Nat. Res. Mod. 5:423-443.

HAMILTON, J.D. 1994. Time Series Analysis. Princeton, N.J.: Princeton University Press.

HARTMAN, R. 1976. The harvesting decision when a standing forest has value. Econ. Inq. 14:52-58.

HEAPS, T. 1984. The forestry maximum principle. J. Econ. Dyn. Contr. 7:131-151.

HYDE, W.F. 1980. Timber Supply, Land Allocation, and Economic Efficiency. Baltimore, MD: Johns Hopkins University Press. 254p.

JACKSON, D.H. 1980. The microeconomics of the timber industry. Boulder, CO: Westview Press. 136p.

KOSKELA, E. AND M. OLLIKAINEN. 2001. Forest taxation and rotation age under private amenity valuation. J. Environ. Econ. Manage. 42(3):374-384.

KUULUVAINEN, J., AND J. SALO. 1991. Timber supply and life cycle harvest of nonindustrial private forest owners: an empirical analysis of the Finnish case. For. Sci. 37(4):1011-1029.

KUULUVAINEN, J., AND O. TAHVONEN. 2001. Testing the forest rotation model:Evidence from panel data. For. Sci. 45:539-551.

LEE, K.J. 1997. Hedonic estimation of nonindustrial private forest landowner amenity values. Ph.D. Diss. Raleigh: North Carolina State University. 80 pp.

LINTNER, J. 1965. The valuation of risk assets and the selection of risky investments in stock portfolios and capital budgets. Rev. Econ. Stat. 47:13-37.

LOHMANDER, P. 1988. Pulse extraction under risk and a numerical forestry application. Syst. Anal. Model. Simul. 5(4):339-354.

LOHMANDER, P. 2000. Optimal sequential forestry decisions under risk. Ann. Oper. Res. 95:217-228.

LU, H.-C., AND J. BUONGIORNO. 1993. Long- and short-term effects of alternative cutting regimes on economic returns and ecological diversity in mixed-species forests. For. Ecol. Manage. 50:173-192.

MARTELL, D.L. 1980. The optimal rotation of a flammable forest stand. Can. J. For. Res. 10:30-34.

MAX, W., AND D.E. LEHMAN. 1988. A behavioral model of timber supply. J. Environ. Econ. Manage. 15(1):71-86.

NAUTIYAL, J. C., AND J. S. WILLIAMS. 1990. Response of optimal stand rotation and management intensity to one-time changes in stumpage price, management cost, and discount rate. For. Sci. 36:212-223.

NEWMAN, D.H. 1988. The optimal forest rotation: A discussion and annotated bibliography. Gen. Tech. Rep. SE-48. USDA Forest Service Southeastern For. Exper. Station, Asheville NC. 47p.

NEWMAN, D.H., AND D.N. WEAR. 1993. Production economics of private forestry: a comparison of industrial and nonindustrial forest owners. Am. J. Agr. Econ. 75(3):674-684.

NORRIS FOUNDATION. 1977-1989. Timber Mart-South. The Daniel B. Warnell School of Forest Resources, University of Georgia, Athens.

NORSTRØM, C.J. 1975. A stochastic model for the growth period decision in forestry. Swed. J. Econ. 77:329-337.

PATTANAYAK, S.K., B.C. MURRAY, AND R.C. ABT. 2002. How joint is joint forest production: an econometric analysis of timber supply conditional on endogenous amenity values. For. Sci. 48(3):476-491.

PELTOLA, J., AND K.C. KNAPP. 2001. Recursive preferences in forest management. For. Sci. 47(4):455-465.

PLANTINGA, A.J., AND R.A. BIRDSEY. 1994. Optimal forest stand management when benefits are derived from carbon. Nat. Res. Mod. 8(4):373-387.

PLANTINGA, A.J. 1998. The optimal timber rotation: an option value approach. For. Sci. 44:192-202.

PRESTEMON, J.P., AND D.N. WEAR. 2000. Linking harvest choices to timber supply. For. Sci. 46(3):377-389.

PROVENCHER, B. 1995. Structural estimation of the stochastic dynamic decision problems of resource users: an application to the timber harvest decision. J. Environ. Econ. Manage. 29:321-338.

PROVENCHER, B. 1997. Structural versus reduced-form estimation of optimal stopping problems. Am. J. Agr. Econ. 79:357-368.

PUKKALA, T. AND J. MIINA. 1997. A method for stochastic multiobjective optimization of stand management. For. Ecol. Manage. 98(2):189-203.

RAUNIKAR, R., J. BUONGIORNO, J.P. PRESTEMON, AND K.L. ABT. 2000. Financial performance of mixed-age naturally regenerated loblolly-hardwood stands in the South Central United States. For. Pol. Econ. 1(3/4):331-346.

REDMOND, C.H., AND F.W. CUBBAGE. 1988. Portfolio risk and returns from timber asset investments. Land Econ. 64:325-37.

REED, W.J. 1984. The effects of the risk of fire on the optimal rotation of a forest. J. Environ. Econ. Manage.11:180-190.

ROUTLEDGE, R.D. 1980. The effect of potential catastrophic mortality and other unpredictable events on optimal forest rotation policy. For. Sci. 26(3):389-399.

SAMUELSON, P.A. 1976. Economics of forestry in an evolving society. Econ. Inq. 14:466-492.

SCARPA, R., J. BUONGIORNO, J.-S. HSUE, AND K.L. ABT. 2000. Assessing the non-timber value of forests: a revealed preference, hedonic model. J. For. Econ. 6(2):83-108.

SHARPE, W.F. 1964. Capital asset prices: a theory of market equilibrium under conditions of risk. J. Finance 19:425-442.

STRANG, W.J. 1983. On the optimal forest harvesting decision. Econ. Inq. 21(4):576-583.

SWALLOW, S.K., AND D.N. WEAR. 1993. Spatial interactions in multiple-use forestry and substitution and wealth effects for the single stand. J. Environ. Econ. Manage. 25:103-120.

TAHVONEN, O. AND S. SALO. 1999. Optimal forest rotations with in situ preferences.J. Environ. Econ. Manage.37(1):106-128.

UUSIVORI, J. 2002. Nonconstant risk attitudes and timber harvesting. For. Sci. 48(3):459-470.

THOMSON, T.A. 1992. Optimal forest rotation when stumpage prices follow a diffusion process. Land Econ. 68:329-42.

WAGNER, J.E., F.W. CUBBAGE, AND C.H. REDMOND. 1995. Comparing the capital asset pricing model and capital budgeting techniques to analyze timber investments. For. Prod. J. 45:69-77.

WASHBURN, C.L., AND C.S. BINKLEY. 1990. Informational efficiency of markets for stumpage. Am. J. Agr. Econ. 72:394-405.

WILLASSEN, Y. 1998. The stochastic rotation problem: a generalization of Faustmann's formula to stochastic forest growth. J. Econ. Dyn. Contr. 22:573-596.

YIN, R., AND D.H. NEWMAN. 1995a. Optimal timber rotations with evolving prices and costs revisited. For. Sci. 41:477-490.

YIN, R., AND D.H. NEWMAN. 1995b. A note on the tree-cutting problem in a stochastic environment. J. For. Econ. 1:181-190.

YIN, R., AND D.H. NEWMAN. 1996. The effect of catastrophic risk on forest investment decisions. J. Environ. Econ. Manage. 31:186-197.

YIN, R., AND D. H. NEWMAN. 1997. When to cut a stand of trees? Nat. Res. Mod. 10(3):251-261.

ZHANG, D.W. 2001. Faustmann in an uncertain policy environment. For. Pol. Econ. 2(2):203-210.

ZINKHAN, F.C., W.R. SIZEMORE, G.H. MASON, AND T.J. EBNER. 1992. Timberland Investments. Timber Press, Portland, OR.

Chapter 5

Forest Production
Costs of Harvesting Timber

Jacek P. Siry, Frederick W. Cubbage, and Erin O. Sills
University of Georgia and North Carolina State University

Basic production economics involves the estimation of production functions, calculation of various types of production costs, comparison of costs with product prices, and determination of profit-maximizing mixes of input use and levels of production. Relationships among inputs, technology, and multiple products determine the productivity and efficiency of firms or organizations.

Efficiency is achieved by minimizing the aggregate costs of inputs to produce a given product or products (or services), by maximizing the value of production of goods or services with a given set of inputs, or by maximizing the net difference between the costs of inputs and prices of outputs. Production economics traditionally uses the market costs of inputs (land, labor, and capital) and the market prices of products (timber, boards, hunting leases, or recreation fees) in determining the most efficient use of resources. In this chapter, we maintain this approach, which assumes that these prices represent the private opportunity costs and values of resources in use or *in situ*. Methods for determining relative values of nonmarket products are considered in section 3 of this book.

Production economics principles are discussed in various textbooks, and this chapter draws from classic references such as Doll and Orazem (1978), Pappas and Hirschey (1990), and Chambers (1988). Numerous studies have applied these principles to estimate profit, cost, or production functions for the forestry and wood products sectors. For example, Newman (1991) used a modified production function to analyze changing southern forest productivity. Smith and Munn (1998) analyzed regional costs of the logging industry in the Pacific Northwest and Southeast. Newman and Wear (1993)

Sills and Abt (eds.), Forests in a Market Economy, 59–76. ©Kluwer Academic Publishers. Printed in The Netherlands.

compared industrial and nonindustrial forest owners using a restricted profit function.

In this chapter, we focus on short-run production relationships, where some or many inputs are fixed, and only one or a few vary. In the long run, all production inputs can vary and adjust to optimal economic levels. Most forestry stand-level analyses include time as a variable, since timber production and ecosystem values take decades to materialize. Nevertheless, these analyses can still be considered short run if they vary only a few production inputs, such as tree spacing or rotation length. A common approach is to fix all inputs other than the one of interest in order to estimate its optimal level in production. Chapters 4, 6, 12, and 13 treat the issue of time more explicitly.

The next section reviews the empirical literature on production and cost functions for timber harvesting in the southern United States. Estimation of profit functions is illustrated in chapters 8 and 9. This chapter continues with a review of the basic principles of production theory. The final section presents an application to aggregate production and average cost functions for harvesting timber in the southern US.

1. HARVESTING COSTS IN THE SOUTH

Estimating the total, average, or marginal costs of forest regeneration, timber harvesting, and forest products manufacturing are typical applications of production economics. Analysts may use total or average production rates (i.e., seedlings planted per day, trees cut and skidded per hour, or logs sawn into various dimensions per day), with input costs and output prices, to estimate average costs of production per hour, day, week, month, or year. They may estimate production or cost functions, particularly as they are influenced by inputs such as timber volumes, equipment employed, harvest method, or tree size. Data from individual firms or cost studies may be aggregated to estimate industry-wide productivity, costs, and trends over time. A variety of approaches and levels of sophistication have been employed in such cost calculations, ranging from simple hand calculations to spreadsheet models to elegant computer simulation models.

Timber harvesting productivity and cost estimation were described by Matthews (1942) and operationalized by researchers and practitioners in the 1970s, notably Plummer (1977) and Stuart (1981). Kluender et al. (1998), Wang et al. (1998), Cubbage (1983) and others have since estimated southern U.S. timber harvesting costs for individual machines, harvesting locations, harvesting systems, and variable inputs. Most of these analysts estimated regression equations for harvesting productivity, then used average machine and labor costs per hour to calculate harvesting cost functions by relevant site variables. Three examples of these approaches follow.

Cubbage (1983) used the Harvesting System Simulator (Stuart 1981) to estimate effects of productivity rates, input costs, and stand volumes on average timber harvesting costs for four southern harvesting systems—bobtail trucks, tree-length cable-skidders, full-tree feller-buncher grapple-skidders, and whole-tree chippers. Average 1980 harvesting costs were $22.46 per cord for the chipper system, $28.68 per cord for the full-tree system, $32.27 per cord for the bobtail truck/chainsaw system, and $36.68 for the tree-length/chainsaw system. Most stand characteristics, productivity rate changes, and input cost changes do not change the relative ranking of these costs. Input cost changes of 20% or more would generally only change total system costs by less than 10%. However, production quotas that limited weekly output would increase average costs more than 10% per cord. Changes in input productivity would have nearly equal effects on average costs. For example, a 25% increase in productivity would reduce average costs by about 25%.

More recently, Wang et al. (1998) used spatially explicit simulation in a geographical information system (GIS) modelling framework to examine the effects of interactions of stand type, harvesting method, and equipment on the productivity of southern timber harvesting. They did not calculate costs, but these productivity estimates (production functions) would provide the basis for subsequent harvesting cost estimates. For example, they found that harvesting planted stands had greater average production rates than harvesting natural stands (11 vs. 7 cords per productive man-hour); feller-bunchers harvested timber faster than timber harvester machines or chainsaws (15, 8, and 4 cords per productive man-hour, respectively); and making clearcuts was more productive than shelterwood or single-tree selection systems (10, 9, and 8 cords per productive man-hour, respectively). For clearcuts, feller-buncher systems produced an average of 435 cords per week, a harvester/forwarder system produced 244 cords per week, and a chainsaw/grapple-skidder system produced 140 cords per week. To determine the cheapest systems and the effects of these variables on average costs, we would need to estimate the weekly costs of running the entire operation.

Kluender et al. (1998) examined the effects of removal intensity and tree size on harvesting costs and profitability. They examined 16 stands, with timber harvesting at intensities ranging from 0.27 to 1.0 (proportion of basal area removed) and with different average diameters (DBH) of the stands. Average tree diameter had a much greater effect on average cost than proportion of basal area removed, which changed costs by only a small percentage for any diameter size. Average costs were as much as $40 per

100 cubic feet (CCF) for 6-inch average tree sizes, dropped rapidly to about $20/CCF for 10-inch trees, and tapered off asymptotically to about $15/CCF for 20-inch trees. Profit functions were the inverse of these costs, with losses occurring until about 8-inch tree sizes, then profits rising rapidly to about $13/CCF for an average 13-inch tree size, and to almost $15/CCF at a 20-inch average tree size.

2. PRODUCTION THEORY

2.1 Production Functions

A production function is a basic input-output relationship that describes a biophysical relationship or production process that converts inputs (x) into outputs (y) in a technically efficient manner. A single output, multiple input production function can be written as:

$$y = f(x) \qquad 5.1$$

Growing timber (y) using land, labor, machinery, and fertilizer and herbicide (x) in a given time could be represented by such a function. Production relationships may be expressed by the total production or total product (TP) (per amount of input) or by single input variation measures, including the average product (AP) and the marginal product (MP). The following equations summarize these basic concepts:

$$AP_{xi} = \frac{y}{x_i}$$
$$MP_{xi} = \frac{\partial y}{\partial x_i} \qquad 5.2$$

where x_i represents the quantity of input i. Figure 5.1 shows a single output–single input production function with the classical sigmoid shape. The total product or output curve represents the varying rates at which an input (e.g., fertilizer, x) is converted to a product (e.g., timber, y). The total product curve first increases at a rapid rate, then slows its growth rate but still increases, and eventually reaches its maximum and declines. This relationship represents universal biological and economic principles stated in the law of diminishing returns. This law posits that the marginal return of any given input will eventually decrease as more and more of that single input is used to produce a single output. The first application of fertilizer will increase southern pine growth to a larger extent than the second and following applications in the same amount. Excessive use of any input, including fertilizer, could eventually cause illness, growth declines, or increased mortality.

Average product (*AP*) is total output divided by the total amount of that input, e.g., timber volume growth per amount of fertilizer applied. Marginal product (*MP*) describes the incremental output produced by the addition of one more unit of input, e.g., timber volume growth resulting from one more fertilizer application. An increasing marginal product indicates that inputs are producing proportionately more output in that range of production ($x_0 < x < x_1$). A decreasing marginal product ($x_1 < x < x_2$) indicates that output is increasing at a decreasing rate with respect to inputs. When marginal product reaches zero ($x = x_2$), it means that extra inputs provide no additional output. And when marginal product becomes negative ($x > x_2$), it implies that extra inputs cause a decrease in the total production.

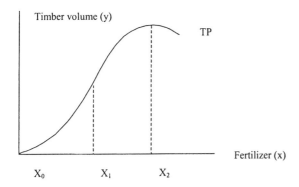

Figure 5.1. Production function

A unit-free measure of marginal input productivity is the elasticity of production, defined as the percentage change of output with respect to a percentage change of input:

$$\varepsilon_i = \frac{\partial y}{\partial x_i} \frac{x_i}{y} \qquad 5.3$$

The elasticity of production indicates whether output is increasing proportionately more or less than an input. For example, an elasticity greater than one could indicate that a 1% increase in fertilization rates increases timber volume growth by more than 1%.

2.2 Cost Functions

Cost functions represent the minimum cost of producing a good or service at given input prices (w). For a lumber mill, a cost function provides the minimum cost of producing a certain amount of lumber, given prices of stumpage, labor, machinery, materials, and energy used in production. Input prices are assumed to be exogenous, i.e., firms are price takers in input markets. The cost function is defined as:

$$c(w, y) = \min wx, \quad s.t. \ y = f(x) \qquad 5.4$$

Since in the short run some inputs are fixed at predetermined levels, they can be divided into variable and fixed inputs. Fixed costs are large costs that cannot be changed in the short run, such as the land area or a pulp mill. Variable costs may include labor, materials, and energy. In the long run, all inputs become variable.

Total costs (TC) are the sum of fixed costs plus total variable costs (TVC). Average and marginal quantities can be calculated for total and variable costs.

$$AC = \frac{TC}{y} \quad or \quad AVC = \frac{TVC}{y}$$
$$MC = \frac{\partial(TC)}{\partial y} \quad or \quad \frac{\partial(TVC)}{\partial y} \qquad 5.5$$

These relationships determine output supply behavior of the firm, as shown in figure 5.2. The firm will continue to operate in the short run as long as it covers its variable costs. This is true where marginal costs exceed average variable costs, because in perfect competition, the firm produces up to the point where its marginal cost equals the price of its output. As long as the price of output (equal to MC) exceeds average variable costs, the firm will continue to operate. The firm could still lose money if it does not also cover fixed costs, but it could at least use some of the net returns from variable short-run production to cover part of its fixed costs. Therefore, the MC curve above the AVC is the firm's supply curve.

Forest Production

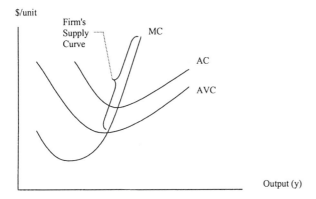

Figure 5.2. Firm's supply curve

In aggregate, those short-run marginal costs represent the costs of production of all firms for a given output and determine the industry's supply, which is the sum of all the output produced by all firms in a given market area. As long as the firms could sell a greater value of output than the price of variable inputs used, they should continue production. For example, lumber mills will continue to operate as long as the price of lumber they produce is higher than average variable costs of stumpage, labor, materials, and energy per unit of lumber output. In the long run, if the firms do not cover fixed costs as well, they will be forced out of business. This implies that in the long run, the price of output must exceed its average costs.

2.3 Revenue Functions

Revenue functions represent the maximum revenue that can be derived from given output prices *(p)* and a set of inputs. For a lumber mill, a revenue function would indicate the maximum revenue that can be derived from producing various lumber dimensions, given their prices and quantities of inputs such as stumpage, labor, machinery, materials, and energy used in production. In the single output case, revenue maximization reduces to the production of the maximum output permitted by a given input set; multiple outputs need to be considered to have a true economic problem, as in the product-product relationships discussed in section 2.6.

The revenue function is:

$$R(p,x) = \max \; py, \quad s.t. \; y = f(x) \qquad 5.6$$

Total revenue (*TR*) is calculated by multiplying the amount of output sold by its price. One can also calculate total revenue variation measures such as average revenue (*AR*), which equals the total revenue divided by output quantity, and marginal revenue (*MR*), which represents the change in the total revenue resulting from the sale of an additional unit of output:

$$AR = \frac{TR}{y}$$
$$MR = \frac{\partial TR}{\partial y}$$

5.7

In perfect competition, firms are assumed to be price takers in output markets. This means that individual firms supply only a small fraction of the entire industry output, and no firm can change the price of its output by producing more or less of that output. For example, this may be the case in a region where timber is supplied by a large number of small timber growers, none of which has a market share large enough to influence the price of timber in that region. It follows that the firm can sell any amount of output without affecting market prices. Since the amount of output does not affect the market price of output, *AR* and *MR* are constant and equal to the price of output:

$$AR = MR = p$$

5.8

Consequently, the firm can influence its revenues only by selling more or less of its output. But output changes imply cost changes; therefore, managers must consider both revenues and costs in order to make optimal economic decisions.

2.4 Profit Functions

Economic theory postulates that the objective of the firm is to maximize profits, which are defined as the difference between the firm's revenue (e.g., revenues from log sales) and costs (e.g., total operating costs of a logging enterprise) (equation 5.9). In making decisions about its activities, the firm faces technological constraints that are implied by a production function and market constraints that are often set by the assumption that the firm is a price taker in output and input markets. Production functions, input prices, output prices, and cost functions provide the means to calculate the profit-maximizing levels of input use and product manufacture.

Forest Production

$$\Pi(p,w) = \max\{pf(x) - wx\}$$
$$= \max\{py - c(w,y)\} \quad\quad 5.9$$

Profit maximization can be thought of as a two-stage process. Profits can be maximized with respect to inputs and their prices or with respect to outputs and their prices. In the first stage, which can be considered as the short run, profits are maximized for a given level of output. Since output and, therefore, revenues are fixed, profits are maximized by minimizing costs. In the second stage, which can be considered as the long run, output levels can vary and are chosen to maximize profits. Maximizing profits with respect to inputs results in the following criteria:

$$pMP_{xi} = w_{xi} \quad or \quad MP_{xi} = \frac{w_{xi}}{p} \quad\quad 5.10$$

Similarly, one can derive the profit-maximizing criterion for outputs as:

$$p - MC = 0 \quad or \quad p = MC \quad\quad 5.11$$

Using equation 5.8, we derive the classic criterion for profit maximization:

$$MC = MR \quad\quad 5.12$$

This condition stating that the marginal revenue of each action should equal its marginal cost determines the optimum mix of inputs and outputs. For firms that are price takers, information on product prices and input productivity can be used with this condition to derive profit-maximizing production levels, which are the most economically efficient use of resources. In our timber-growing example, in order to achieve maximum profits, one must consider and compare the increased revenues from accelerated timber volume growth with the costs incurred by applying more fertilizer. Profit-maximizing conditions will be met when marginal timber revenues equal marginal costs of applying fertilizer.

2.5   Input-Input Relationships

The relationship between two inputs and one output is determined by the productivities of the individual inputs and their interaction. An isoquant represents the output of one product that can be obtained with varying amounts of two inputs, for example, the volume of timber that can be harvested with varying combinations of labor and machinery. Inputs are typically substitutes, and thus in our example, a given volume of timber can be produced with various combinations of labor and machinery. A logger may decrease machinery use by replacing it with more labor and still be able to harvest the same amount of timber. The rate of substitution of one input for another is expressed in terms of the marginal rate of technical substitution (MRTS):

$$MRTS_{x_1 \to x_2} = \frac{\partial x_2}{\partial x_1} = -\frac{MP_{x_1}}{MP_{x_2}} \qquad 5.13$$

Figure 5.3 presents the classical two-input (labor and machinery) and one-output (timber) isoquant, with slope measured by MRTS. The isoquant has a negative slope and is convex towards the origin, implying a diminishing MRTS. As labor replaces machinery, more and more labor is needed to replace an additional piece of machinery; that is, MP_{x2} diminishes while MP_{x1} increases. The elasticity of input substitution is a unitless measure that describes how inputs substitute to produce a given level of output at a given point:

$$\sigma = \frac{\partial x_1}{\partial x_2} \frac{x_2}{x_1} \qquad 5.14$$

Input prices must be considered in order to determine the economically optimal level of input use. These costs are graphically represented by isocost lines (equation 5.15), which describe various combinations of inputs that cost the same fixed amount. Note that the slope of an isocost line is equal to the ratio of input prices $(-w_1/w_2)$.

$$C = w_1 x_1 + w_2 x_2 \qquad 5.15$$

To produce a given level of output using the least expensive combination of outputs, the firm should operate on the lowest possible (closest to origin) isocost line. This isocost line will be tangent to the isoquant that represents the desired output level:

$$\frac{\partial x_2}{\partial x_1} = -\frac{w_1}{w_2} \qquad 5.16$$

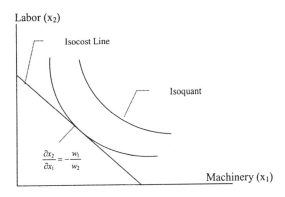

Figure 5.3. Input-input relationships

2.6 Product-Product Relationships

The production possibilities frontier (PPF) represents the maximum amounts of two outputs that can be produced from a given amount of a single input. For example, consider pulpwood and sawtimber outputs from a fixed area during a given time period. PPFs could be constructed to show the different combinations of outputs possible, given varying single inputs such as planting densities, fertilization regimes, or pruning. A controversial and as of yet poorly quantified example is production of outdoor recreation and timber in a given area.

Products may be complementary, independent, or competitive in nature. Timber and deer habitat are often complementary products because many timber management practices improve deer habitat. Independent products are not affected by each other's production. Timber and waterfowl might be an example of independent products in a landscape where they do not use the same resources (wetland vs. forest land). But the provision of most forest products involves tradeoffs, implying that in order to produce more of one product (timber), the amount of another product must be reduced (outdoor recreation), given fixed input levels.

Figure 5.4 presents the classical concave PPF with two competitive products (timber and recreation) and one input (forest land). The technological or biological rate of substitution of one product for another, given a fixed level of input, is termed the marginal rate of product substitution, or *MRPS*:

$$MRPS_{y_1 \to y_2} = \frac{\partial y_2}{\partial y_1} \qquad 5.17$$

The concavity of the PPF implies increasing opportunity costs: the more timber is produced, the more recreation has to be given up to obtain an extra unit of timber and vice versa. Any level of timber production will reduce recreation values. But timber production on selected sites, which may be heavily stocked with timber but not particularly suited for recreational uses, will reduce recreation values to a lesser extent than widespread timber management across all sites.

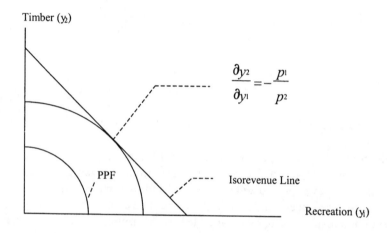

Figure 5.4. Output-output relationships

In order to determine the optimal combination of outputs, information about their prices or values has to be incorporated. Isorevenue lines (equation 5.18) represent various combinations of timber and recreation that generate the same revenues. Their slope ($-p_1/p_2$) reflects product prices, which for most forest management decisions remain constant over the range of resources being managed. (See chapter 19 for methods for calculating the price or value of forest recreation opportunities.)

$$R = p_1 y_1 + p_2 y_2 \qquad 5.18$$

If prices capture all values of forest outputs, then the logical goal of forest management is to generate the highest possible profits. Assuming fixed input costs, this goal translates into operating on the highest possible (furthest from the origin) isorevenue line. The PPF corresponding to the fixed input available represents the maximum feasible production, and thus the highest isorevenue line is the one tangent to that PPF:

$$\frac{\partial y_2}{\partial y_1} = -\frac{p_1}{p_2} \qquad 5.19$$

Finally, most forestry operations involve multiple outputs and multiple inputs. In the case of two outputs and two inputs, the efficiency conditions are that the *MRPS* (slope of the PPF for the two outputs) must be equal across the two inputs, and that the *MRTS* (slope of the isoquant for the two inputs) must be equal across the two outputs. These marginal conditions generalize to the N output–N input case.

2.7 Estimation

Production, cost, revenue, and profit functions are closely related because costs, revenues, and profits are all constrained by the underlying technology, as represented by the production function. The shape of cost functions, for example, is always the inverse of the corresponding production functions. If the total production is increasing at an increasing rate (increasing *MP*), then the total costs are increasing at a decreasing rate (decreasing *MC*). If the *MP* is decreasing, then the *MC* is increasing. These relations allow one to compute cost functions from production functions, or production functions from cost functions. Functions possessing this quality are called dual.

The knowledge of production functions and input prices allows us to derive cost functions that are directly linked to the basic production technologies. Since the technology constrains optimizing behavior of the firm, one should be able to reverse this process and derive production functions from cost functions, because the firm's responses should reveal the underlying technology. Indeed, cost functions can be estimated directly using econometric methods, and then the underlying production functions can be derived. Direct estimation of cost functions is often easier, because all inputs can be combined into a single cost measure by using their prices. This is not possible with production quantities, and data on the complete set of production relationships is usually difficult to obtain. The choice between using production functions and input prices versus direct cost estimation depends on the data available, the need for examining production processes, and the economic measures analyzed and assumptions made.

When production and cost functions are estimated for individual harvesting systems or firms, independent variables often include tract size, stand volume, tree size, equipment configuration, production rates, or input prices. More aggregate analyses tend to estimate basic relationships between total production and input levels (e.g., capital and labor) or between total cost and production levels. We turn next to a case study of aggregate production and cost functions for timber harvesting.

3. PRODUCTION AND COST OF HARVESTING SOUTHERN TIMBER

Cubbage and others (Carter and Cubbage 1994, Carter et al. 1994, Cubbage and Carter 1994, Cubbage et al. 1989) extended the harvesting system cost estimation, reviewed in section 1, to examine trends in several southern timber harvesting systems. In this section, we summarize the findings of this research on aggregate cost and production functions. Chapter 7 reports stochastic frontier regressions estimated with the same data. The data are from two surveys of timber harvesting companies in the southern United States and include information on 3,680 logging crews in 1979 and 463 in 1987. These crews used eight different configurations of harvesting equipment, including six shortwood systems and two longwood systems. Survey respondents provided detailed information on the inputs and output of their logging system. These are used to calculate average production per crew and average cost per cord, as reported in table 5.1.

The average harvesting cost is the sum of equipment and labor costs per week divided by weekly production in cords of wood harvested (equation 5.5). The average costs per cord were generally much less for the longwood systems, in both 1979 and 1987. Average costs actually rose for most shortwood systems between 1979 and 1987, while average costs per cord dropped by almost $8 per cord for the highly mechanized longwood systems, from $34 per cord to $26 per cord. Capital-intensive systems were most cost efficient in both periods and gained cost advantages by 1987. This rapid increase in cost efficiency was reflected in a large shift to longwood systems, from 58.7% of all production in 1979 to 90.4% in 1987. The declining efficiency of small-scale, labor-intensive shortwood systems continues today. We therefore focus on the increasingly dominant longwood harvesting systems.

In both 1979 and 1987, there were just over 1,000 cable-skidder longwood operations in the South. Their share of production fell from 23.5% to 19.7% over that time period, due to a shift to grapple-skidders. Here, we evaluate changes between 1979 and 1987 in the production and cost functions of longwood operations with cable-skidders, using the survey data described above.

Forest Production

Table 5.1. Southern U.S. timber harvesting production and costs per week by technology class, 1979 and 1987 (Carter et al. 1994)

Technology class	1979 production			1987 production		
	Cords per class	Cords per crew	Cost[a] ($/cord)	Cords per class	Cords per crew	Cost[a] ($/cord)
Shortwood						
Manual bobtail	20,813	20.8	43.7	3,233	26.9	43.5
Manual bigstick	124,777	27.0	44.0	26,365	26.8	51.4
Manual/skidder	35,806	36.3	52.9	13,655	38.8	60.5
Forwarder	27,629	70.2	44.8	7,858	86.1	48.6
Cable-skidder	36,334	69.4	52.0	8,494	59.8	54.9
Grapple-skidder	25,763	189.3	32.2	16,189	160.5	31.1
Total (shortwood)	270,763	34.4	45.4	75,794	41.7	48.9
Longwood						
Cable-skidder	145,222	143.1	36.3	155,290	127.8	30.0
Grapple-skidder	229,958	254.9	32.3	556,233	294.7	25.4
Total (longwood)	384,180	190.7	34.0	711,523	227.7	26.4
Total (all classes)	654,943	66.1	38.7	787,317	154.3	28.8

[a] Average weighted by total production levels; indexed to 1988 dollars.

Following equation 5.1, total production (cords per week) is estimated as a function of inputs (asset value of machinery and number of employees) in an OLS regression. Assuming that the survey respondents are profit-maximizing logging firms, we expect diminishing marginal returns to inputs, that is, the relationship between inputs and output should fall in the range of x_1 to x_2 in figure 5.1. Two functional forms are considered for estimation of this relationship: log-log, which allows for diminishing returns, and linear, which may also approximate the relationship in that limited range of the production function. In addition to the two inputs, the specifications include a dummy variable (D) indicating the survey year for the observation, with 0 = 1979 and 1 = 1987. Table 5.2 reports the results.

Table 5.2. Production function for longwood cable-skidder system

Variable	Log-Log[a]		Linear	
	Coeff.	St. Err.	Coeff.	St. Err.
Intercept	0.32		-26.7	
Assets	0.84	6.44**	0.02	7.31**
Employees	0.43	3.49**	11.5	5.01**
D for 1987	-0.24	-0.3	11.6	-0.99
D • Assets	0.11	0.52	0.02	0.21
D • Employees	-0.1	-0.49	7.1	1.78*
R^2	0.48		0.57	

Sample size = 480   Dependent variable is cords per week.
[a] Assets, employees, and cords per week are logged.
** = significant at 1% level   * = significant at 10% level

Both assets and employees have significant positive coefficients in the regressions, consistent with basic production theory. In the log-log functional form, the coefficients on assets and employees are partial output elasticities; their values in the range of 0 to 1 are consistent with diminishing marginal returns in both 1979 and 1987.[1] Overall, the estimation results suggest that there has not been any structural change in this technology class between 1979 and 1987. The only exception is in the linear specification, where the significant coefficient on the last variable suggests that the productivity of employees has increased substantially.

As described in section 2.2, the average cost of a firm is a function of its level of output. Thus, average cost is regressed on weekly production, the dummy (D) for 1987, and an interaction term. Table 5.3 reports the results of log-log and quadratic average cost functions for the longwood cable-skidder system.

Table 5.3. Average cost functions for longwood cable-skidder system

	Log-log[a]		Quadratic	
Variable	Coeff.	St. Err.	Coeff.	St. Err.
Intercept	5.51		67.28	
Cords/wk	-0.38	-17.5**	-0.25	-8.06**
Cords/wk ^{2}			0.0004	6.0**
D for 1987	0.64	4.28**	1.47	0.34
D • words/wk	0.18	-5.91**	-0.08	-1.85*
D • cords/wk ^{2}			0.00	1.47
R^2	0.7		0.45	

Sample size = 480
[a] Dependent variable and cords/wk are logged.
** = significant at 1% level * = significant at 10% level

The log-log functional form suggests diminishing marginal costs across the full range of firms surveyed, while the quadratic functional form indicates that cost initially declines and then increases with scale of production.[2] The latter result is consistent with theory, as illustrated by figure 5.2. The log-log regression indicates that between 1979 and 1987, costs have increased overall (significant positive coefficient on D), as has the response of costs to production levels (significant coefficient on interaction term). The quadratic regression provides more ambiguous results on structural change, with an insignificant coefficient on D and a small and marginally significant coefficient on one interaction term. The log-log specification appears to be a better fit to the data, given its substantially higher R^2. Based on that specification, we can conclude that there has been structural change in this technology class such that the scale of the operation has become even more important in controlling costs.

4. CONCLUSION

The basic production economics techniques discussed in this chapter are used for estimating costs, revenues, and profits for market goods and services. The aggregate timber harvesting analysis illustrates how production economics can be applied in forest operations, using survey data to estimate the basic shape and structural changes in production and average cost functions. The results suggest that there has not been significant change for this technology class, except for the increased influence of the scale of operations on total cost. This is a simplified analysis, insofar as factors other than production levels affect average costs, and factors other than capital and labor inputs affect productivity. Richer specifications of cost functions would be useful for firms planning or estimating harvesting or other forest production costs, for forest planning and budgeting, and for forest policy analyses. The challenge is to obtain the data required for these specifications. Relatively simple aggregate cross-sectional analyses, as illustrated in this chapter, provide estimates of regional costs for forestry production activities, indicators of changes across time, and validity checks for cost estimates from individual firms or from simulation approaches.

5. LITERATURE CITED

CARTER, D., AND F. CUBBAGE. 1994. Technical efficiency and industry evolution in southern pulpwood harvesting. Can. J. For. Res. 24: 217-224.

CARTER, D., F. CUBBAGE, B. STOKES, AND P. JAKES. 1994. Southern pulpwood harvesting productivity and cost changes between 1979 and 1987. Res. Paper NC-318. USDA Forest Service, North Central For. Exper. Station, St. Paul, MN. 33 p.

CHAMBERS, R. 1988. Applied Production Analysis: A Dual Approach. Cambridge University Press. Cambridge, U.K. 331 p.

CUBBAGE, F. 1983. Simulated effects of productivity rates, input costs, and stand volumes on harvesting costs. For. Prod. J. 33(2): 50-56.

CUBBAGE, F., AND D. CARTER. 1994. Productivity and cost changes in southern pulpwood harvesting, 1979 and 1987. So. J. Appl. For. 18(2): 83-90.

CUBBAGE, F., P. WOJTKOWSKI, AND S. BULLARD. 1989. Cross-sectional estimation of empirical southern United States pulpwood harvesting cost functions. Can. J. For. Res. 19: 759-767.

DOLL, J., AND F. ORAZEM. 1978. Production Economics: Theory with Applications. John Wiley & Sons. New York, NY. 406 p.

KLUENDER, R., D. LORTZ, W. MCCOY, B. STOKES, AND J. KLEPAC. 1998. Removal intensity and tree size effects on harvesting cost and profitability. For. Prod. J. 48(1): 54-59.

MATTHEWS, D. 1942. Cost Control in the Logging Industry. McGraw-Hill Book Co., New York, NY. 374 p.

NEWMAN, D. 1991. A modified production function analysis of changing southern forest productivity. Can. J. For. Res. 21(8): 1278-1287.

NEWMAN, D. AND D. WEAR. 1993. Production Economics of Private Forestry – a comparison of industrial and nonindustrial forest owners. Am. J. Ag. Econ. 75(3): 674-684.

PAPPAS, J., AND M. HIRSCHEY. 1990. Managerial Economics, Sixth Edition. The Dryden Press, Chicago, IL. 826 p.

PLUMMER, G. 1977. Harvesting cost analysis. P. 65-79 *in* Logging Cost and Production Analysis. Timber Harvesting Rep. No. 4. LSU/MSU Logging and Forestry Operations Center, Long Beach, MS.

SMITH, P., AND I.A. MUNN. 1998. Regional Cost Function Analysis of the Logging Industry in the Pacific Northwest and Southeast. For. Sci. 44(4): 517-525.

STUART, W. 1981. Harvesting analysis technique: A computer simulation system for timber harvesting. For. Prod. J. 31(11): 45-53.

WANG, J, W. GREENE, AND B. STOKES. 1998. Stand, harvest, and equipment interactions in simulated harvesting prescriptions. For. Prod. J. 48(9): 81-86.

---

[1] In the log-log functional form, the partial output elasticity for assets in 1979 is just the coefficient on the natural log of assets. For 1987, the elasticity is the sum of the coefficients on ln (assets) and on D•ln (assets).

[2] The initial decline in cost is indicated by the negative coefficient on production (and D• production), while the eventual increase is indicated by the positive coefficient on the squares of these terms. The minimum average cost occurs at 320 cords per week in 1979 and 330 cords per week in 1987.

Chapter 6

Financial Analysis of Timber Investments

F. Christian Zinkhan and Frederick W. Cubbage
The Forestland Group LLC and North Carolina State University

Timber is part of the investment portfolio of a wide range of investors. Traditional forestry investors include farmers who own forest land and the large forest products firms that have purchased forest land to grow timber, usually to supply large pulp and paper mills. Over the last few decades, many new investors have inherited, purchased, or otherwise acquired timberland. These are often passive investors, not actively involved in timber management. However, like all economic actors, they are either directly or indirectly concerned with the returns to timber production. Both passive and active timberland investors have alternative investment vehicles for their scarce capital. Common analytical frameworks can be applied to timber and nontimber assets to help investors assess these alternative asset classes.

This chapter reviews these frameworks for various types of timberland investments, starting with traditional capital budgeting techniques as applied to timber investments. It then introduces the use of modern financial theory to analyze timber investments and continues with a review of new vehicles for timber investments and their use by sophisticated new timberland owners.

1. MODERN FINANCIAL ANALYSIS OF TIMBER INVESTMENTS

Capital budgeting techniques have been the principal means of analysis of timber investments. These techniques discount the values of costs and returns to calculate present values, land expectation values, cost/benefit ratios, or internal rates of return. The forestry literature has many examples of capital budgeting techniques used to determine the value of timber and timberland investments. Each of these capital budgeting techniques has advantages and disadvantages (Brealey and Myers 1991, Clutter et al. 1983, Davis and Johnson 1987). We review these techniques and the use of modern portfolio theory (MPT) here, drawing from a prior review by Wagner et al. (1995).

1.1 Capital Budgeting Techniques

The most often used capital budgeting criteria for forestry investments are the net present value (*NPV*), land expectation value (*LEV*), and internal rate of return (*IRR*). The respective formulas follow:

$$NPV = \sum_{t=0}^{T} B_t(1+i)^{-t} - \sum_{t=0}^{T} C_t(1+i)^{-t} \qquad 6.1$$

$$LEV = \frac{\sum_{t=0}^{T} B_t(1+i)^{-t} - \sum_{t=0}^{T} C_t(1+i)^{-t}}{(1-(1+i))^{-T}} \qquad 6.2$$

The *IRR* is defined as that discount rate that equates the present value of the benefits with the present value of the costs:

$$\sum_{t=0}^{T} B_t(1+IRR)^{-t} = \sum_{t=0}^{T} C_t(1+IRR)^{-t} \qquad 6.3$$

where B_t = a benefit at time t, i = annual discount rate, C_t = a cost at time t, T = lifetime of project or rotation length.

The *NPV* converts a series of periodic income flows to a single number that can be used to compare mutually exclusive investment alternatives over the same investment horizon at a given discount rate (cost of capital).[1] For single investment decisions, one would accept an investment that has a positive *NPV* if enough capital were available. If the *NPV* were negative, one would reject that investment. In order to compare *NPV*s of repeatable projects (rotations) of different lengths, one would have to convert all those investments to the same time horizon, such as the least common denominator of all time horizons.

The *LEV* calculates the present value of an infinite series of projects (rotations). This provides a simple means to convert investments with different time horizons to one simple common denominator of infinity. *LEV* is applied just like *NPV* in making investment decisions, with positive *LEVs* inferring investment acceptability and negative *LEVs* suggesting project rejection.

The *IRR* is defined as the discount rate that makes the present value of the benefits of a project exactly equal to the present value of the costs of a project. *IRR* indicates the annual rate of return that an investment would generate. For individual investments, the *IRR* is usually compared with some given hurdle rate or with other potential investments. Projects with *IRRs* greater than the hurdle rate are considered acceptable given adequate capital.

Most investment or project decisions compare multiple investments with a limited capital budget or constraint. For selecting among many exclusive projects, one would choose the maximum *NPV* or *LEV*, or largest *IRR*.

Net present value is generally recommended as being superior in most textbooks (e.g., Brealey and Myers 1991, Copeland and Weston 1988), and *LEV* is the best extension of this approach for very long-lived forestry investments of unequal time lengths. For investors who can clearly determine their discount rate, *NPV* and *LEV* provide the best means to maximize profits. Selecting the projects with the highest total *LEV* with a given limited budget will generate the most net returns for that amount of capital. *IRR* is generally considered a theoretically inferior capital budgeting criterion, but its use persists for many practical reasons. It is easy to understand, explain, and compare with other investment vehicles. It avoids problems of project scale or length in making comparisons. And often owners do not know their discount rate, so *IRR* provides a means to compare investments heuristically with the implied cost of capital.

Choice of the discount rate is crucial for *NPV* and *LEV* analyses and decisions. The discount rate represents an organization's opportunity cost of capital for an investment. For private firms or investors, this implies the alternative rate of return that the investor could receive in some other investment of similar risk. This is often calculated as the weighted average of debt (loans) and equity (stock) for private firms. For public organizations, the cost of capital is usually determined by the government or by an international lending agency. It too should represent some average of debt financing such as the cost of government borrowing. Next we turn to one approach for estimating the discount rate for an investment.

1.2 Capital Asset Pricing Model

The traditional finance and relatively new forestry literature provides various overviews of the capital asset pricing model (CAPM) and its applications to forestry (Brealey and Myers 1991, Conroy and Miles 1989,

DeForest et al. 1991, Mills 1988, Redmond and Cubbage 1988, Wagner and Rideout 1991, Zinkhan 1988b).

According to the CAPM, an asset's total financial risk is the sum of two components: diversifiable and nondiversifiable risk. Nondiversifiable (systematic) risk reflects an asset's price movements caused by changes in the macroeconomy. Diversifiable (nonsystematic) risk reflects factors unique to the particular asset that are independent of the macroeconomy (e.g., research and development, markets for the subject's products, firm management, or capital and labor productivity of the organization).

An investor can eliminate diversifiable risk by creating a portfolio containing assets from various sectors in the economy (Elton and Gruber 1995, Haugen 1987). Thus, the investor is rewarded with higher returns only for the nondiversifiable risk. The CAPM is used to estimate an asset's nondiversifiable risk. Therefore it values an asset with respect to the market. Sharpe (1964) defined CAPM as:

$$E(R_{at}) = R_{ft} + \beta_a E(R_{mt} - R_{ft}) \qquad 6.4$$

where R_{at} = nominal rate of return of asset a in time t, R_{ft} is the nominal rate of return of a risk-free asset in time t, β_a is the index of nominal nondiversifiable risk of asset a (also referred to as the *beta* coefficient), R_{mt} is the nominal rate of return of the market portfolio in time t, and E is the expected value operator.

Equation 6.4 defines the expected, or *ex ante*, rate of return for any asset. Since *ex ante* returns are not observable, Jensen (1969) showed that CAPM parameters could be estimated from *ex post* returns:

$$R_{at} - R_{ft} = \alpha_a + \beta_a(R_{mt} - R_{ft}) + \varepsilon_t \qquad 6.5$$

where

$$\beta_a = Cov(R_a, R_m)/Var(R_m) \qquad 6.6$$

$$\alpha_a = E(R_{at} - R_{ft}) - \beta_a E(R_{mt} - R_{ft}) \qquad 6.7$$

where $Cov(R_a, R_m)$ is the covariance between the returns of an asset and the market portfolio, $Var(R_m)$ is the variance of the returns to the market portfolio, α is the *alpha* coefficient, and ε_t is the stochastic residual error.

The lefthand side of equation 6.5 defines the excess returns to the asset over what would be expected for an asset with that risk. The term in parentheses on the righthand side of equation 6.5 defines the excess returns to the market portfolio.

The *beta* coefficient in equations 6.5 and 6.6 defines the percent movement of an asset's returns with that of the market portfolio. If the *beta* coefficient is greater than one, the asset's returns are more risky than the market. If the *beta* coefficient is less than one, the asset's returns are less sensitive to the market and are thus less risky than the market. The *beta* coefficients also may be negative, although this is uncommon with traditional financial instruments. A negative *beta* implies the returns of the asset are countercyclical to those of the market portfolio. Such an asset would reduce the overall risk of an investor's portfolio dramatically.

The *alpha* coefficient in equations 6.5 and 6.7 is the Jensen Performance Index (1969). The *alpha* coefficient has an expected value of zero because the CAPM implies that an asset's returns are determined solely by its nondiversifiable risk. Therefore, a statistically significant and positive (or negative) *alpha* value indicates that returns are greater (or less) than necessary to compensate for a given level of nondiversifiable risk.

The CAPM also may be used to estimate a risk-adjusted discount rate (i_{at}) for an asset in a given period t:

$$i_{at} = R_{ft} + [E(R_{mt}) - R_{ft}]\beta_a \qquad 6.8$$

The risk-adjusted discount rate reflects the compensation required by the investor for an asset's unique nondiversifiable risk. This provides a unique risk-adjusted discount rate that could be used in capital budgeting analyses of pending investments in an asset if its future cash flows were uncertain. Fortson (1986) reviewed five techniques for selecting a discount rate, with applications to forestry investments. He noted that using a single discount rate for all projects in cost/benefit analysis was inappropriate and that CAPM could be used to calculate the appropriate unique discount rate for various investments.

Traditional CAPM analyses estimate the regression coefficients (*alpha* and *beta*) using historical returns from stocks or other financial instruments. The periodic returns on the stock—rate of price change plus dividends (R_{at})—are modeled as a function of total market returns (R_{mt}) as represented by a market proxy such as the Standard & Poor's 500 Index (S&P 500). The CAPM has been applied to forestry investments directly by using timber price appreciation and annual growth rate returns as a proxy for R_{at}. Other studies have used aggregate timberland performance indexes or simulated timber investment returns compared to a market index.

Without a capital constraint, an investor should allocate funds to all projects offering a positive *alpha*. When viewed on an *ex ante* basis, a positive alpha suggests an expected return greater than the return required relative to the *beta* risk within the CAPM. In concept, *alpha* is similar to the popular economic value added measures of performance that have been applied to timber investments (see Zinkhan 1996).

Due to the logic and simplicity of its derivation, the CAPM has gained broad acceptance in financial markets. However, there has been criticism (see Jagannathan and McGrattan 1995 for a review). The most significant criticism is the lack of a positive relationship between stocks' *betas* and actual returns found by Fama and French (1992). Factors other than beta may need to be considered in asset pricing models. Additional research is needed, especially with respect to assets such as timber, which are not traded in organized markets with ready buyers and sellers.

1.3   Timber Investment Analyses Using CAPM and Capital Budgeting Techniques

The preceding discussion describes the fundamentals of traditional forestry finance and MPT as applied to forestry investments. A number of research studies in the 1980s and 1990s examined the application of modern financial techniques to forestry investments and compared them with traditional forestry outcomes (e.g., Conroy and Miles 1989, Cubbage and Redmond 1985, DeForest et al. 1991, Mills 1988, Wagner and Rideout 1991, Washburn and Binkley 1990, and Zinkhan 1988b). Many of these are summarized in Wagner et al. (1995) and paraphrased here for illustration.

Both traditional capital budgeting and CAPM techniques have been used to analyze timber investments by modern investors. When used together, they can complement each other both in terms of the amount of information they provide about timber investments and the comparison of those investments with other asset classes. Traditional forestry investments generally favor species or projects with high stumpage values, short rotations, and modest initial and annual costs, in order to maximize the present value of future returns. The CAPM differs somewhat because it relates stock or timber returns to those of a market proxy. Thus, a substantial increase in value (e.g., rapidly increasing stumpage prices) or dividends (e.g., rapid growth rates) relative to the level of nondiversifiable risk will tend to make for desirable investments per the CAPM.

Wagner et al. (1995) analyzed typical forestry investments using capital budgeting techniques. Based on the capital budgeting criterion and uniform expected rates of stumpage appreciation, Douglas fir plantation investments had the highest present values of six species examined at $2,500 per acre (using an 8% real discount rate). Sawtimber rotations of southern pine had the next highest NPVs and LEVs at about $600 per acre. Southern pine pulpwood returns were negative at a price of $15 per cord (-$30/acre), but had the third highest NPV/LEV at a price of $26.90 per cord ($200 per acre). Hardwood sawtimber regimes had LEVs of about $120 per acre. Internal rates of return for these management regimes were about 12% for those with positive NPVs and 7.5% for the rest.

Based on the sum of *ex post* rates of annual stumpage price appreciation (generally for a 37-year period ending in 1988) and estimates of growth rates, Wagner et al. (1995) reported that Douglas fir timber and eastern hardwoods had the highest measured annual returns, at 13.8% and 13.3%, respectively. Southern pine sawtimber rotations on plantations (12.7%) and pine pulpwood rotations (10.9% to 12.1%) had lower average annual returns. For all species, returns were comparable to market returns on the S&P 500—within a percentage point or two—during the same time period.

CAPM analyses indicated that all but one of the species management regimes had positive *alpha* values, indicating superior returns compared to the market for the level of nondiversifiable risk assumed. In addition, they all had negative *beta* values, ranging from −0.18 to −0.34, indicating that they were countercyclical to the returns of the market proxy. This suggests that timber investments can be an important part of an overall portfolio of investments by improving the relative risk and return performance of that portfolio. In fact, DeForest et al. (1991) showed that timber assets would be useful to add as a moderate component of a portfolio in order to enhance overall risk/return performance.

1.4 Timber Investments and the Efficient Frontier

A basic issue for investors with diversified portfolios is whether to include any timber assets. If an investor does include timber assets, a second issue is how much of the portfolio to allocate to timber assets. The efficient frontier is an MPT analytical tool that can be used to help address these issues. The efficient frontier is used primarily to make strategic asset allocation decisions at the portfolio level. Then, if an allocation is made to timber, the timberland investment manager should seek out investments that offer either positive NPVs (using the CAPM to establish the cost of equity capital) or positive alphas.

Rational investors seek to maximize the expected return of their portfolios relative to a given level of volatility. Harry Markowitz developed a conceptual and quantitative approach for identifying risk/return efficient portfolios of assets a half century ago (1952). In comparing two portfolios with the same level of return volatility, a rational investor will prefer the one with the higher return. Or, when comparing two portfolios with the same level of expected return, a rational investor will prefer the one with the lower volatility. An inferior, or inefficient, portfolio is one dominated from a risk/return perspective by at least one other portfolio. Considering all possible mixes of investment assets and applying mathematical programming, Markowitz demonstrated how to map out all risk/return efficient portfolios in the form of an efficient frontier and thus eliminate from consideration all other (inferior) portfolios. Sharpe (1964) later derived the CAPM using the efficient frontier as a building block, along with the assumption of the availability of a risk-free investment and a market portfolio.

Previous publications (e.g., Caulfield 1998, Zinkhan 1990, and Zinkhan et al. 1992) have illustrated the potential improvement in the tradeoff between risk and return by adding timberland to the mix of financial asset-dominated portfolios. Figure 6.1 is representative of the conclusions of these asset allocation studies.

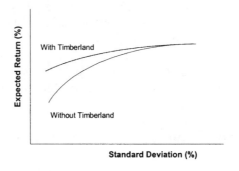

Figure 6.1. Efficient frontiers with and without timberland

When timberland is included as an asset class alternative, the efficient frontier of portfolios that maximizes expected return (or, minimizes risk) for a given level of risk (or, expected return) shifts upward. Such a shift in the efficient frontier suggests that an investor should consider adding some timberland to the portfolio mix.

The inputs needed to estimate an efficient frontier are shown in figure 6.2 and table 6.1. The National Council of Real Estate Investment Fiduciaries (NCREIF) Total Timberland Index is used as a proxy for timberland. This index is based on cash flow and appraisal data reported by several timberland investment management organizations (TIMOs) participating with the NCREIF in this project. TIMOs are firms that raise capital from institutional and other investors and invest it in timberland. In addition to

Financial Analysis of Timber Investments

this composite timberland index for the entire nation, NCREIF also reports subindexes for three regions: the Southeast, the West, and the Northeast.

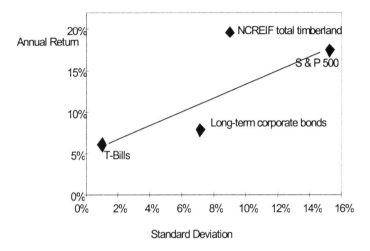

Figure 6.2. Timberland relative to Capital Market Line, 1987 to 1999

The actual return of the NCREIF Total Timberland Index exceeded that of the three financial assets for the period 1987 to 1999 (figure 6.2). For the period 1987 to 1999, the annual return of the NCREIF Western Index (26.0%) exceeded that of the NCREIF Southeastern Index (13.8%). Pressure on privately owned timber in the West increased in the early 1990s due to sharp cutbacks in harvests on federal lands. This pressure caused dramatic increases in timber prices and timberland returns. As the industry reacted by shifting its focus toward the South, pressure was applied to privately owned forest resources in that region. Given its dramatic returns since 1987, timberland plotted well above the capital market line that connects the risk/return positions of Treasury bills and stocks. Some of the return premium achieved on timberland may be attributable to compensation for relative illiquidity; however, a comparable premium relative to the capital market line is not expected in the long-term future.

Table 6.1. Correlation coefficient matrix, 1987 to 1999

	T-Bills	Long-term corporate bonds	S&P 500	NCREIF Total Timberland Index
T-Bills	1.00	0.19	0.03	0.05
LT Corp Bonds		1.00	0.07	0.17
S&P 500			1.00	-0.06

As a benchmark for portfolio asset allocation comparisons, institutional investors and their consultants often refer to a standard portfolio. The standard portfolio is defined as one with the following asset mix: 60% large capitalization common stocks, 30% corporate bonds, and 10% Treasury bills. The risk and expected return of a portfolio with this mix of financial assets, but with 10% also allocated to timberland, is compared to that of the standard portfolio in figure 6.3. These risk/expected-return combinations are based on *ex post* quarterly results for the period 1987 to 1999.

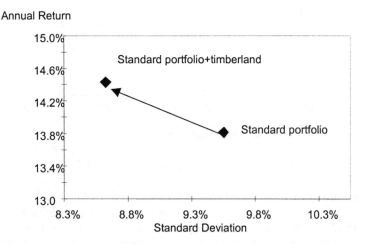

Figure 6.3. Improvement in risk/return efficiency through addition of timberland to a standard portfolio, 1987 to1999

As shown in figure 6.2, the volatility of timberland's returns as reflected by the NCREIF Total Timberland Index was modest relative to large capitalization stocks. It has been noted elsewhere that appraisal-based return series such as the NCREIF Total Timberland Index smooth out fluctuations in the series, and thus standard deviation measures are underestimated (e.g., Zinkhan 1990). Other reasons for the relatively low volatility of returns associated with timberland include the steadiness of the biological growth and ingrowth—the movement of timber into more valuable merchantability classes—components, the rather stable demand for a diversity of timber products, and the long-term nature of the investment.

The inverse relationship between timberland returns and the returns of common stocks is reflected by the negative correlation reported in table 6.1. The volatility reduction highlighted in figure 6.3 is partially attributable to this negative correlation. Based on its superior returns and countercyclical relationship with the S&P 500, it should not be surprising that the NCREIF Total Timberland Index had a positive alpha and a negative beta for this study period, when equation 6.5 is estimated with the data described in table 6.1 and figure 6.2 (see figure 6.4).

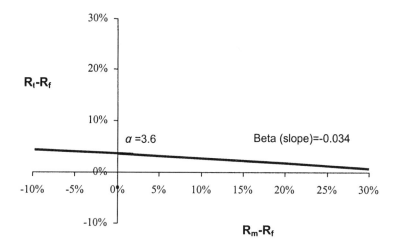

Figure 6.4. Regression line, excess returns on S&P 500 versus excess returns on NCREIF Timberland Index, 1987 to 1999

1.5 Timber Investments and Option Pricing

Traditional capital budgeting approaches to assessing forest resource management alternatives ignore the importance of managerial flexibility. Yet forest managers' options to abandon, delay, expand, or contract certain practices in response to changing operating or market conditions have value. Failure to recognize this value can lead to nonoptimal land use conversion, poor forest rotation decisions, and other problems. Therefore, investors should consider integrating option pricing that recognizes the value of managerial flexibility into the capital budgeting framework. This discussion is based on Zinkhan (1996) and is paraphrased here.

The term option means more than a decision alternative. Options, whether financial or real, derive their value from the value of some underlying asset, such as common stock or timber. They represent the right to do something, such as to purchase stock at a fixed price, rather than an obligation to do something. As a right rather than an obligation, an option will never have a value less than zero. Traded contracts, such as stock-based call and put options, are the primary form of option that financial economists investigate. A call (put) option on an asset provides the buyer with the right to buy (sell) the asset at the exercise price prior to expiration of the option contract.

Economists have been moderately successful in applying traded options valuation principles to such real options as the right to expand one's manufacturing plant in the future. Examples of strategic options embedded within real forest assets are the flexibility of harvest timing, the option to abandon a silvicultural investment prior to completion, and the option to convert from forestry to alternative land uses. Financial economists have developed models such as the Black-Scholes option pricing model (Black and Scholes 1973) and the binomial option pricing model (Rendleman and Bartter 1979) to value traded options. Traditional discounted cash flow models cannot be used to value true call or put options, since the cost of capital is elusive given that the risk of the underlying asset (such as the stock acquired when the option is exercised) is constantly changing (Brealey and Myers 1991). With the risk automatically changing as the option's expiration date is approached, the CAPM cannot be used to estimate the option's cost of capital.

To value an option on an asset using the binomial option pricing model, the analyst assumes that the asset price follows a binomial generating process. For an option on an asset not paying an interim dividend of some sort, the analyst must input the following five variables into the model: the underlying asset's current price, the exercise price, the maturity of the option, the risk-free rate of interest, and the asset return volatility. For a call option, there is a positive relationship between all but one of these factors and the magnitude of the option value. Only the exercise price has an inverse relationship with call option value. With respect to the volatility of the underlying asset's returns, greater expected volatility leads to higher upside prices upon exercise of the option. Although greater expected volatility also increases the possibility that the asset price will plunge far below the exercise price, the option buyer is comforted by the fact that he can opt not to exercise the option under such a condition.

Zinkhan (1991) applied the Black-Scholes option pricing model to the valuation of cutover southern pine timberland. The Black-Scholes option pricing model is a special case of the binomial model where the number of intervals in which there are two possible changes in asset pricing approaches infinity during a given time period. The subject land had potential for conversion to farmland. The base assumptions yielded a land-use conversion option value that was 16.1% of the land expectation value for timber production. When the expected level of the volatility of the change in the gap between timberland and farmland prices was reduced from the base-case level of 41.5% to 20.8% in the analysis, the value of the option fell from 16.1% to 7.6% of the land expectation value for timber production. The valuation of such embedded options represents an additional stage of the capital budgeting process.

Contractual options in forestry are rather common. They include long-term cutting contracts, buyout options in landowner assistance programs, options to buy timberland properties, and development rights sold by timberland owners to developers. These options too can be valued with option pricing models. With more sophisticated investors now participating in timberland markets, the need for assessing and valuing financial contracting opportunities with option pricing models is greater. In the next section, new types of participants in timberland markets are discussed.

2. EVOLUTION OF TIMBERLAND AS AN ASSET CLASS

The preceding discussion summarized the traditional means for analyzing timberland investments and compared those with approaches from modern financial theory. These analytical approaches and applications have helped facilitate large-scale investments by new classes of timberland owners in the United States in the last two decades. The balance of this chapter describes how the new investor classes are developing and how continuous advances in financial analyses being applied to forestry investments contribute to this trend.

Timberland is now recognized as a legitimate investment alternative by many institutional investors. Starting in 1981 with the launch of the first pooled timberland fund for institutional investors, such entities have now allocated almost $8 billion to this asset class. Institutional investors are organizations with fiduciary obligations, including such entities as pension funds, universities, foundations, and trusts. With this level of interest having been revealed by sophisticated investors, approaches for assessing and valuing timberland now often mimic processes used in financial markets. Investment managers expect the TIMOs that develop and manage timberland funds for institutional investors to provide analysis similar to that generated by managers of common stock, bonds, and commercial real estate. Also, investors have started to demand investment vehicles that are similar in structure to those observed in conjunction with more established asset classes.

With the enactment of the Employee Retirement Income Security Act (ERISA) by the U.S. Congress in 1974 and the spread of investment practice consistent with MPT, pension funds and other institutional investors started to allocate capital to asset classes other than publicly traded common stocks and bonds. Alternative assets considered by institutional investors included, among others, commercial real estate and private equity. ERISA required corporate pension funds to diversify their investments in order to reduce the chance of large losses. MPT, with the concepts developed by Markowitz (1952) as the centerpiece, encouraged investment managers to include assets with return series poorly correlated with the core of financial assets. Initially viewed as a subset of the real estate asset class, timberland was selected for its diversification potential by a number of institutions in the early 1980s.

Diversification was not the only criterion considered by institutional investors when selecting timberland. Relative to other alternative investments such as cropland and managed commodities, timberland offers relatively stable returns. With biological growth and ingrowth generating a majority of the expected returns on most timberland investment vehicles, return volatility tends to be modest. Although *expected* returns tend to be less than that achieved by U.S. equities, they generally exceed expected returns on higher quality bonds. Expected returns tend to be especially high when assessed relative to the modest volatility of timberland returns. Also, when performing due diligence on timberland, many of the pioneering institutional investors were intrigued with the opportunity to purchase assets in markets that are less informationally efficient than developed securities markets. A market is relatively inefficient when information about an asset is not rapidly distributed to market participants and then reflected in its price. Less than perfect informational efficiency is a necessary condition for finding investments trading for a price less than net present value. Obviously, the market for a 50,000-acre timberland tract is not as deep as the market for an S&P 500 common stock, leading to a less efficient market.

Simultaneously with the initial interest expressed by institutional investors in timberland, some forest products companies started to consider divestitures of their forest resources. Rinehart (1985) noted that industry could use the freed-up capital to finance the expansion of processing facilities. This symbiotic relationship between institutional investors and forest products companies led to a well-documented trend toward separation of timberland ownership and timber processing (Yin et al. 1998). At about the same time that some institutional investors commenced significant capital commitments to timberland, several forest products companies, including International Paper Company, established captive master limited partnerships (MLPs) to hold their forest assets. Stated motivations for the creation of these publicly traded limited partnerships—categorized as a form of securitization in which a rather illiquid real asset is converted into a traded security—ranged from defensive maneuvering against hostile takeovers to providing Wall Street with the opportunity to directly, and more fully, value timberlands (Zinkhan 1988a). Although financial markets reacted positively to the announcements of this sort of restructuring, the captive MLPs—and such variations as private letter stock—proved to be only an interim measure.

After first concentrating on the disposal of nonstrategic forest assets and the creation of captive timberland-owning entities, forest products companies later turned toward the disposition of some of their core holdings. Many of these large-scale sales were to TIMOs and were funded by institutional investors. The growth of assets under management by TIMOs grew from less than $1 billion in 1990 to about $8 billion in 2001. Institutional investors own less than 4% of the value of domestic privately owned timberland. Although this institutional resource is dwarfed by the holdings of industrial and nonindustrial landowners, we expect the institutional sector to continue to grow and dominate transactions. With respect to transactions, an estimated 16.2 million acres of industrial timberland, or about 23% of industry's land base in the United States, was sold or included in some form of securitization during the period 1996 to 1998 at an average price per acre of $700. Due to the structure of some of the transactions, a portion of the securitized acreage effectively remains under the control of the originating forest products company.

Some of the TIMOs have allocated a portion of the funds provided by institutional investors to timberland outside of the United States. Early investments were made in New Zealand and Chile, with special interest in fast-growing radiata pine plantations. More recently, investors have shown interest in diversifying into other regions and forest types. Sellers of timberland seem to recognize the need for a premium return when political risk is present. With markets for timber often limited, many investments are structured as joint ventures involving a partner experienced in timber processing and forest products marketing.

The TIMOs generally manage their institutional assets in either separate accounts or pooled funds. A separate account holds the timberland of one major investor in a single portfolio. To achieve some degree of diversification, a capital commitment of $25 to $50 million is generally required for the initiation of a separate account. In contrast, a pooled fund collects capital from numerous entities and allocates it to a portfolio of properties. A pooled fund offers investors committing at least $1 to $3 million the opportunity to participate in a rather large, diversified portfolio of timberland. In this fashion TIMOs serve an important economic function: denomination intermediation. Whereas the pooled fund structure provides an investor with greater diversification than otherwise, the separate account can offer greater control to the institutional investor and lower management fees. With respect to investment duration, the investor tends to have greater discretion with separate accounts than pooled funds. Pooled funds usually have base lifetimes of 10 years, extendable under certain conditions by the TIMO to 15 or 20 years.

Often, industry seeks to maintain access to the timber on the sold property for at least a limited duration. This is typically structured through a timber supply arrangement between the buyer and seller, in which timber is sold at the floating fair market value over time. More complex structures have been created that enable the forest products company to just partially monetize their timberland assets. With a joint venture structure, the forest products company contributes timberland, and the TIMO contributes capital. The forest products company either redirects this capital for internal purposes, or it is used by the joint venture to expand its acreage under management. Typically, the forest products company contracts out its forestry personnel to the joint venture for a fee. International Paper Company recently developed an interesting new vehicle to partially monetize its forest assets. That company has sold packages including the standing timber component of some of its intermediate-aged, Southern U.S. pine plantations to TIMOs. International Paper retains the land, stipulates the maximum period over which the buyer can conduct harvesting operations, regenerates the forests, and even offers forest management services.

The securitization of timberland has been a popular topic at forest economics meetings (e.g., Zinkhan and Jenkins 1998; Caulfield and Flick 1999). Examples of securitization are the previously mentioned timberland MLPs and the creation of letter stock by Georgia-Pacific Corporation's Timber Company. More recently, Plum Creek Timber Company converted its MLP structure into the first publicly traded timberland real estate investment trust (REIT). Without an income tax being applied at the MLP or REIT levels, both entities are income-tax efficient for taxable investors such as individuals. However, tax-exempt institutional investors generally avoid MLPs because of unique tax issues. Because REITs do not pose such difficulties to tax-exempt investors, institutional investors can effectively acquire them. Thus, a timberland REIT can potentially attract a deep market for its shares, involving both institutional and retail buyers. In addition to the greater liquidity offered by publicly traded securities, securitization provides the individual investor with the opportunity to participate in timberland with a very modest capital allocation. However, one significant challenge to the expansion of securitization of timberland markets is the apparent steep short-term cost of capital required by participants in the REIT markets. Specifically, REIT investors generally require relatively high and consistent dividend yields. This requirement is not necessarily consistent with maximizing the value of a well-recognized harvest-timing option for landowners: allowing timber to grow on the stump when prices are poor.

Timberland is now a popular investment alternative that competes against other real as well as financial assets. One reason for its success in attracting new sources of capital is its performance relative to other investment alternatives in a CAPM context. This success in attracting capital could lead to frictions between timberland owners and timberland investment managers.

Separation of timberland ownership and management results in potential conflicts of interest. Financial economists developed agency theory (Jensen and Meckling 1976) to analyze these conflicts of interest. In agency theory, timberland owners are the principals and the managers (TIMOs or forestry consulting firms) are their agents. To minimize conflicts of interest, timberland owners should consider compensation packages that reward the managers for behaving in ways that improve the economic welfare of the owners. A system involving graduated incentive fees, which offer managers greater rates of compensation as returns to owners increase, is a common measure for minimizing the conflicts between timberland owners and TIMOs. To ensure an effective structure, the owner needs to both choose targets that will encourage long-term—and not just short-term—performance and offer potential incentives that are adequate to influence behavior. Of course, the owners should weigh the cost of extra compensation and monitoring costs against any anticipated, improved marginal returns.

3. CONCLUSIONS

In a global context, timberland assets can vary across such factors as the legal structure holding the investment, quality and intensity of management, species mix, age, site productivity, political risk, susceptibility to fire and other perils, and location relative to timber markets. Thus, the implications of research findings regarding the investment performance of timberland need to be considered in this context.

Institutional investors generally recognize the portfolio diversification benefits of timberland. However, they have expressed a need for further pragmatic research relating to the following three topics:

- Accurate historical measures of losses from natural perils so that a greater variety of insurance products can be developed.
- Analysis of the relationship of timberland returns to those generated by other alternative assets such as private equity, oil and gas, managed commodities, and hedge funds.
- The assessment of timberland returns relative to pricing factors other than the market risk of the CAPM. Portfolio managers often construct portfolios with specific sensitivities to such factors as inflation, interest rate risk, and fluctuations in aggregate economic activity.

Finally, investors are curious about the relative degree to which forest managers can influence performance through various actions. Both simulations and case study methodologies can be used to further explore the sensitivity of financial performance to managerial inputs.

4. LITERATURE CITED

BLACK, F. AND M. SCHOLES. 1973. The pricing of options and corporate liabilities. J. Polit. Econ. 81:637-654.
BREALEY, R.A., AND S.C. MYERS. 1991. Principles of Corporate Finance. McGraw-Hill, New York.
CAULFIELD, J.P. 1998. A fund-based timberland investment performance measure and implications for asset allocation. So. J. of Appl. For. 22:143-147.
CAULFIELD, J.P., AND W.A. FLICK. 1999. Prospects and challenges with securitized timberland. in Proceedings of the 1999 Southern Forest Economics Workshop, I.A. Munn et al., (eds.). Department of Forestry, Mississippi State University, Starkville, MS.
CLUTTER, J.L., J.C. FORTSON, L.V. PIENAAR, G.H. BRISTER, AND R.L. BAILEY. 1983. Timber Management: A Quantitative Approach. McGraw-Hill, New York.
CONROY, R., AND M. MILES. 1989. Commercial forestland in the pension portfolio: The biological beta. Fin. Analysts J. 45:46-54.
COPELAND, T.E., AND J.F. WESTON. 1988. Financial Theory and Corporate Policy. Addison-Wesley Publishing Co, New York.
CUBBAGE, F.W., AND C.H. REDMOND. 1985. Capital budgeting practices in the forest products industry. For. Prod. J. 35:55-60.
DAVIS, L.S., AND K.N. JOHNSON. 1987. Forest Management. McGraw-Hill Book Co., New York.
DEFOREST, C.E., F.W. CUBBAGE, C.H. REDMOND, AND T.G. HARRIS. 1991. Hedging with trees: Timber assets and portfolio performance. For. Prod. J. 41:23-30.
ELTON, E.J., AND M.J. GRUBER. 1995. Modern Portfolio Theory and Investment Analysis. John Wiley and Sons, New York.
FAMA, E., AND F. FRENCH. 1992. The cross-section of expected stocks returns. J. Fin. 47:427-466.
FORTSON, J.C. 1986. Factors affecting the discount rate for forestry investment. For. Prod. J. 36:67-72.
HAUGEN, R.A. 1987. Modern Investment Theory. Prentice-Hall, Englewood Cliffs, NJ.
JAGANNATHAN, R., AND E.R. MCGRATTAN. 1995. The CAPM debate. Federal Reserve Bank of Minneapolis Quarterly Review 19:2-17.
JENSEN, M. 1969. Risk, the pricing of capital assets and the evaluation of investment portfolios. J. Bus. 42:167-247.
JENSEN, M., AND W.H. MECKLING. 1976. Theory of the firm: Managerial behavior, agency costs and ownership structure. J. Fin. Econ. 3:305-360.
MARKOWITZ, H.M. 1952. Portfolio selection. J. Fin. 7:77-91.
MILLS, W.L. 1988. Forestland: Investment attributes and diversification potential. J. For. 86:19-24.
REDMOND, C.H., AND F.W. CUBBAGE. 1988. Portfolio risk and returns from timber asset investments. Land Econ. 64:325-337.
RENDLEMAN, R., AND B. BARTTER. 1979. Two-state option pricing. J. Fin. 34:1093-1110.
RINEHART, J.A. 1985. Institutional investment in U.S. timberlands. For. Prod. J. 35:13-18.
SHARPE, W.F. 1964. Capital asset prices: A theory of market equilibrium under conditions of risk. J. Fin. 14:425-442.

WAGNER, J.E., F.W. CUBBAGE, AND C.H. REDMOND. 1995. Comparing the capital asset pricing model and capital budgeting techniques to analyze timber investments. For. Prod. J. 45:69-77.

WAGNER, J.E. AND D.B. RIDEOUT. 1991. Evaluating forest management investments: The capital asset pricing model and the income growth model. For. Sci. 37:1591-1604.

WASHBURN, C.L., AND C.S. BINKLEY. 1990. On the use of period-average stumpage prices to estimate forest asset pricing models. Land Econ. 66:379-393.

YIN, R., J. CAULFIELD, M.E. ARONOW, AND T.G. HARRIS. 1998. Industrial timberland: Current situation, holding rationale, and future development. For. Prod. J. 48:43-48.

ZINKHAN, F.C. 1988(a). The stock market's reaction to timberland ownership restructuring announcements: a note. For. Sci. 34:815-819.

ZINKHAN, F.C. 1988(b). Forestry projects, modern portfolio theory, and discount rate selection. So. J. Appl. For. 12:132-135.

ZINKHAN, F.C. 1990. Timberland as an asset for institutional portfolios. Real Est. Rev. 19:69-74.

ZINKHAN, F.C. 1991. Option pricing and timberland's land-use conversion option. Land Econ. 67:317-325.

ZINKHAN, F.C. 1996. Economic value added and forest management. The Compiler 14:14-16.

ZINKHAN, F.C., AND H.R. JENKINS. 1998. Much ado about REITs. *in* Proceedings of the Southern Forest Economics Workshop, K.L Abt and R.C. Abt, (eds.). North Carolina State University, Raleigh,NC.

ZINKHAN, F.C., W.R. SIZEMORE, G.H. MASON, AND T.J. EBNER. 1992. Timberland Investments. Timber Press, Portland, OR.

---

[1] The discount rate i is a discrete annual rate; that is, it applies once a year. The equivalent continuous discount rate, r, as used in chapter 4, is ln (1+i). Most forestry projects and investments are calculated initially with a before-tax, real (without inflation) discount rate. Effects of inflation or taxes can be important and are added as necessary.

Chapter 7

Timber Production Efficiency Analysis

Douglas R. Carter and Jacek P. Siry
University of Florida and North Carolina State University

The term efficiency is widely used in economics to describe an optimal allocation of resources and is promoted as one logical economic goal for individuals, firms, and societies. In the simplest sense, efficiency means making the most out of what you have—minimizing waste of scarce resources. In this chapter we describe how one might go about thinking of both technical and economic, or productive, efficiency in a theoretical framework, as well as provide an overview of empirical techniques used to estimate productive efficiency levels of individual firms. The technical efficiency of southern U.S. logging firms is examined as a case study.

Why would anyone be interested in this sort of exercise? The answer lies in understanding the motivations, and limitations, of economic agents. For example, firms or other decision-making units are, as assumed in several other chapters of this text, profit maximizers or cost minimizers (at selected output levels). Firms capable of attaining these goals over time are more likely to survive in a competitive market, while others may fall by the wayside. Central to the notion of profit maximization is the selection by the firm of optimal levels of both inputs and outputs, given an existing production technology. Also central is the ability of the firm to transform those same inputs into outputs in the most efficient manner possible.

The fact of the matter is that not all firms or other decision-making units are capable of meeting these goals equally well. The reasons for this may contain valuable information. At the micro level, we may be able to determine which factors are keeping some firms from achieving their stated objectives in relation to their competitors. Or, from a broader perspective, one may conclude that scarce resources are simply being wasted—an overall efficient use of resources is not being achieved, which is arguably suboptimal from a societal point of view. We may also be interested in determining whether or not government regulation or control of resources is less or more efficient than that attainable in the private sector. Production efficiency analysis provides us with a set of tools for estimating efficiency differentials in these and other scenarios.

A simple illustrative example of the basic idea is given in figure 7.1. Standard production economics analysis begins with the production function $y=f(x)$ where y is output and x is input. Empirical estimation of the production function is accomplished by gathering data on observed input-output relationships. One method for estimating the function is to use standard regression techniques to estimate $\overline{f(x)}$ (see chapter 5). However, this is an average production function by construction. Some firms are observed doing better than $\overline{f(x)}$, and some less. Consequently, $\overline{f(x)}$ does not represent the best-practice technology, since some firms are observed doing better than the average. By contrast, $f(x)'$ could be considered to represent the true production function. Indeed, Chambers (1988) notes that the definition of the production function "...specifically excludes[s] the possibility of technical inefficiency." Technicalities aside, $f(x)'$ is commonly referred to as the frontier production function.

Production efficiency analysis has grown prolifically over the past 25 years. Efficiency studies have been performed in almost every conceivable economic sector, from agriculture to banking, from transportation to utilities. Studies have been conducted at all levels, from the individual firm to the country. Those interested in learning more will find that excellent references have been published recently, including Coelli et al. (1998), Färe et al. (1994), Fried et al. (1993), and Kumbhakar and Lovell (2000).

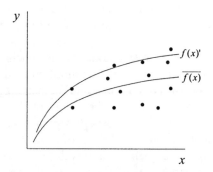

Figure 7.1. Average and frontier production functions

1. DEFINING THE PRODUCTION TECHNOLOGY AND TECHNICAL EFFICIENCY

We begin our formal treatment by assuming that production takes on the following characteristics. Producers use a purely positive N vector of inputs denoted $x = (x_1,...,x_N) \in R_{++}^N$ to produce a purely positive M vector of outputs denoted $y = (y_1,...,y_M) \in R_{++}^M$.

All feasible input-output vector relationships are described by the graph of the production technology or production possibilities set, PT, such that

$$PT = \{(y,x): x \text{ can produce } y\} \qquad 7.1$$

Two concepts of technical efficiency are represented in figure 7.2. Point a represents a case where production is technically inefficient for the one input, one output case from either an *input-reduction* or *output-augmentation* point of view. From the input-reduction viewpoint, too much input x' is being used to produce y'. Indeed, y' could be produced efficiently with x^*. Alternatively, the output-augmentation viewpoint is that y^* could be produced with x', rather than the currently produced y'.

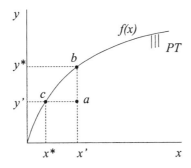

Figure 7.2. Input-reduction and output-augmentation

With multiple inputs and outputs, we define input and output sets and input and output isoquants. The input sets of the production technology, which describe the sets of all input vectors that are feasible for each output vector y, are defined as

$$L(y) = \{x : (y,x) \in PT\} \qquad 7.2$$

Alternatively, the output sets of the production technology describe the sets of all output vectors that are feasible for each input vector x and are defined as

$$P(x) = \{y : (y,x) \in PT\} \qquad 7.3$$

Only a subset of each input or output set would be recognized as efficient, given the above definitions.

A critical category of efficient vector relationships is defined via input and output isoquants. An input isoquant

$$\text{Isoq } L(y) = \{x : (y,x) \in PT, \lambda x \notin L(Y), \lambda < 1\} \qquad 7.4$$

gives the set of input vectors capable of producing output vector y. However, if the input vector x is radially contracted by λ, a scalar, then that input vector is no longer capable of producing output vector y. Thus, the isoquant represents a vector that is radially efficient. Similarly, an output isoquant

$$\text{Isoq } P(x) = \{y : (y,x) \in PT, \theta \ y \notin P(x), \theta > 1\} \qquad 7.5$$

gives the set of radially efficient output vectors that can be produced with each input vector x. Figure 7.3 illustrates this concept using the input isoquant, while figure 7.4 does the same using the output isoquant.

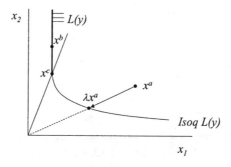

Figure 7.3. Input-oriented technical efficiency using input isoquants

In figure 7.3, the input vector x^a is currently being used to produce y. However, a more efficient use of inputs is possible at position λx^a, i.e., an equiproportionate reduction of inputs by λ. Thus, we say that x^a is input-oriented technically efficient by λ. For a single output, it follows that the definition for input-oriented technical efficiency is given by the function

Timber Production Efficiency Analysis

$$TE_I(y,x) = \min\{\lambda : y \leq f(\lambda x)\}, \qquad 0 < TE_I(y,x) \leq 1 \qquad 7.6$$

while an output-oriented technical efficiency function is given by (figure 7.5)

$$TE_O(y,x) = [\max\{\theta : \theta y \leq f(x)\}]^{-1}, \qquad 0 < TE_O(y,x) \leq 1 \qquad 7.7$$

Distance functions, introduced by Shephard (1953, 1970) are further used to describe multi-input, multi-output technical efficiency relationships.[1]

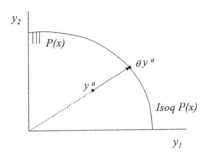

Figure 7.4. Output-oriented technical efficiency using output isoquants

The equiproportionate reduction in inputs, or augmentation of outputs, concept of technical efficiency was introduced by Debreu (1951) and Farrell (1957) and is widely used as the conceptual basis for efficiency measurement in empirical work. However, it does have some noticeable limitations. Take, for example, vector x^b in figure 7.3. Although this input set sits on the input isoquant, it is not very efficient in relation to vector x^c since it uses excess units of x_2. Also, there is no assurance that an equiproportionate reduction in inputs is the best path to a particular efficiency objective. Nonradial measures of efficiency may be warranted. Many nonradial measures have been examined in the literature, and most are linked to applications that make use of mathematical programming techniques. However, they are not without their own significant limitations.

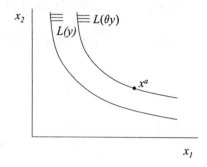

Figure 7.5. Output-oriented technical efficiency using input isoquants

2. THE MEASUREMENT AND DECOMPOSITION OF COST EFFICIENCY

Technical efficiency (in the sense of Debreu-Farrell) is a necessary, but not sufficient, condition for achieving cost minimization or profit maximization. In addition to technical efficiency, consideration must be given to input and output prices and their influence on optimal input-output allocations. In this section, we focus our efficiency analysis discussion on simple (single output) cost minimization. Extensions to revenue and profit maximization can be found in Kumbhakar and Lovell (2000).[2]

In addition to previously defined input and output vectors, we assume that firms face a vector of input prices $w = (w_1,...,w_N) \in R_{++}^N$. For a single output, the cost frontier is a function

$$c(y,w) = \min_x \{w^T x : y \leq f(x)\} \qquad 7.8$$

The cost frontier defines the minimum expenditure required to produce a single output at input prices w. The ratio of minimum to observed cost is termed cost efficiency and given by the function

$$CE(y,x,w) = c(y,w)/w^T x, \qquad 0 < CE(y,x,w) \leq 1 \qquad 7.9$$

Cost efficiency is made up of two components. The first relates to the firm's ability to make maximum use of inputs, which we have already termed technical efficiency, and the second relates to the firm's ability to use inputs in the correct proportion as indicated by their relative prices, or allocative efficiency. Figure 7.6 decomposes cost efficiency into these separate elements.

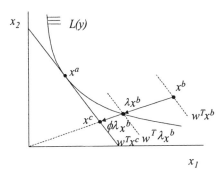

Figure 7.6. Measurement and decomposition of cost efficiency

Assume a firm is currently using x^b to produce y. The cost-minimizing use of inputs is found elsewhere, however, at x^a, with cost $w^T x^a$, where the firm not only produces on the isoquant but also uses the correct proportion of inputs as dictated by their relative prices w. We note that the allocation at x^c costs the same as it does at x^a, i.e., $w^T x^c = w^T x^a$. Thus, the overall level of cost inefficiency can be decomposed into two components, an input-oriented technical component, λ, and an allocative component, ϕ. The technical component is given as before, i.e.,

$$TE_I(y,x) = \min\{\lambda : y \leq f(\lambda x)\}, \qquad 0 < TE_I(y,x) \leq 1 \qquad 7.10$$

The allocative component is the ratio ϕ of the current cost of inputs at λx^b, which is equal to $w^T \lambda x^b$, to the least cost input allocation at x^c, which is equal to $w^T x^c$. Thus, allocative efficiency is the ratio of cost efficiency to input-oriented technical efficiency, or

$$AE_I(y,x,w) = CE(y,x,w)/TE_I(y,x) \qquad 0 < AE_I(y,x,w) \leq 1 \qquad 7.11$$

It follows that the only way to achieve maximum cost efficiency, i.e., $CE=1$, is for both technical and allocative efficiency to also be equal to one.

3. EMPIRICAL METHODS

Stochastic frontier analysis (SFA) and data envelopment analysis (DEA) are the two major techniques for productive efficiency measurement. The former is a parametric regression-based method, and the latter is a nonparametric mathematical programming method. While DEA is a bounding technique and does not need any modification for efficiency measurement, the classical regression model that estimates average values had to be modified for efficiency measurement purposes (figure 7.1).

SFA was proposed independently by Aigner et al. (1977), Battese and Corra (1977), and Meeusen and van den Broeck (1977). They modified the classical regression model by proposing a composed error specification

$$y_i = f(X_i;\beta)e^{\varepsilon_i} \qquad 7.12$$

where

$$\varepsilon_i = v_i - u_i \qquad 7.13$$

The symmetric component, normally distributed, $v_i \sim N(0,\sigma_v^2)$, represents any stochastic factors beyond the firm's control affecting the ability to produce on the frontier. The asymmetric component, $u_i \sim |N(0,\sigma_u^2)|$, $u_i \geq 0$, is in this case assumed to be distributed as a half-normal and can be interpreted as pure technical inefficiency. Technical efficiency in this case is the ratio of observed to maximum or frontier output

$$TE_i = e^{-u_i} = \frac{y_i}{f(X_i;\beta)e^{v_i}} \qquad 7.14$$

which is analogous to the concept of output-oriented technical efficiency described in section 1.

Assuming independence between u_i and v_i, the log likelihood function for the composed error model can be given as

$$\ln \ell(y \mid \beta,\gamma,\sigma^2) = \frac{N}{2}\ln(2/\pi)$$
$$- N\ln\sigma + \sum_{i=1}^{N}\ln\left(1 - \Phi\left(\frac{\varepsilon_i \gamma}{\sigma}\right)\right) - \frac{1}{2\sigma^2}\sum_{i=1}^{N}\varepsilon_i^2 \qquad 7.15$$

where

(i) $\varepsilon_i = y_i - f(X_i; \beta)$
(ii) $\sigma^2 = \sigma_u^2 + \sigma_v^2$ 7.16
(iii) $\gamma = \sigma_u / \sigma_v$

and $\Phi(\bullet)$ is the cumulative distribution function of a standard normal random variable. Solutions for the consistent and asymptotically efficient maximum likelihood estimators can be found by solving for the parameter vector $(\beta, \sigma^2, \gamma)$ that maximizes the log likelihood function.

Two significant calculations can be further derived from the parameter estimates. The average or mean technical efficiency for the data sample is given by (Lee and Tyler 1978)

$$\overline{TE} = E(e^{-u}) = 2e^{\sigma_u^2/2}(1-\Phi(\sigma_u)), \qquad 0 \le \overline{TE} \le 1 \qquad 7.17$$

while the observation- or firm-specific technical efficiency can be calculated as (Jondrow 1982)

$$TE_i = e^{-\left[\frac{\sigma_u \sigma_v}{\sigma}\left(\frac{\phi(\varepsilon_i \gamma/\sigma)}{1-\Phi(\varepsilon_i \gamma/\sigma)} - \left(\frac{\varepsilon_i \gamma}{\sigma}\right)\right)\right]}, \qquad 0 \le TE_i \le 1 \qquad 7.18$$

where $\phi(\bullet)$ is the probability distribution function of a standard normal random variable.

SFA is a parametric estimation technique. Therefore, the choice of functional form for the regression has important consequences because misspecification will result in biased estimates. Two popular choices include the Cobb-Douglas and translog functional forms. The first, while simpler to estimate, is a rather inflexible and restrictive formulation. For example, the Cobb-Douglas production function imposes fixed input elasticities and returns to scale as well as unitary elasticities of substitution. While the translog functional form relaxes these restrictions, it frequently does not satisfy theoretical properties and is more prone to multicollinearity and degrees of freedom problems.

In contrast to SFA, DEA uses linear programming methods to develop a nonparametric piecewise frontier enveloping the data. DEA calculates a maximal performance index for each firm relative to all other firms in the sample, with the sole requirement that each firm lie on or below the frontier. Firms that lie on the frontier represent the best production practices, and thus, by determination, are efficient. For those that lie below the frontier,

DEA can identify the sources and levels of efficiency relative to efficient firms.

The objective is to measure the performance of each firm relative to the best practice in the sample of n firms. This is done by developing weights for each firm's inputs and outputs that solve the following problem (Coelli et al. 1998)

$$\max_{u,v} \frac{u' y_i}{v' x_i}$$

s.t.

$$\frac{u' y_j}{v' x_j} \leq 1, \quad j = 1, 2, \ldots, n$$

$$u, v \geq 0$$

7.19

where (x_i, y_i) is the input-output vector of the firm being evaluated and (x_j, y_j) is the input-output vector of the jth firm in the sample. The solution to this problem is a set of weights that maximizes the weighted output-to-input ratio for the firm being evaluated subject to the constraint that no firm has a ratio larger than unity. This nonlinear problem, which has an infinite number of solutions, can be transformed into the linear programming multiplier problem by imposing the constraint $v'x_i=1$, which yields

$$\max_{\mu, \upsilon} \mu' y_i$$

s.t.

$$\upsilon' x_i = 1$$

$$\mu y_j - \upsilon x_j \leq 0 \quad j = 1, 2, \ldots, n$$

$$\mu, \upsilon \geq 0$$

7.20

The model was originally proposed by Charnes et al. (1978) and was named the CCR model after them. This model has an output-augmenting orientation and assumes constant returns to scale.

Alternatively, one can also solve a dual linear envelopment program

$$\min_{\theta, \omega} \theta$$

s.t.

$$-y_i + Y\omega \geq 0$$

$$\theta x_i - X\omega \geq 0$$

$$\omega \geq 0$$

7.21

where θ is a scalar, ω is an $n \times 1$ vector of constants, X is the input set and Y is the output set. The program involves fewer constraints and is therefore easier to solve. The program radially contracts the firm's input vector while remaining within the feasible input set. The value of θ represents the technical efficiency score for the firm. If radial input contraction is not possible, then $\theta = 1$, and the firm is technically efficient. Otherwise, radial contraction is possible, and the firm is not efficient.

The CCR model is appropriate for the firms that operate at an optimal scale. Banker et al. (1984) extended the model to account for variable returns to scale. This was achieved by adding a convexity constraint to equation 7.21. The constraint takes the form $n'\omega = 1$, where n is a vector of ones. This model allows the estimation of technical efficiency measures free of scale effects, thus avoiding problems encountered when applying the CCR model to firms that are not operating at an optimal scale.

The choice between SFA and DEA depends on available data and research objectives. SFA imposes a specific functional form and a specific error structure, and therefore accounts for noise and can be used to conduct hypothesis tests. In contrast, DEA imposes neither a specific functional form nor a distributional structure for the efficiency term. Even though measurement error, noise, and outliers may influence the shape and position of every type of frontier, their effect on a DEA frontier is more pronounced. These differences in data, behavioral assumptions, and estimation methods determine the relative strengths and weaknesses of the two approaches.

Once performance is measured, sources of performance differences can be identified in order to develop management and policy prescriptions for improving performance. Performance differences may be caused by factors such as geographic location, demographic and social conditions, ownership, market structure, uncertainty, regulatory policies, managerial experience, training, and environmental conditions and regulations. Environmental variables play an important role in forest operations. These variables include (1) natural production conditions such as average temperature, rainfall, and soils; (2) forest resource conditions such as growing stock, age structure, and species composition; and (3) policy conditions such as conservation and protection measures. Once potential factors that may affect performance are identified, they can be incorporated into a second-stage regression to determine the sources of performance differences. The results of this analysis can help make and adjust policies and managerial decisions. For more on methods used to determine the sources of performance differences see, for example, Coelli et al. (1998) and Lovell (1993).

While there are relatively few applications of efficiency measurement in the forestry literature, their number has increased in recent years. The earliest efficiency measurement study in forestry was by Rhodes (1986), who applied DEA to investigate the performance of U.S. national parks. The analysis was designed to compare the overall performance of different parks and better meet challenges such as budget cuts, environmental threats, deteriorating facilities, and expansion of responsibilities. The study found that several parks were consistently inefficient.

Kao led a number of studies that employed DEA to evaluate the management efficiency of Taiwan's forest districts (Kao 2000; Kao and Yang 1991, 1992; Kao et al. 1993). The measurement was carried out for 13 forest districts from 1978 to 1992 and evaluated alternative approaches to their consolidation into a few larger districts, which can benefit from scale economies. These studies measured technical forest management efficiency and decomposed it into pure technical and scale components. The forests were assumed to provide multiple outputs, such as timber, recreation, and soil conservation. Average growing stock was used as a proxy for soil conservation, based on the assumption that higher average growing stock level implies less soil erosion. The forests produced these outputs using financial capital (budget), initial stocking, labor, and land. The studies found a small improvement in performance following the consolidation of the forest districts. Likewise, Viitala and Hänninen (1998) used DEA to assess efficiency levels in 19 state-funded, regional forestry boards in Finland. These authors use DEA because of its ability to accommodate multiple outputs even without price information (a typical situation among public forestry organizations), less restrictive assumptions, and easy decomposition of efficiency scores into pure technical and scale components.

In comparison, Siry and Newman (2001) used SFA to examine the technical efficiency of forest districts in Poland. The study focused on how the transition toward a market economy impacts the performance of public forest management. Operating in an increasingly competitive market environment, the forest districts initiated reforms to improve their economic performance. These reforms focused on management decentralization, reduction of labor costs, creation of smaller forest districts, sale of state-owned equipment, and privatization of management operations (contracting private enterprises to perform logging, thinning, and other forest management operations). A stochastic frontier production function was estimated to evaluate the efficiency of timber production and management policies. The function relates timber produced in forest districts to inputs such as land, growing stock, employees, roads, personnel vehicles, and logging equipment. The empirical results provide evidence of substantial technical inefficiency, which has not been effectively addressed by creating smaller districts or reducing the workforce. The results provide strong support for the continued privatization of forest operations. In another SFA study, Grebner and Amacher (2000) analyzed the impacts of policy shifts and privatization on the efficiency of timber production in New Zealand.

They found that cost efficiency decreased in the short run as the economy adjusted to the changes.

SFA and DEA have also been applied to efficiency and/or technical change measurement in the private forest industry sector. Using SFA, Carter and Cubbage (1994, 1995) estimated the technical efficiency and technical change of the logging industry in the U.S. South. Brännlund et al. (1995) and Hetemäki (1996) examined the impact of environmental regulations on the profitability of Swedish and Finnish paper mills, respectively. While the former uses DEA, the latter is based on SFA. Similar studies of the effects of pollution control on the performance of U.S. pulp and paper producers include Pittman (1981, 1983) and Färe et al. (1989, 1993). Even though some of these works may not employ frontier analysis techniques, their common finding is that pollution control and environmental regulation do have a significant impact on the performance of pulp and paper producers. In addition, Yin (1998, 2000) employed DEA to measure the cost efficiency of linerboard producers in North America and further decomposed cost efficiency scores into their technical and allocative components.

These studies indicate that efficiency measurement is highly data-dependent, and common forestry concerns such as long production periods still apply. If these concerns can be satisfactorily addressed, then efficiency measurement provides information about the direction of change as well as useful managerial inputs. Efficiency measurement helps to identify technological comparisons, key constraints, and means of increasing efficiency. Measurement techniques accommodate multiple outputs and can be used for the evaluation of nonmonetary outputs.

4. TECHNICAL EFFICIENCY IN SOUTHERN U.S. PULPWOOD HARVESTING

The southern U.S. is one of the most important timber-producing regions in the world. The region supplies 17% of global industrial roundwood production. Much of the roundwood harvested in the South is used in the production of pulp and paper products. The economic importance of the logging industry is not inconsequential—approximately half of the cost of delivered wood to the mill is due to harvesting and transportation, the other half arising from the cost of stumpage. Reducing costs through improved efficiency in this sector can therefore have a significant impact on the overall profitability of forestry in the South.

The southern pulpwood logging industry has experienced substantial structural shifts in the past (1979 to 1987). Due to massive capital-labor substitution, the number of employees decreased 30%, while capital assets increased 12%. At the same time, average production per harvesting crew increased from 66 to 154 cords per week, a 133% increase. Because total production in the South did not increase at the same rate, industry consolidation resulted. The number of firms declined 52% over the period.

This evolution suggests that profit maximization was an important motivator and that associated technical efficiency would be important as well. In an effort to understand how technical efficiency and technical change may have influenced the evolution of this industry, these were estimated for the southern U.S. pulpwood logging industry over the period 1979 to 1987 using stochastic frontier production function analysis. Firm-specific characteristics were then used in an auxiliary regression to help explain observed technical efficiency differentials between individual firms.[3]

4.1 Methods and Data

Technical efficiency in southern pulpwood harvesting is modelled using the SFA methodology discussed above. In this study, we do not attempt to measure allocative efficiency, due to a lack of reliable price data. Our simple production function uses a translog specification and includes measures of capital (K) and labor (L), as well as a dummy variable for primary species harvested, either hardwood (S1) or mixed (S2). Softwood species (S3) is the base dummy. Our output variable (Y) is measured as average total weekly production in cords. Three regressions are estimated: one using only the 1979 data, another using only the 1987 data, and a pooled regression that includes a time variable (t) which measures Hicks neutral technical change on an average annual basis over the 8-year period.

Data for this analysis stem from two comprehensive American Pulpwood Association (APA) surveys of southern U.S. pulpwood producers in 1979 and 1987. Surveys gathered information on individual firm characteristics such as owner age, owner education, years in business, equipment type and age, number of employees and crews, weekly production or harvest, and type of harvest (i.e., pine, hardwood, or mixed). Capital stock is defined as the replacement value of assets. Thus depreciation of equipment over time is included. Labor is reported as the number of employees plus the owner. Further discussion of the survey is available in Cubbage and Carter (1994).

4.2 Results

Stochastic frontier regression results are given in table 7.1. It is difficult to glean anything meaningful directly from the regression estimates due to the translog model form. However, we can say that capital and labor inputs are significant in determining total production, and firms that harvested primarily hardwoods produced less than those that harvested softwoods. The estimate on t shows that technical change was 1.8% per year. Thus the frontier function was shifting outward over time. The stochastic nature of the function is demonstrated by the fact that around 10% of the observed input-output combinations actually lie above the estimated frontier functions.

Table 7.1. Results of stochastic frontier regressions

Variable	1979	1987	Pooled
Constant	2.688*	2.811*	3.128*
Ln K	-0.200*	-0.138	-0.466*
Ln L	1.050*	0.729*	1.033*
$.5(\ln K)^2$	0.214*	0.235*	0.298*
$.5(\ln L)^2$	0.037	0.184	0.058
Ln K ln L	-0.123*	-0.121	-0.136*
S1	-0.343*	-0.282	-0.355*
S2	0.021	-0.221*	-0.039*
t	—	—	0.018*
σ^2	0.666*	0.647*	0.658*
γ	2.343*	3.061*	2.307*
Pseudo R^2	0.74	0.75	0.73
η	0.846	0.904	0.842
TE	0.600	0.595	0.603
% residuals > 0	10.5	8.9	11.7

*indicates two-tailed statistical significance to the 0.05 level

With regard to technical efficiency, all regressions estimate average industry technical efficiency to be around 60%. This did not change significantly from 1979 to 1987. The frequency distributions of technical efficiency (table 7.2) indicate what percentage of firms fall within different efficiency categories. Table 7.2 indicates relatively smooth distributions. Most firms are between 60% and 70% technically efficient. The proportions that fall above 80% and below 40% are relatively low.

In an effort to explain technical efficiency differentials among logging firms, we regress technical efficiency scores multiplied times 100 (i.e., TE_i * 100) against various factors hypothesized to influence them. These included owner age, years in business or experience, number of crews, and total production per week. Squared terms for these factors account for nonlinear effects. Finally, a dummy variable accounts for technology choice: firms are categorized as either shortwood or longwood producers. Because of the substantial shift to longwood systems, we expected that shortwood systems were comparatively technologically inefficient.

Table 7.2. Frequency distributions for firm-specific technical efficiency

Ranges	1979	1987	Pooled
$0 \leq TE_i < .10$	.0043	.0045	.0038
$.10 \leq TE_i < .20$	.0262	.0349	.0242
$.20 \leq TE_i < .30$	.0418	.0541	.0381
$.30 \leq TE_i < .40$	.0789	.0590	.0856
$.40 \leq TE_i < .50$	.1215	.1527	.1329
$.50 \leq TE_i < .60$	.1852	.1715	.1807
$.60 \leq TE_i < .70$	.2297	.2105	.2230
$.70 \leq TE_i < .80$	.2231	.1783	.2118
$.80 \leq TE_i < .90$	.0845	.1300	.0973
$.90 \leq TE_i \leq 1.00$	.0047	.0045	.0025

Results of these regressions are found in table 7.3. The age of the owner (1979, 1987) and years of experience (1979 only) are important explanatory factors impacting the level of technical efficiency. This is evidence of differences in managerial ability among firms influenced by working experience and/or perhaps better decision-making skills over time.

The level of output is also important, suggesting that economies of scale influence efficiency levels. These effects were more pronounced in 1979 than 1987. The regression indicates that a 10-cord increase in total weekly production increases technical efficiency between 0.7% (1987) and 1.9% (1979). Holding output constant, firms relying on shortwood systems are less efficient than those that use longwood systems by between 9% (1987) and 16.7% (1979). Longwood systems represent a technological advance, and these regressions may explain the rapid adoption of longwood systems. Finally, owner-operators who have more than one crew face a reduction in efficiency for each additional crew used, between 2.5% (1987) and 3.1% (1979). These firms were less technically efficient at the crew level than those that could produce more output within a single crew. One explanation is diminished owner-operator control as the number of crews increases.

It is notable that the auxiliary regressions have low explanatory power. Only about 20% of the observed differences in technical efficiency are explained by them. It is likely that other factors are important in determining relative firm-efficiency levels. These might include the harvesting location, local regulations, or quotas. Our original frontier model specification also lacks adequate specificity for the natural resource input, which is likely to be highly variable from location to location. Still, these auxiliary regressions do indicate that statistically significant relationships exist between aspects of managerial ability and technical efficiency.

Table 7.3. Factors influencing technical efficiency

Variable	1979	1987
Constant	44.89*	-4.665*
Age	0.475*	2.779*
Age^2	-0.006*	-0.030*
Experience	0.229*	-0.124
Experience^2	-0.007*	0.004
Crews	-3.087*	-2.520*
Output	0.189*	0.066*
Output^2	-2E-04*	-2E-05*
Shortwood	-16.74*	-8.972*
R^2	0.19	0.23

* indicates two-tailed statistical significance to the 0.01 level

The average technical efficiency level of the industry did not change over the estimation period. One possible explanation for this constancy in observed mean efficiency levels is that firms were chasing a moving frontier, since technical change was meaningful over the period (1.8% per year). In such a situation, persistent technical inefficiency might be expected, whereas one might expect the average level of technical efficiency to rise over time in an environment where technical change is constant. New data on southern logging systems should allow this hypothesis to be tested.

Finally, because inefficiency impacts producer costs and profits, increasing the efficiency level of the industry can reduce the costs of pulpwood production. Increasing efficiency could reduce industry costs of production by up to one minus the mean level of technical efficiency (or 40%) if all firms produced on the technically efficient frontier. If allocative inefficiency (not measured) is also a substantive factor, then economic (cost) efficiency would be even less than the technical efficiency scores reported here (see section 2), and potential gains could be larger from more efficient operations. These savings can have the desired effect of making forestry more profitable, which has broad societal implications. Training programs aimed at increasing harvesting efficiency could thus yield positive returns both to loggers, in increased profitability, and in terms of societal welfare by enhancing the comparative advantage of forest land use.

5. LITERATURE CITED

AIGNER, D.J., C.A.K. LOVELL, and P. SCHMIDT. 1977. Formulation and estimation of stochastic frontier production function models. J. Econometrics 6: 21-37.

BANKER, R., A. CHARNES, and W. COOPER. 1984. Some models for estimating technical and scale efficiencies in data envelopment analysis. Manage. Sci. 30: 1078-1092.

BATTESE, G.E., and G.S. CORRA. 1977. Estimation of a production frontier model: With application to the pastoral zone of eastern Australia. Austra. J. Agri. Econ. 21: 169-179.

BRÄNNLUND, R., R. FÄRE, and S. GROSSKOPF. 1995. Environmental regulation and profitability: An application to Swedish pulp and paper mills. Environ. Res. Econ. 6: 23-26.

CARTER, D.R., and F.W. CUBBAGE. 1994. Technical efficiency and industry evolution in southern U.S. pulpwood harvesting. Can. J. For. Res. 24(2): 217-224.

CARTER, D.R., and F.W. CUBBAGE. 1995. Stochastic frontier estimation and sources of technical efficiency in southern timber harvesting. For. Sci. 41(3): 576-593.

CHAMBERS, R.G. 1988. Applied production analysis: A dual approach. Cambridge University Press. New York, NY. 331 p.

CHARNES, A., W. COOPER, and E. RHODES. 1978. Measuring the efficiency of decision making units. Eur. J. Oper. Res. 2: 429-444.

COELLI, T., D.S.P. RAO, and G.E. BATTESE. 1998. An introduction to efficiency and productivity analysis. Kluwer Academic. Boston, MA. 275 p.

CUBBAGE, F.W., and D.R. CARTER. 1994. Productivity and cost changes in southern pulpwood harvesting, 1979 to 1987. So. J. Appl. For. 18(2): 83-90.

DEBREU, G. 1951. The coefficient of resource utilization. Econometrica 19(3): 273-292.

FÄRE, R., S. GROSSKOPF, C.A.K. LOVELL, and C. PASURKA. 1989. Multilateral productivity comparisons when some outputs are undesirable: A nonparametric approach. Rev. Econ. Stat. 71: 90-98.

FÄRE, R., S. GROSSKOPF, C.A.K. LOVELL, and S. YAISAWARNG. 1993. Derivation of shadow prices for undesirable outputs: A distance function approach. Rev. Econ. Stat. 75: 374-380.

FÄRE, R., S. GROSSKOPF, and C.A.K. LOVELL. 1994. Production frontiers. Cambridge University Press, New York, NY. 296 p.

FARRELL, M.J. 1957. The measurement of production efficiency. J. Royal Stat. Soc. Series A 120: 257-281.

FRIED, H.O., C.A.K LOVELL, and S.S. SCHMIDT (Eds.). 1993. The measurement of productive efficiency: Techniques and applications. Oxford University Press. New York, NY. 426 p.

GREBNER, D.L., and G.S. AMACHER. 2000. The impacts of deregulation and privatization on cost efficiency in New Zealand's forest industry. For. Sci. 46(1): 40-51.

HETEMÄKI, L. 1996. Essays on the impact of pollution control on a firm: A distance function approach. Finnish Forest Research Institute. Res. Pap. 609. 166 p.

JONDROW, J., C.A.K. LOVELL, I.S. MATEROV, and P. SCHMIDT. 1982. On the estimation of technical inefficiency in the stochastic frontier production function model. J. Econometrics 19: 233-238.

KAO, C., and Y. YANG. 1991. Measuring the efficiency of forest management. For. Sci. 37(5): 1239-1252.

KAO, C., and Y. YANG. 1992. Reorganization of forest districts via efficiency measurement. Eur. J. Oper. Res. 58: 356-362.

KAO, C., P. CHANG, and S. HWANG. 1993. Data envelopment analysis in measuring the efficiency of forest management. J. Enviro. Manage. 38: 73-83.

KAO, C. 2000. Measuring the performance improvement of Taiwan forests after reorganization. For. Sci. 46(4): 577-584.

KUMBHAKAR, S.C., and C.A.K. LOVELL. 2000. Stochastic frontier analysis. Cambridge University Press. New York, NY. 333 p.

LEE, L., and W.G. TYLER. 1978. The stochastic frontier production function and average efficiency. J. Econometrics 7: 385-389.

LOVELL, C.A.K. 1993. Production frontiers and productive efficiency. P. 3-67 *in* The measurement of productive efficiency, H.O. Fried, C. A. K. Lovell, and S. S. Schmidt (eds.). Oxford University Press. New York, NY.

MEEUSEN, W., and J. VAN DEN BROECK. 1977. Efficiency estimation for Cobb-Douglas production functions with composed error. Int. Econ. Rev. 18: 435-444.

PITTMAN, R. 1981. Issues in pollution interplant cost differences and economies of scale. Land Econ. 57(1): 1-17.

PITTMAN, R. 1983. Multilateral productivity comparisons with undesirable outputs. Econ. J. 93: 883-891.

RHODES, E. 1986. An exploratory analysis of variations in performance among U.S. national parks. P. 47-71 *in* Measuring efficiency: An assessment of data envelopment analysis, R. Silkman (ed.). Jossey-Bass. San Francisco, CA.

SHEPHARD, R.W. 1953. Cost and production functions. Princeton University Press, Princeton, NJ. 104 p.

SHEPHARD, R.W. 1970. Theory of cost and production functions. Princeton University Press, Princeton, NJ. 308 p.

SIRY, J.P., and D.H. NEWMAN. 2001. A stochastic production frontier analysis of Polish state forests. For. Sci. 47(4): 526-533.

VIITALA, E., and H. HÄNNINEN. 1998. Measuring the efficiency of nonprofit forestry organizations. For. Sci. 44 (2): 298-307.

YIN, R. 1998. DEA: A new methodology for evaluating the performance of forest products producers. For. Prod. J. 48(1): 29-34.

YIN, R. 2000. Alternative measurements of productive efficiency in the global bleached softwood pulp sector. For. Sci. 46(4): 558-569.

---

[1] We do not examine this topic further here. Most of the functions described here—e.g., input and output sets, isoquants, technical efficiency functions, cost functions—must satisfy certain mathematical properties, which can be readily found in the referenced literature (see, for example, Kumbhakar and Lovell 2000, Lovell 1993).

[2] Issues regarding appropriate scale are best addressed using profit functions.

[3] This case study is based upon previous work found in Carter and Cubbage (1994), Cubbage and Carter (1994), and Carter and Cubbage (1995).

Chapter 8

Aggregate Timber Supply
From the Forest to the Market

David N. Wear and Subhrendu K. Pattanayak
USDA Forest Service and Research Triangle Institute

Timber supply modeling is a means of formalizing the production behavior of heterogeneous landowners managing a wide variety of forest types and vintages within a region. The critical challenge of timber supply modeling is constructing theoretically valid and empirically practical aggregate descriptions of harvest behavior. Understanding timber supply is essential for assessing tradeoffs between forest production and the environment, for forecasting timber market activity and timber prices, and for evaluating the level and distribution of costs and benefits of forest policies. It follows that timber supply modeling is an essential interface between forest production economics and policy and decision making. This chapter examines timber supply modeling, focusing especially on issues regarding aggregation of timber stocks (some of this chapter is based on Wear and Parks 1994). A section on general theory is followed by a discussion of various contemporary modeling approaches. The explicit aggregation of forest capital and description of capital structure in the analysis of timber supply remain as core research issues. We conclude with an empirical example that explores these topics.

1. THEORY

Timber supply models summarize the production behavior of forest managers in a market setting. Their conceptual foundation is the biological/physical production possibilities of timber growing and inventory adjustment, as well as information on the objectives of forest landowners.

Sills and Abt (eds.), Forests in a Market Economy, 117–132. ©Kluwer Academic Publishers. Printed in The Netherlands.

When sector-level timber supplies are to be examined, heterogeneous forest land and owners with heterogeneous objectives must be aggregated.

1.1 Timber Production Function

Underlying any economic study of production is a production function that translates inputs into outputs (see chapter 5). For timber supply, the inputs should include the age of the forest, (a) the level of forest management effort (E), and the quality of the land (q) (Binkley 1987). Merchantable timber volume per unit area (V) is given by the yield function:

$$V = v(a, E; q) \qquad 8.1$$

The marginal physical product of age and management effort is positive and decreasing in the relevant ranges of age and effort. Provided that the forest manager's objective function and discount rate can be specified, then the forest yield function can be used to define if and when a forest stand would be harvested. For example, consider a manager who faces prices (p) for timber and (w) for management effort (in this case, effort used to reforest the land after harvest). When the land is maintained indefinitely in forest use, the manager will maximize profit by selecting harvest ages (a) and levels of effort (E) to optimize:

$$\pi^F = \max\{a, E\} \sum_{j=0}^{\infty} \{pv(a, E; q)e^{-ra} - wE\}e^{-raj} \qquad 8.2$$

The optimum profit obtained, π^F, is the present net value for an infinite sequence of identical harvest ages. This formulation provides a valuation for forest land of quality (q) when there are no trees present at the beginning of the manager's planning horizon. The manager's problem can easily be modified to account for standing timber inventories; however, when profit from timber enterprise is the only argument in the objective function (cf. Hartman 1976), the solution for optimum age (a^*) is unaffected by the manager's starting inventory of timber. With this definition of profit, the manager recognizes that there is an opportunity cost to holding old trees rather than faster-growing young trees, and that this opportunity cost influences the harvest timing decision.

As long as the manager's optimum timber profits are positive and greater than the value of land in alternative uses, then the manager's solution to 8.2 will identify profit-maximizing harvest dates, harvest volumes, and levels of regeneration effort. The optimum harvest age is obtained where the marginal benefits from delaying the harvest are just equal to the marginal opportunity costs of the delay (see chapter 4).

When forest management decisions are guided by utility rather than profit maximization (i.e., objectives include more than marketable timber products), the forest management problem may be more complex than the problem described by equation 8.2. For example, nonpriced amenity services in the manager's objective function, or forest-level constraints, may bind on local decisions (see Pattanayak et al. 2002 for a recent summary). However, even when these questions are addressed in the manager's problem, similar decision rules result (i.e., harvest occurs where marginal benefits and costs of delaying harvest are balanced). For subsequent discussion here, we posit that a decision rule exists which defines the economically optimal harvest age for each forest owner and quality class. If we define the manager's current expectation of future market prices as p^e, the manager's optimum harvest age under these expectations, $(a*)$, is given by

$$a*(p, p^e; q) = a : MBD^e(a, E; q) = MOC^e(a, E; q) \qquad 8.3$$

The optimum harvest age depends on current market signals (p) and market expectations (p^e). This optimum age is not necessarily the same as that given by the timber-only solution and may vary over time as price expectations are revised. Equation 8.3 is a long-run solution when all elements of p^e are equal to p.

1.2 Aggregate Supply

The core challenge of modeling and evaluating timber supply is constructing some meaningful aggregation of the individual stand harvest decision to define the relationship between aggregate harvest quantity and price. Neoclassical models of supply build on the assumption of a typical producer and, accordingly, develop from a prototypic production function such as equation 8.1. However, timber inventories are heterogeneous (they can be viewed as rather complex capital stocks), and timber is produced from forests allocated to a variety of uses, most of which yield joint products. This indicates that each forest type has a different production function or, more conveniently, that some quality variable shifts a common production function. Modeling timber supply is therefore a nontrivial undertaking. One way to develop an aggregate supply model is to start with the aggregation of individual harvest choices.

$$S^{SR} = \sum_{i=1}^{I} \sum_{j=1}^{J} A_{i,j} \{v(a, E; q)\} * H_{i,j} = g(p, p^e)$$

$$\text{where } H_{i,j} = \begin{bmatrix} 1 & \text{if harvest occurs at } p, p^e \\ 0 & \text{otherwise} \end{bmatrix}$$

8.4

where $A_{i,j}$ is the area of forest in age class i and quality class j; $H_{i,j}$ is a binary variable that describes the harvest decision; and, because H depends on the age of the stand as described above, the aggregate function (g) is conditional on the age and quality distribution of the forest. Supply is conditioned on the current distribution of quality and age classes and implies that a common decision model describes the behavior of all landowners.

Generalizing supply to address variable forest conditions requires including some description of forest inventory as an argument in the supply function. One way to consider the influence of forest inventory in g is simply to include a variable that measures total inventory quantity (I).

$$I = \sum_{i=1}^{I} \sum_{j=1}^{J} \{v(a, E; q)\} \text{ and}$$
$$S^{SR} = g(p, p^e, I)$$

8.5

This approach is common in applied analysis but provides no logical link to the mechanism of the harvest decision defined by equation 8.2—i.e., the aggregate quantity of inventory says nothing about the age and quality distribution of inventory which influences the harvest choices that define supply. To illustrate, equation 8.5 implies that a gain in the sapling volume of the forest would have the same impact on aggregate output as would a gain in sawtimber volume. To address this shortcoming, this simplified model might be expanded to include arguments for separate vintages of timber that are the components of I.

Another way to motivate the development of an aggregate timber supply is to specify an aggregate production or transformation function. In this case, the focus shifts from the stand as the fundamental unit of production to the forest as a whole. The intertemporal aggregate transformation function is defined as:

$$T(S, I_{t-1}, I_t, X, Y) = 0$$

8.6

where S is timber output; I_{t-1} and I_t are vectors of inventory volumes for timber of various qualities at the beginning and the end of the time step; X is

Aggregate Timber Supply

a vector of other inputs; and Y is a vector of other outputs. This approach accommodates the view of timber as both input (starting growing stock) and as output (harvested timber and inventory at the end of the period) and accommodates the joint production nature of forest production. Collapsing S and I into the vectors X and Y provides a compact representation:

$$T(X,Y) = 0 \qquad 8.7$$

If the production model in 8.7 has certain desirable attributes related to its curvature and the separability of inputs and outputs, then a dual function can be used to summarize the behavior of profit-maximizing entities operating with the technology described by the production model (see Chambers 1988). This dual function is generally either a cost function or profit function but could be a revenue function, depending on the degree of flexibility allowed by the production technology. If, for example, we assume that output prices or quantities are fixed for the time step analyzed, then the producer can only minimize cost so that the cost function is the appropriate dual function. If both output and input quantities can be fully adjusted to optimal levels, then a profit function is the appropriate dual function.

Consider the profit function:

$$\pi^{LR} = h(p,w) = 0 \qquad 8.8$$

where p is a vector of output prices corresponding to elements of the output vector Y, and w is a vector of input prices corresponding to elements of the input vector X. Because all inputs and outputs are fully variable—that is, they can be adjusted to their optimal levels without adjustment costs—a long-run profit function is defined (indicated by the superscript LR). If, on the other hand, some of the inputs cannot be optimally adjusted, then a constrained or short-run profit function would apply. In such a case, the observed profit is a function of output prices, the prices of fully variable inputs, and the level of those inputs that cannot be fully adjusted—generally labelled quasi-fixed inputs and included as a separate vector Z here. The inventory of timber is generally quasi-fixed in the case of forest production because the starting age/species distribution heavily constrains what the forest composition can be at the end of the period.

$$\pi^{SR} = h(p,w;Z) \qquad 8.9$$

Output supply can be directly derived from the profit function by taking the partial derivative of profit with respect to price (Hotelling's lemma).

$$\frac{\partial \pi^{SR}}{\partial p_i} = S_i = g_i(p, w; Z) \qquad 8.10$$

Building an aggregate supply model using this approach requires (1) defining the full complement of inputs and outputs relevant to the transformation function (equation 8.7), (2) defining meaningful input and output aggregates for empirical work, and (3) determining whether inputs/outputs are fully variable over the time step considered (so price enters the function) or quasi-fixed (so quantity enters the function). In addition, in forestry we face concerns regarding the aggregation of different owner groups. Typically, separate functions are defined for forest industry and nonindustrial private forest owners.

Developing supply models from the summation of individual choice models and from an aggregate production model leads to the same conceptual endpoint. Supply should be a function of price and of input costs and is conditioned on the state of the forest inventory. The core issues in empirical work are (1) determining a meaningful description of forest inventory and (2) addressing dissimilar forest owners.

2. MODELING APPROACHES

Aggregate timber supply models have been developed using a variety of approaches which develop either from aggregation of individual or representative forest owner models or from aggregate production models. In general, these can be grouped into normative and positive approaches, but it is important to emphasize that timber market modeling, especially when applied to the long run, in application is rooted as much in data management as it is in theoretical modeling. As a result, timber supply models are often developed as hybrids of these two approaches.

2.1 Normative Approaches

One approach to modeling timber supply is to apply stand-level management optimization models (equations 8.2 and 8.3) for all forest types within the region being studied. The approach requires assumptions regarding market structure (e.g., perfect competition), a forest inventory (i.e., a definition of forest land by quality classes), and some definition of the biological production functions (e.g., empirical growth and yield tables). Given this information, the analyst can calculate the optimal rotation and derive the implied annual contribution of each forest type to total production.

Aggregate Timber Supply 123

Supply relationships can then be examined by mechanistically simulating the production response to changes in timber prices.

2.1.1 Static Engineering Models

The original applications of normative models to timber supply defined the optimal rotation for each quality class of forests and then summed up the average annual harvest implied for each forest class to define supply (e.g., Vaux 1954). These are long-run models because they do not explicitly address the age structure of forest capital (they assume a long-run adjustment to an optimal age distribution). They address *maximum potential* timber output, in that they provide no practical mechanism for describing the actual behavior of timberland owners. These models provide an extremely rich supply specification and, as shown by Hyde (1980) and Jackson (1980), can provide tractable comparative statics for forest sector policies related, for example, to public land management and timber taxes. An application of engineering models to the supply effects of public timber management is developed in chapter 12.

2.1.2 Intertemporal Optimization

Another class of engineering models directly simulates short-run harvest and inventory adjustments by focusing on intertemporal optimization in timber management (e.g., Berck 1979). This is accomplished by linking the anticipated effects of production on prices through time, recognizing that increasing harvests in one period can increase scarcity in subsequent periods. Accordingly, these models must explicitly account for the age distribution of the existing forest and its influence over production possibilities in the short run, as well as on the evolution of the age distribution through time. Furthermore, the connection between prices and quantities and therefore the market clearing mechanism must be specified.

The market problem can be solved using an optimization method that derives from Samuelson's (1952) finding that the competitive market solution occurs where the sum of consumer and producer surpluses is maximized. So, given a demand function for Y

$$p = \alpha + \beta Y \qquad 8.11$$

and an aggregate supply function

$$p = g^{-1}(Y) \qquad 8.12$$

Samuelson's intertemporal objective function is a simple quadratic function of Y:

$$Z = \sum_{i=1}^{T} (\alpha_i - \beta_i Y_i) Y_i - c(Y_i)(1+r)^{-i} + M_{T+1}(1+r)^{-(T+1)} \qquad 8.13$$

where i indexes time and M_{T+1} is a terminal value assigned to standing inventory at the end of the planning period (comparable to the bareland value in the Faustmann stand-level model). With information on the biological production functions, cost functions ($c(Y)$), and the demand relationships (which now vary by time period), it is possible to solve 8.13 by maximizing Z with respect to harvests (Y).

This technique allows a detailed specification of timber inventories and other technical inputs. In this way it is similar to the engineering approaches. It departs from a purely normative assessment by incorporating econometric demand models in a market-simulating objective function. This specificity allows the direct analysis of a wide variety of questions about optimal investment levels and paths under varying conditions for various classes of ownerships. Another important aspect of this modeling approach is that, unlike econometric models, it provides a framework for simulating production in new policy environments and for unprecedented changes in environmental factors (e.g., climate change).

Gilless and Buongiorno (1987) have applied the methodology to U.S. pulp and paper industries. Sallnas and Eriksson (1989) provide an interesting approach that explicitly builds the market solution from the set of solutions to individual stand-level problems using a decomposition technique. They also allow for noise, or departure from the technically optimal solutions, for individual forest categories using an entropy constraint. Perhaps the most extensive application of the mathematical programming approach is found in a model of the U. S. rural land-based sectors called the Forest and Agricultural Sector Optimization Model (FASOM) (Adams et al. 1996) that focuses on the land use interface between agriculture and timber production. All of these models are constructed as static simulations, solving the problem on a period-by-period basis. It is also possible to approach the problem using dynamic optimization techniques such as optimal control or dynamic programming. Sedjo and Lyon (1990) provide an optimal control analysis of global timber markets. Sohngen and Sedjo (1998) compare and contrast the performance of the two different approaches.

While these models hold great advantage for generating hypotheses, evaluating nonmarginal changes, and making detailed long-run forecasts,

Aggregate Timber Supply 125

normative models have no role in testing hypotheses regarding supply behavior. They are also highly sensitive to specification error in defining the behavior of forest managers. Positive timber supply models provide an alternative approach that can address these issues.

2.2 Positive Approaches

The structure of a statistical model of economic behavior is generally derived from the economic theory of rational behavior and then estimated using historical observations of production and consumption decisions. These models offer tools for testing economic hypotheses and have been applied in a wide variety of forms to timber markets. Statistical models of timber supply also define the core of many forecasting models in use today.

2.2.1 Individual Choice Models

One body of timber supply models is developed from observations on individual harvesting decisions. These individual choice models examine directly the implied marginal conditions between harvesting and delaying harvest on a particular stand for a particular owner (e.g., equation 8.3). Studies constructed at this level generally assume that the manager maximizes utility. Accordingly, positive harvest choice models are estimated using discrete choice methods (e.g., logit and probit models) fit to cross-sectional observations of harvest and delay choices and landowner and forest quality characteristics (see Binkley 1981 and Dennis 1990).

By framing these decisions within a household production problem, these models can recognize tradeoffs between forestry and other household consumption decisions and between timber products (which may be sold to provide income and wealth) and other services (e.g., amenities) from forests that may be consumed by the household. The earlier empirical applications of the household production logic do not, however, explicitly model the nontimber outputs. Chapter 14 explicitly estimates harvesting and amenity choices. Studies that employ the household production framework provide insights into provision of wood products from a forested landscape with variable forest ownership characteristics and variable forest conditions. They are therefore useful for inferring the choice of variables and form of aggregate supply models.

Recent research has begun to explore the derivation of timber supply directly from these individual choice models. The conceptual bridge between individual and aggregate response is an estimate of how much of the inventory is represented by the observed individual. Hardie and Parks (1991) developed this bridge in their study of forest regeneration, using an area-based sampling frame of forest inventories conducted by the USDA Forest Service. This innovation allows aggregate behavioral responses to be built up directly from the individual survey plots. Prestemon and Wear (2000) applied this approach to define the aggregate softwood supply from the Coastal Plain of North Carolina. In effect, they directly applied the aggregation shown in equation 8.4 but with H representing harvest probabilities (shares) rather than binary choices. While promising for small areas, insights into supply may ultimately be limited by the lack of social data associated with owners of the plots.

2.2.2 Aggregate Supply Models

Aggregate timber supply models have also been developed from aggregate production models. Nearly all aggregate timber market models specify the same form for timber supply. Supply is modelled as a function of price (p) and standing forest inventory (I):

$$S = g(p; I, Z) \qquad 8.14$$

where S is timber supply and g is a function which is generally consistent with equation 8.10. This model derives implicitly from a forest production function, where I represents the accumulated capital inputs to forestry (i.e., the timber inventory, which results from time, effort, land quality, and possibly other capital inputs). Z is a vector of other supply shifters that may or may not be included in the supply model. Timber supply should be positively related to I and positively related to price. Because of price endogeneity, supply is usually jointly estimated with demand (i.e., variables influencing demand are needed to identify the supply equation). They are typically estimated using simultaneous equation techniques with time series data (Adams and Haynes 1980, Daniels and Hyde 1986, Newman 1987).

While these supply formulations have proved very useful for market analysis, they are not always explicitly tractable to theories of production behavior (Binkley 1987, Wear 1991). For example, these models cannot distinguish the effects of various structures of forest capital that might be represented by the aggregate quantity of timber inventory. When the age distribution and species composition of the forest capital is relatively constant over time, this may not be a problem. However, the misspecification may be serious when these qualities vary substantially (i.e., when simulated over long time periods).

Recent research into empirical aggregate supply models focuses on incorporating more inventory detail or explicitly deriving supply equations. Pattanayak et al. (2002) incorporate age distribution information in an aggregate supply model. Newman and Wear (1993) explicitly derive timber supply equations for the Coastal Plain of the southeastern United States from an aggregate short-run profit function directly comparable to equation 8.10. The model was estimated using cross-sectional observations of output, price, cost, and inventory variables. Yin and Newman (1997) apply a dynamic profit function approach.

2.3   Timber Supply: An Empirical Analysis

The southeastern United States is arguably the most important and the most active timber market in the world. Here, forest investment and harvesting have shifted timber production from an extractive to an agricultural endeavor. Timber harvests have increased steadily since the 1950s as forest inventories accumulated on lands previously dedicated to the production of cotton and other row crops. In 1997 the South contributed 58% of the wood products produced in the United States (Powell et al. 1993). Forest products industries have shifted their processing capital to the South from other regions, and forest production is an important part of many rural economies. In this section we develop a model of southern timber markets using historical time series data. The intent of the analysis is to illustrate some of the concepts developed in the theory section.

2.3.1   Defining Timber Supply

To develop an equation for timber supply, we construct the dual to the production technology described in equation 8.7:

$$\pi^{SR} = h(p, w; Z, t) \qquad 8.15$$

where p is a vector of output prices, w is a vector input prices, Z is the vector of quasi-fixed inputs to production, and t indexes profit consistent with the dating of technology in equation 8.7. In our production model we have forest capital inputs that are unlikely to be optimally adjusted in the short run (i.e., within the annual time step of the data used for this analysis), so assume that these should be classified as quasi-fixed inputs (i.e., there are no elements of w in equation 8.16. The supply of an output is derived from the profit function using Hotelling's lemma:

$$S_i = \frac{\delta \pi^{SR}}{\delta p_i} = g_i(p; Z, t) \qquad 8.16$$

so that the righthand side includes output prices (*p*) quasi-fixed input quantities (*Z*), and a time index (*t*). We apply a quadratic profit function to the case with two variable outputs (sawtimber Y_s and pulpwood Y_p), two quasi-fixed inputs (natural forests Z_n and planted forests Z_a), and a quasi-fixed output (forest growth Z_g):

$$\pi^{SR} = a_0 + a_t t + \sum_i a_i p_i + \sum_j a_j Z_j + \frac{1}{2} \sum_i \sum_k b_{ik} p_i p_k + \frac{1}{2} \sum_j \sum_l b_{jl} Z_j Z_l + \sum_i \sum_j b_{ij} p_i Z_j + \sum_i c_i t p_i \qquad 8.17$$

Summations of *i* and *k* are over *s* and *p*, and summations of *j* and *l* are over n, *a*, and *g*. The supply equations are therefore given by Hotelling's lemma as follows:

$$S_S = \frac{\delta \pi^{SR}}{\delta p_s} = a_s + b_{ss} p_s + b_{sp} p_p + b_{sn} Z_n + b_{sa} Z_a + b_{sg} Z_g + c_s t$$

$$S_P = \frac{\delta \pi^{SR}}{\delta p_p} = a_p + b_{pp} p_p + b_{ps} p_s + b_{pn} Z_n + b_{pa} Z_a + b_{pg} Z_g + c_p t$$

$$8.18$$

The forest quantity variables are measured as capital indices that require weighting each age/vintage class of forest area by its implied rental price using methods outlined in Wear (1994). These define a Tornqvist index of forest capital similar to indices of capital built from stocks of buildings, machinery, and other long-lived capital assets used in manufacturing (see generally, Caves et al. 1982).

2.3.2 Estimating the Supply Equations

We estimate these supply equations using annual observations on dependent and independent variables for the period 1965 to 1994. Forest capital measures are specified as described above. Timber quantities are estimated from output of final goods (e.g., lumber) using technical conversion factors. Prices of timber are indexed by softwood sawlog and softwood pulpwood prices reported by the state of Louisiana.

Price of timber products is endogenous to the sector, so we need additional information to identify the supply equations. This is accomplished by specifying demand equations for the two products. Following Newman (1987), we specify demand as follows:

$$D_i = g_i(p, p_{fi}, w_i, Y_{i,t-1}, t) \qquad 8.19$$

where p_{fi} are the final product prices for the two sectors (final goods prices are defined as sector-specific producer price indices for lumber and wood products and paper and allied products sectors reported by the Bureau of Economic Analysis). Lagged production of i is represented by $Y_{i,t-1}$ and w_i is the wage rate for labor in the specified sector. For the pulpwood demand equation, we also include as an explanatory variable the quantity of residues from solidwood sectors used in paper production. This material has grown as an important substitute for pulpwood over the estimation period. Equation 8.19 is therefore consistent with derived demand using the profit function for the respective wood-using sector.

To address price endogeneity and the joint presence of autocorrelation and a lagged endogenous variable on the righthand side, we apply a three-stage least-squares estimator to the system of equations defined by 8.18 and 8.19. The initial Durbin-Watson and Durbin's h statistics indicated that autocorrelation corrections for the supplies of sawtimber and pulpwood were necessary. In addition, we impose symmetry on the profit function by requiring $b_{ps} = b_{sp}$. Note that the time trend was dropped from the estimation of the supply equations. Estimation results are shown in table 8.1.

Of 25 estimated coefficients, 18 are significant at the 10% level. All significant price/quantity coefficients have the expected signs: positive price coefficient for sawtimber supply and negative coefficients for both pulpwood and sawtimber demands. Evaluation of the coefficients for the capital measures indicates that pulpwood is responsive to planted capital but not to natural capital. Sawtimber supply is significantly responsive to both forms of capital. Sawtimber supply elasticities with respect to these measures of capital indicate that supply is much more responsive to natural capital (78.46) versus planted capital (0.059), consistent with expectations.

2.4 Contemporary Research Issues

The appropriate approach to modeling timber supply depends on the objectives of the analysis. Because they model production in a mechanistic fashion, engineering approaches, especially those driven by a dynamic optimization framework, hold advantage for analyses of the long-run supply

Table 8.1. Parameter estimates for the equation system defining pulpwood and sawtimber markets in the U.S. South

Parameter	Coefficient estimate	Approx. std.err.	T-ratio	Prob. > |T|	
Pulpwood supply					
Intercept	954.211	1552.9	0.61	0.540	
Price (pulpwood)	-1.253	17.819	-0.07	0.944	
Price (sawtimber)	41.995	23.177	1.81	0.484	
Planted forest capital	0.378	0.535	0.71	0.079	*
Natural forest capital	-87.054	1631.8	-0.05	0.958	
Growth	0.143	0.0434	3.29	0.002	**
Adjusted R^2 = .93					
Pulpwood demand					
Intercept	-881.388	241.6	-3.65	0.001	**
Quantity (pulpwood)	-0.007	0.002	-3.47	0.001	**
Price (paper)	-0.123	0.045	-2.73	0.010	**
Residues	0.011	0.002	-6.72	0.000	**
Lagged pulpwood	0.002	0.002	1.29	0.205	
Wages (paper)	1.552	0.560	2.77	0.009	**
Time trend	0.461	0.124	3.70	0.001	**
Adjusted R^2 = .63					
Sawtimber supply					
Intercept	-5374.75	2087.9	-2.57	0.015	**
Price (sawtimber)	1.374	0.774	1.78	0.085	*
Price (pulpwood)	41.995	23.177	0.71	0.484	
Planted forest capital	100.276	28.875	3.47	0.001	**
Natural forest capital	6454.966	2057.7	3.14	0.004	**
Growth	-0.052	0.063	-0.83	0.411	
Adjusted R^2 = .63					
Sawtimber demand					
Intercept	8959.024	2343.2	3.82	0.001	**
Quantity (sawtimber)	-0.282	0.0580	-4.85	0.000	**
Price (lumber)	2.737	0.3427	7.99	0.000	**
Lagged sawtimber	0.170	0.0532	3.19	0.003	**
Wages (lumber)	47.204	12.4658	3.79	0.001	**
Time trend	-4.619	1.2202	-3.79	0.001	**
Adjusted R^2 = .91					

* = significance at the 5% level
** = significant at the 1% level

and structural changes in forest production (e.g., related to climate change). However, engineering approaches are susceptible to specification error and, because they are not fit to observations, may not be well suited for short- and medium-run analysis of market dynamics and policy effects.

In contrast, positive econometric models of supply are calibrated to observed producer behavior and provide a mechanism for testing various hypotheses regarding behavior and impacts. However, econometric models are not without potential specification problems. While individual choice models have a high degree of specificity regarding the structure of supply, aggregate models of supply often are based on *ad hoc* aggregations of timber capital. The theory discussed above suggests that the structure as well as the amount of forest capital should have a bearing on timber supply estimates.

Ongoing research focuses on the explicit aggregation of forest capital in timber supply models. One approach is to conduct modeling at finer scales, even at the level of the individual, and then aggregate the outcomes to simulate supply responses (e.g., Prestemon and Wear 2000). Another approach is to expand aggregate timber supply models to include additional information on the structure of timber inventory on the righthand side of the supply equation (e.g., Pattanayak et al. 2002). This can be accomplished by including additional variables—e.g., by expanding the vector of inventory attribute variables, Z, in equation 8.16—or by aggregating inventory in a way that is consistent with capital theory—i.e., consistent with the contribution of each component of the capital stock to production (as developed in the example in this chapter). This is an attempt to make the inventory shifter a measure of capital rather than purely a biological factor. One promising avenue of research is hybridizing normative and positive supply analyses. This approach allows empirical findings to inform long-run timber market projections.

3. LITERATURE CITED

ADAMS, D.M., AND R.W. HAYNES. 1980. The 1980 softwood timber assessment market model: Structure, projections, and policy simulations. For. Sci. Mono. 22, 64 p.

ADAMS, D.M., R.J. ALIG, B.A. MCCARL, J.M. CALLAWAY, AND S.M. WINNETT. 1996. An analysis of the impacts of public timber harvest policies on private forest management in the U.S. For. Sci. 42(3): 343-358.

BERCK, P. 1979. The economics of timber: A renewable resource in the long run. Bell J. Econ. 10(2):447-462.

BINKLEY, C.S. 1981. Timber supply from nonindustrial forests. Bulletin No. 92, Yale University, School of Forestry and Environmental Studies, New Haven, CT.

BINKLEY, C.S. 1987. Economic models of timber supply. P.109-136. *in* The Global Forest Sector: An analytical perspective. Kallio,M., D.P. Dykstra, and C.S. Binkley (eds.). John Wiley and Sons, New York. 703 p.

CAVES, D.W., L.R. CHRISTENSEN, AND W.E. DIEWERT. 1982. The economic theory of index numbers and the measurement of input, output, and productivity. Econometrica 50(6):1393-1414.

CHAMBERS, R. 1988. Applied production analysis: A dual approach. Cambridge University Press, New York. 331 p.

DANIELS, B., AND W.F. HYDE. 1986. Estimation of supply and demand elasticities for North Carolina timber. For. Ecol. Manage. 14:59-67.

DENNIS, D.F. 1990. A probit analysis of the harvest decision using pooled time series and cross-sectional data. J. Environ. Econ. Manage. 18:176-187.

GILLESS, J.K., AND J. BUONGIORNO. 1987. PAPYRUS: A model of the North American pulp and paper industry. For. Sci. Monograph No. 28. 37 p.

HARDIE, I.W., AND P.J. PARKS. 1991. Individual choice and regional acreage response to cost-sharing in the South, 1971-1981. For. Sci. 37(1):175-190.

HARTMAN, R. 1976. The harvest decision when the standing forest has value. Econ. Inquiry 14: 52-58.

HYDE, W.F. 1980. Timber supply, land allocation and economic efficiency. Johns Hopkins Press, Baltimore, MD. 224 p.

JACKSON, D.H. 1980. The Microeconomics of the Timber Industry. Westview Press, Boulder, CO., 136 p.

NEWMAN, D.H. 1987. An econometric analysis of the southern softwood stumpage market: 1950-1980. For. Sci. 33:932-945.

NEWMAN, D.H., AND D.N. WEAR. 1993. The production economics of private forestry: A comparison of industrial and nonindustrial forest owners. Am. J. Agri. Econ. 75:674-684.

PATTANAYAK, S.K., B.C. MURRAY, AND R.C. ABT. 2002. How joint is joint forest production? An econometric analysis of timber supply and amenity values in the U.S. South. For. Sci. 48(3):479-491.

POWELL, D.S., J.L. FAULKNER, D.R. DARR, Z.ZHU, AND D.W. MACCLEERY. 1993. Forest resources of the United States, 1992. Gen. Tech. Rpt. RM-234. USDA Forest Service Rocky Mountain Forest and Range Experiment Station, Fort Collins, CO. 132 p.

PRESTEMON, J.P., AND D.N. WEAR. 2000. Linking harvest choices to timber supply. For. Sci. 46(3):377-389.

SALLNAS, O., AND L.O. ERIKSSON. 1989. Management variation and price expectations in an intertemporal forest sector model. Nat. Res. Model. 3:385-398.

SAMUELSON, P.A. 1952. Spatial price equilibrium and linear programming. Am. Econ. Rev. 42:283-303.

SEDJO, R.A., AND K.S. LYON. 1990. The long-term adequacy of the world timber supply. Resources for the Future.Washington DC. 230 p.

SOHNGEN, B., AND R. SEDJO. 1998. A comparison of timber market models: Static simulation and optimal control approaches. For. Sci. 44(1)2436.

VAUX, H.J. 1954. Economics of young growth sugar pine resources. Bulletin No. 78, Division of Agricultural Sciences, University of California, Berkeley.

WEAR, D.N. 1991. Evaluating technology and policy shifts in the forest sector: Technical issues for analysis. P. 147-155. *in* Proceedings of the 1991 Southern Forest Economics Workshop, S.J. Chang (ed.). University of Kentucky, Washington DC.

WEAR, D.N. 1994. Measuring investment and productivity in timber production. For. Sci. 40(1):192-208.

WEAR, D.N., AND P.J. PARKS. 1994. The economics of timber supply: An analytical synthesis of modeling approaches. Nat. Res. Model. 8(3):199-223.

YIN, R., AND D.H. NEWMAN. 1997. Long-run timber supply and the economics of timber production. For. Sci. 43(1):113-120.

Chapter 9

Timber Demand
Aggregation and Substitution

Robert C. Abt and SoEun Ahn
North Carolina State University

Forest economics' existence as a distinct subdiscipline is usually attributed to the unique characteristics of the forest resource (Gregory 1987). The issues of longer production period, opportunity cost of land, and trees as factory and product, along with Faustmann's subsequent insight, are generally the beginning and end of the forest economics section of broader resource economics courses. Thus, the focus of forest economics has been understandably supply-sided. Once the assessment of forest resources goes beyond stand management to include prices and markets, however, demand analysis plays an equally important role.

Demand for forest resources falls into the standard categories of demand derived from production processes of goods or consumer demand for amenities and other noncommodity goods. The focus of this chapter is on derived demand for forest product commodities. Applied derived demand analysis in forestry has followed the development of analytical techniques. The development of flexible functional forms and applications of duality in production theory in the 1970s were followed by a flurry of applications in forestry in the 1980s. This early literature tends to be populated by cost- or profit-function-based studies that center in a particular industry. Studies have been conducted for specific industries and regions (deBorger and Buongiorno 1985, Nautiyal and Singh 1985, Singh and Nautiyal 1986), across regions (Abt 1987, Martinello 1987), and across market levels (Haynes 1977, Merrifield and Singleton 1986, Stier 1980). Although the results from these studies vary depending on applications, there seem to be some indications that 1) the results are likely to be sensitive to the level of aggregation over firms, regions, and markets; 2) wood and labor are substitutes, at least, in the aggregate; 3) factor own-price elasticities are inelastic; and 4) technical change is labor-saving/capital-using.

This chapter concentrates on aggregation, input substitution, and technical change in timber derived demand analysis. Aggregation is a ubiquitous issue in applied economics. Its role in forest commodities is interesting because fixed-factor assumptions, which may be tenable at the firm level, are commonly applied to aggregate-level policy questions where there is little empirical support for the assumption. Input substitutability between wood and labor is particularly important since the employment impact of various forest policies is a key question on both public and private lands. The size and direction of these impacts are directly affected by assumptions about input substitution in the aggregate production function.

Section 1 reviews the derivation of timber demand from cost or profit functions, followed by an examination of the implications of the heterogeneity of individual mill technologies on aggregate production behavior. Sections 3 and 4 address the decomposition of derived demand to isolate the substitution and output effects and investigates the implications of harvest control policies on the labor demand. Section 5 discusses productivity measurement and summarizes the literature on technical change. The final section assesses our current state of knowledge and provides ideas for future research.

1. DERIVED DEMAND

Figure 9.1 is a schematic diagram of product flows in the U.S. South and describes the link between derived demand for timber and the demands for end-use products. For instance, the demand for the roundwood (logs) is derived from demand for housing, furniture, and other solid wood products. Sawtimber and pulpwood are traditional terms that describe forest product markets with separate measurement systems (board feet versus cords), separate markets (housing versus paper), and separate facilities (sawmills versus pulp mills).

To derive the factor demands, consider a production function of a mill, which produces lumber (y). For simplicity, assume that the mill uses only two variable inputs, roundwood (x_w) and labor (x_l) for its production.

$$y = f(x_w, x_l) \qquad 9.1$$

Then, the mill's cost function[1] is defined as:

Timber Demand

$$c(p_w, p_l, y) = \min p_w x_w + p_l x_l \quad \text{such that } \bar{y} = f(x_w, x_l) \tag{9.2}$$

where p_w and p_l are the prices of roundwood and labor, given to producers.

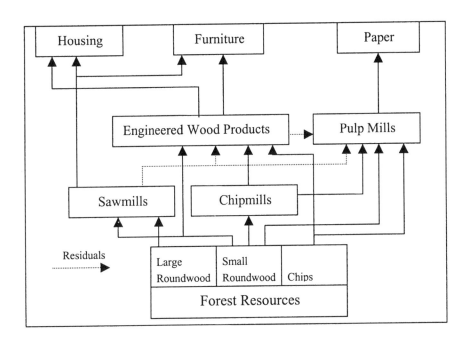

Figure 9.1. Schematic of product flows and derived demand for timber

Thus, the cost function gives the input combination, which minimizes cost to produce a given output. If the cost function is differentiable, then by Shepherd's lemma, the unique cost-minimizing derived demand for each input exists as follows:

$$x_w(p_w, p_l, y) = \frac{\partial c(p_w, p_l, y)}{\partial p_w} \tag{9.3}$$

$$x_l(p_w, p_l, y) = \frac{\partial c(p_w, p_l, y)}{\partial p_l} \tag{9.4}$$

These results imply that derived demands can be characterized by examining empirical properties of cost functions.

Alternatively, derived demands can be obtained from the mill's profit function instead of the cost function. The mill's profit function is defined as:

$$\pi(p_y, p_w, p_l) = \max p_y y - p_w x_w - p_l x_l \qquad 9.5$$

where p_y is the price of output, which is also exogenous to producers. Thus, the profit function gives the combination of output and inputs, which maximizes profit of a firm, given the prices of output and inputs. If the profit function is differentiable, then by Hotelling's lemma, the unique profit-maximizing supply and derived demand functions are represented as:

$$y(p_y, p_w, p_l) = \frac{\partial \pi(p_y, p_w, p_l)}{\partial p_y} \qquad 9.6$$

$$x_w(p_y, p_w, p_l) = -\frac{\partial \pi(p_y, p_w, p_l)}{\partial p_w} \qquad 9.7$$

$$x_l(p_y, p_w, p_l) = -\frac{\partial \pi(p_y, p_w, p_l)}{\partial p_l} \qquad 9.8$$

Profit-maximizing derived demands can be written as:

$$x_w(p_y, p_w, p_l) = x_w(p_w, p_l, y(p_y, p_w, p_l)) \qquad 9.9$$

$$x_l(p_y, p_w, p_l) = x_l(p_w, p_l, y(p_y, p_w, p_l)) \qquad 9.10$$

In words, the profit-maximizing input demands equal the cost-minimizing input demands evaluated at the profit-maximizing output. The empirical recovery, however, of mill-level derived demands using the derivatives of cost or profit functions has not had much success due to the lack of data at mill level and the aggregate nature of policy decisions. See Chambers (1988) for further discussion of aggregation over firms.

Perhaps one of the biggest challenges associated with aggregation over firms is to understand the relationship between firm- and industry-level behavior. If individual firms are exact replicas of each other, the aggregation is straightforward. The aggregate output and input demands can be found by multiplying individual functions by the number of firms in the industry. However, if we allow heterogeneity of individual firms, which is more likely to be the case in reality, things get complicated quickly. Houthakker (1955-1956) showed that if all firms within an industry used fixed-proportion technologies and capacity was pareto distributed, then the aggregate production function is Cobb-Douglas. Fisher (1993) examined the theoretical requirements for the existence of aggregate production functions. His analysis suggests that the strict theoretical conditions for aggregation are unlikely to exist. In practice he notes that empirical analysis of aggregate production is likely to be most useful in aggregates that are more homogeneous and stable. Below we analyze empirical aggregation over individual mills in the lumber industry and examine the effects of mill heterogeneity on industry aggregate production structure.

2. AGGREGATE PRODUCTION STRUCTURE

There is an intuitive appeal to the proposition that the production technology between input (roundwood) and output (lumber) in individual sawmills is fixed.[2] The simple process of cutting cylinders into rectangles inherently limits the possibility of gains in efficiency from substitution. This is the basis for log scales (e.g., Doyle and Scribner) used in timber mensuration, which explicitly assume that mill output can be predicted as a function of wood input alone. However, even if individual mills exhibit a fixed-proportion technology, aggregate production may exhibit significant responses to aggregate output and input price changes due to the heterogeneity of mills in an industry.

To investigate the behavioral differences between mills and industry, consider the decomposition of aggregate input demand response to aggregate output changes. The relationship between firm and industry elasticities is conveniently explored using logarithmic forms and the chain rule.[3] Assume that industry output is a function of the output of individual mills. Each mill is assumed to use a fixed-proportion technology between input and output.[4]

$$\frac{d \ln X_w}{d \ln Y} = \sum_i b_i \frac{d \ln x^i_w}{d \ln y^i} \frac{d \ln y^i}{d \ln Y} \qquad 9.11$$

where X_w is aggregate roundwood demand, x^i_w is roundwood demand for a mill i, Y is aggregate lumber production, y^i is lumber production for mill i, and $b_i = x^i_w / X_w$ is share of roundwood demand of mill i to the aggregate demand.

The lefthand side (LHS) of equation 9.11 is the percentage change in aggregate input demand with respect to percentage change in aggregate output. The righthand side (RHS) can be interpreted as decomposing the aggregate input-output effect into individual effects that are weighted by the individual mill's share of input demand to the aggregate (b_i) and the individual mill's output response to the change in the aggregate output ($d\ln y^i/d\ln Y$). Since all mills have a fixed-proportion technology, the second term is equal to one ($d\ln x^i_w/d\ln y_i$) for all i and equation 9.11 can be rewritten as:

$$\frac{d \ln X_w}{d \ln Y} = \sum_i b_i \frac{d \ln y^i}{d \ln Y} \qquad 9.12$$

Substituting input share weights with output share weights ($y^i / Y = q_i$) in equation 9.12 gives:

$$\frac{d \ln X_w}{d \ln Y} = \sum_i q_i \frac{x^i_w / y^i}{X_w / Y} \frac{d \ln y^i}{d \ln Y} \qquad 9.13$$

The first term in the RHS of equation 9.13 is the relative ratio of an individual mill's proportion between input and output to the aggregate counterpart. It measures the roundwood input intensity of an individual mill relative to the industry. Following Mork (1978), the LHS of equation 9.13 could be interpreted as the expectation of the product of two terms in the RHS. Using $E(AB) = E(A)E(B) + Cov(A,B)$:

$$\frac{d \ln X_w}{d \ln Y} = \left(\sum_i q_i \frac{x^i_w / y^i}{X_w / Y} \right) \left(\sum_i q_i \frac{d \ln y^i}{d \ln Y} \right) + Cov\left(\frac{x^i_w / y^i}{X_w / Y}, \frac{d \ln y^i}{d \ln Y} \right) \qquad 9.14$$

The first two terms of equation 9.14 reduce to one, leaving:

$$\frac{d \ln X_w}{d \ln Y} = 1 + Cov\left(\frac{x^i_w / y^i}{X_w / Y}, \frac{d \ln y^i}{d \ln Y} \right) \qquad 9.15$$

Timber Demand

The covariance term represents the relationship between the relative input intensity of individual mills and their relative output response.

The hardwood lumber industry has been characterized as having many small mills that use wood intensively. These small mills go in or out of business depending on the demand for lumber. This implies that the more wood-intensive mills may be more responsive to the changes in aggregate output, leading to a positive covariance. Therefore, equation 9.15 indicates that the aggregate input elasticity with respect to aggregate output is always greater than one even if it is assumed to be one at the individual mill-level. Although the fixed-proportion technology may be acceptable at mill level production, the same does not hold for the aggregate.

A similar exposition can be applied to the aggregate own- and cross-price elasticities of roundwood demand. The following equations decompose the own- and cross-price elasticities of the aggregate roundwood demand into weighted averages of individual mills. Weights are the individual mill's share of input demand and the individual mill's output response to a change in prices.

$$\frac{d \ln X_w}{d \ln p_w} = \sum_i b_i \frac{d \ln x_w^i}{d \ln y^i} \frac{d \ln y^i}{d \ln p_w} \qquad 9.16$$

$$\frac{d \ln X_w}{d \ln p_l} = \sum_i b_i \frac{d \ln x_w^i}{d \ln y^i} \frac{d \ln y^i}{d \ln p_l} \qquad 9.17$$

where p_w is the price of roundwood and p_l is the price of labor. Applying a similar derivation technique, equations 9.16 and 9.17 can be represented as equations 9.18 and 9.19, respectively:

$$\frac{d \ln X_w}{d \ln p_w} = \left(\sum_i q_i \frac{d \ln x_w^i}{d \ln p_w}\right) + Cov\left(\frac{x_w^i / y^i}{X_w / Y}, \frac{d \ln y^i}{d \ln p_w}\right) \qquad 9.18$$

$$\frac{d \ln X_w}{d \ln p_l} = \left(\sum_i q_i \frac{d \ln x_w^i}{d \ln p_l}\right) + Cov\left(\frac{x_w^i / y^i}{X_w / Y}, \frac{d \ln y^i}{d \ln p_l}\right) \qquad 9.19$$

The first term in the RHS of equation 9.18 is the weighted average of an individual mill's own-price elasticity of roundwood demand, where weights are the output shares. The covariance term defines the relationship between the relative input intensity of individual mills to the aggregate and the individual mill's output responses to the change in roundwood price. Mills that are wood intensive are likely to be more responsive to changes in roundwood price. This suggests that the covariance term is positive. The first term in the RHS of equation 9.19 is the weighted average of the individual mill's cross-price elasticity, where weights are the output shares. The covariance term characterizes the relationship between the relative input intensity of individual mills to the aggregate and the output responses of individual mills to the changes in labor price. If labor-intensive mills are more responsive to the changes in labor price than in wood price, then this indicates a negative covariance.

In sum, the above two equalities show that the aggregate own- and cross-price elasticities are not equal to the weighted average of individual mills' price elasticities unless the covariance terms are zero. The covariance terms—relationship between input use intensity and mill sensitivity to input price changes—may be zero in the short run but would be expected to be positive over the long run. Given the long-run nature of forest production and forest policies, these effects are likely to be policy relevant.

The above analysis of aggregation indicates that derived demand studies are likely to be sensitive to aggregation across space and products as well as aggregation over individual firms within an industry. Independent tests of regional differences (Abt 1987), output separability (Wear 1987), and input separability (Lander 1999) indicate that aggregation across regions or product space masks significant heterogeneity in production structure. Given the region-specific nature of the forest resource, commodity definitions, and markets, results are likely to be particularly sensitive to aggregation across product and geographic space.

3. SUBSTITUTION AND OUTPUT EFFECTS

The previous section shows that observed aggregate lumber industry behavior may not be representative of individual mills. Empirical aggregate production studies may be most important as reduced form relationships between policy-relevant variables. Numerous studies have examined the input substitutability between roundwood and labor. A common measure employed in the literature to measure the substitutability between inputs i and j is the Allen elasticity of substitution given by:

$$\sigma_{ij} = \frac{C(\mathbf{p}, y) C_{ij}(\mathbf{p}, y)}{C_i(\mathbf{p}, y) C_j(\mathbf{p}, y)} \qquad 9.20$$

Timber Demand

where C is total cost, **p** is a vector of input prices, and the subscripts on C are partial derivatives with respect to the prices of inputs i and j. It can be shown that the elasticity of input substitution (σ_{ij}) is related to cross-price elasticity of factor demand for ith factor with respect to the price of jth factor (ϵ_{ij}) and the cost share of factor j (S_j).

$$\sigma_{ij} = \frac{\varepsilon_{ij}}{S_j} \qquad 9.21$$

The empirical evidence that roundwood and labor are substitutes in the aggregate is substantial (table 9.1). The policy implications of substitution, however, have not been widely recognized. Analyzing the impact of harvest changes on the regional economy through changes in employment has been a core component of forest resource assessment. Examples include assessments of national harvest constraints (Adams and Haynes 1980) and studies of regulations on private lands such as best management practices (Lickwar 1992, Aust et al. 1996).

Table 9.1. Summary of elasticity of substitution from previous studies

Source	Approach	Industry[a]	Region	σ_{LW}[b]
Humphrey and Moroney (1975)	Cost function	SIC24	U.S.	0.61
Humphrey and Moroney (1975)	Production function	SIC24	U.S.	1.24
Merrifield and Haynes (1983)	Production function	Lumber and plywood	Pacific Northwest	-0.91
Abt (1984)	Cost function	SIC242	Appalachian	0.80
			Southern	0.95
			Western	0.60
deBorger and Buongiorno (1985)	Cost function	Paper	U.S.	0.42
deBorger and Buongiorno (1985)	Cost function	Paperboard	U.S.	0.39
Martinello (1985)	Cost function	Pulp and paper	Canada	0.21
Nautiyal and Singh (1985)	Cost function	Lumber	Canada	0.60
Singh and Nautiyal (1986)	Cost function	Lumber	Canada	0.24
Wear (1987)	Cost function	SIC24	U.S.	0.71
Lander (1999)	Cost function	SIC242	U.S. South	0.87
Lander (1999)	Profit function	SIC242	U.S. South	0.67

[a]SIC24—Standard Industrial Classification for solid wood products industries; SIC242—Standard Industrial Classification for sawmills and planing mills
[b]Allen elasticity of substitution between labor and wood calculated at mean values of data

The traditional approach to investigating the employment effects of harvest change is based on a two-stage fixed-proportion assumption. Input-output models are prominent in these analyses, where in the first stage proportional effects of harvest change on employment and output are assumed. Input-output models are then used in the second stage to calculate the multipliers that summarize links to other sectors in the region. Although the fixed-proportion assumption may approximate total impact in regions with single or few mills for the short run, this approach becomes inappropriate as the region or time scale expands.

To analyze this policy question, consider the response of the lumber industry to an input price increase induced by supply reduction in a region. Industry response can be modeled as two distinct adjustments. Figure 9.2 illustrates this process. In the short run, industry adjusts its input combination, reducing roundwood use and increasing labor use, given the roundwood price increase. Recall that this substitution is likely to depend more on the mix of mills than on within-mill adjustments. This is the movement along the same isoquant from the initial point (W_0, L_0) to (W_1, W_1) on Y_0 in figure 9.2. The industry will also reduce output because of the higher cost associated with increased roundwood price. This causes inward shift of the isoquant $(Y_0 \rightarrow Y_1)$, and the industry settles at new input combination (W_2, L_2) and reduced output level (Y_1). The first stage of adaptation is due to substitution between inputs, and the second stage of adaptation is attributable to output adjustment resulting from input price change.

The total effect of a roundwood price increase on labor demand is the sum of substitution and output effects. Our policy interest is usually in the net effect, so understanding the interaction of substitution and output effects is critical. Under reasonable circumstances, the substitution effect (expected to be positive) and output effect (expected to be negative) may completely offset each other, leading to an ambiguous sign of the total effect on labor demand.

Applications of the cost function characterize input use response to changes in input price or output, but do not measure the output response to a change in price. These compensated (Hicksian) factor demands are valid for measuring the substitution effect, but without the output effect they have limited policy relevance. Similarly, fixed-proportion studies have limited policy relevance by ignoring the substitution effect.

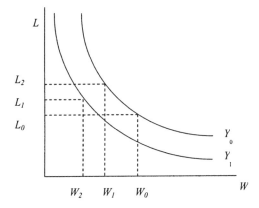

Figure 9.2. Adaptation of lumber industry to increase in roundwood price

Therefore, uncompensated (Marshallian) derived demands, traditionally obtained from the profit function, are more useful because this framework determines optimal output level endogenously, capturing the effect of both substitution and output responses. An alternative method to isolate substitution and output effects is to define an indirect production function, where output is maximized given the expenditure, and output-maximizing factor demands are obtained using Roy's identity.

4. AN APPLICATION OF A DIFFERENTIAL DEMAND SYSTEM TO U.S. LUMBER INDUSTRY

Holt and Fulcher (2001) recently proposed a differential factor demand system as an alternative method to recover both Hicksian (compensated or output-fixed) and Marshallian (uncompensated or expenditure-fixed) factor demand elasticities by utilizing the dual relationship between indirect production and cost functions. They employed the established relationship between compensated and uncompensated factor demands, analogous to Slutsky decomposition in consumer theory, in the development of a differential factor demand system.[5] What makes this method noteworthy is its capability to derive both Hicksian and Marshallian demand elasticities from the cost function. This method provides an explicit way to separate substitution and output effects with the cost function by relating uncompensated demand to compensated demands. In this section, we apply their differential factor demand system to the southern U.S. lumber industry.

If the objective is to maximize output subject to an expenditure constraint and using Roy's identity, output-maximizing input demands can be represented as:

$$X_i^u = X_i^u(p_1,...,p_n,C,t) \qquad 9.24$$

with inputs $i = 1,...,n$, total cost $C = \Sigma p_i X_i$, and t is a technical change index.

Following Holt and Fulcher (2001), the total derivative of 9.24 may be rearranged to:

$$d \ln X_i = \sum_j \varepsilon_{ij}^u d \ln p_j + \varepsilon_{iC} d \ln C + \varepsilon_{it}^u dt \qquad 9.25$$

where $\epsilon^u_{ij} = \partial \ln X^u_i / \partial \ln p_j$ is the ij^{th} Marshallian factor price elasticity, $\epsilon_{iC} = \partial \ln X^u_i / \partial \ln C$ is the ith expenditure elasticity, and $\epsilon^u_{it} = \partial \ln X^u_i / \partial \ln t \mid_{t=1}$ is the ith technical change bias elasticity.

Note that at the optimal output level, compensated factor demands are obtained by Shepherd's lemma from the dual cost function $C(p_1,...,p_n,y,t)$ as:

$$X_i^c(p_1,...,p_n,y,t) = X_i^u(p_1,...,p_n,C(p_1,...,p_n,y,t),t) \qquad 9.26$$

Differentiating both sides of equation 9.26 first by p_j and second by t, and expressing the results in elasticity form produces:

$$\varepsilon_{ij}^u = \varepsilon_{ij}^c - S_j \varepsilon_{iC} \qquad 9.27$$

and

$$\varepsilon_{it}^u = \varepsilon_{it}^c - \varepsilon_{iC} \varepsilon_{Ct} \qquad 9.28$$

where $S_j = p_j X_j / C$ is the jth expenditure share, and $\varepsilon_{Ct} = \partial \ln C / \partial \ln t \mid_{t=1}$ is the elasticity of cost with respect to technical change. Equation 9.27 is the Slutsky decomposition of the uncompensated elasticity of factor demand. Substituting equations 9.27 and 9.28 into 9.25 gives an estimable differential input demand system:

$$S_i d \ln X_i = \beta_i \left[d \ln \tilde{C} - \varepsilon_{Ct} dt \right] + \sum_{j=1}^n \theta_{ij} d \ln p_j + \alpha_i dt \qquad 9.29$$

where $d \ln \widetilde{C} = \sum_{j=1}^{n} S_j d \ln X_j$, $\beta_i = S_i \varepsilon_{iC}$, $\theta_{ij} = S_i \varepsilon_{ij}^c$, and $\alpha_i = S_i \varepsilon_{it}^c$.

We used U.S. lumber industry (SIC242) data from Lander (1999). The data is pooled across three southern states and includes three variable inputs (wood, production labor, and nonproduction labor) and one fixed input (capital) for the years 1964-1988. Converting the system to discrete-change data and incorporating the fixed input (capital) into the system, a final short-run differential factor demand system is estimated as:

$$\widetilde{S}_{it} \Delta \ln X_{it} = \beta_i \left[\Delta \ln \widetilde{C}_t - \varepsilon_{Ct} - \varepsilon_{Ck} \Delta \ln X_{kt} \right]$$
$$+ \sum_{j=1}^{n} \theta_{ij} d\Delta p_{jt} + \alpha_i + \eta_{ik} \Delta \ln X_{kt} + \upsilon_{it}$$
9.30

and

$$\Delta \ln \widetilde{C}_t = \sum_{i=1}^{n} \widetilde{S}_{it} \Delta \ln X_{it} = \varepsilon_{Cy} \Delta \ln y_t + \varepsilon_{Ct} + \varepsilon_{Ck} \Delta \ln X_{kt} + \mu_{it}$$
9.31

where i includes indices for wood (w), production labor (l), and nonproduction labor (n); k = capital, t = 1964-1988; $\widetilde{S}_{it} = 0.5(S_{it}+S_{it-1})$ is the two-year moving average of the ith expenditure share; and Δ denotes a first-difference operator. In equations 9.30 and 9.31, υ_{it} and μ_{it} are error terms that are normally distributed and have zero mean. After imposing cross-equation and symmetry restrictions onto the system and deleting one equation (nonproductive labor equation in our application) to avoid a singular covariance matrix, equations 9.30 and 9.31 are jointly estimated using seemingly unrelated regressions. The parameters in the omitted equation are recovered by utilizing cross-equation and symmetry restrictions.

The parameter estimates from this estimation of a differential input demand system are reported in table 9.2. The estimated system R^2 is 0.709 implying that the model fits the data reasonably. Most parameter estimates (15 out of 18) are statistically significant at 5% significance level. The negative sign of ϵ_{Ct} suggests that technical progress has reduced short-run cost over the time in the lumber industry. The estimate of scale elasticity derived from (1/ ϵ_{Cy}) is 1.122 and statistically significant at 5% level.

The economic implications of the coefficients in table 9.2 are clearer using the elasticities calculated at sample means (table 9.3). Estimated expenditure and output elasticities are positive and significant for all inputs. As expected, compensated own-price elasticities are negative for all inputs. Wood and production labor both exhibit inelastic own-price elasticities near

–0.5 (wood –0.5183; production labor –0.5427). The positive signs of compensated cross-price elasticities among inputs imply that inputs are substitutes for one another. However, the larger magnitude of ϵ^c_{lw} than ϵ^c_{nw} (0.42 vs. 0.20) indicates that production labor is more sensitive to wood prices than nonproduction labor.

For the own-price effect, the substitution and output effects reinforce a reduction in wood demand ($W_0 \rightarrow W_1 \rightarrow W_2$ in figure 9.2). Examining the compensated and uncompensated own price elasticities (ϵ^c_{ww} and ϵ^u_{ww}), the total effect is –0.8774 (ϵ^u_{ww}) as compared to –0.5183 (ϵ^c_{ww}) without the output effect. The estimated Allen elasticity of substitution between wood and productive labor is 1.1481, which is larger than the estimates in the previous studies reported in table 9.1.

Table 9.2. Parameter estimates from differential factor demand system

Parameter	Estimate	Standard Error	T-Statistic	P-Value
ϵ_{Cy}	0.8910	0.0873	10.2059	0.000
ϵ_{Ct}	-0.0346	0.0127	-2.7252	0.006
ϵ_{Ck}	0.4466	0.2037	2.1919	0.028
α_w	-0.0036	0.0046	-0.7821	0.434
α_l	-0.0094	0.0038	-2.4802	0.013
α_n	-0.0216	0.0078	-2.7531	0.006
β_w	0.3591	0.0241	14.9305	0.000
β_l	0.2411	0.0176	13.7335	0.000
β_n	0.3998	0.0286	13.9603	0.000
η_{wk}	0.0044	0.0729	0.0609	0.951
η_{lk}	0.0456	0.0611	0.7457	0.456
η_{nk}	0.3965	0.1257	3.1548	0.002
θ_{ww}	-0.1887	0.0192	-9.8381	0.000
θ_{wl}	0.1187	0.0158	7.5027	0.000
θ_{wn}	0.0699	0.0148	4.7231	0.000
θ_{ll}	-0.1542	0.0184	-8.3961	0.000
θ_{ln}	0.0354	0.0120	2.9421	0.003
θ_{nn}	-0.1054	0.0187	-5.6441	0.000

Note: w, l, n, and k represent indices for wood, productive labor, nonproductive labor, and capital, respectively.

As described in section 3 above, it is the net impact of both output and substitution effects that is most policy relevant. While the output-compensated elasticity ϵ^c_{lw} (= 0.4179) explains the pure substitution effect (the movement from (W_0, L_0) to (W_1, L_1) on Y_0 in figure 9.2) associated with an increase in wood price, the magnitude of the uncompensated effect ϵ^u_{lw} (= 0.1090) summarizes the net impact of both (the movement from (W_0, L_0) to (W_2, L_2) in figure 9.2). As discussed earlier and shown by the estimates in table 9.3, given marginal changes in prices, the output effect can be dominated by the substitution effect.

Table 9.3. Elasticity estimates of factor demands

Parameter	Estimate	Std. err.	T-statistic	P-value
\multicolumn{5}{c}{Compensated price elasticities of input demand}				
ϵ^c_{ww}	-0.5183	0.0527	-9.8381	0.000
ϵ^c_{wl}	0.3262	0.0435	7.5027	0.000
ϵ^c_{wn}	0.1921	0.0407	4.7231	0.000
ϵ^c_{lw}	0.4179	0.0557	7.5027	0.000
ϵ^c_{ll}	-0.5427	0.0646	-8.3961	0.000
ϵ^c_{ln}	0.1248	0.0424	2.9421	0.003
ϵ^c_{nw}	0.1987	0.0421	4.7231	0.000
ϵ^c_{nl}	0.1007	0.0342	2.9421	0.003
ϵ^c_{nn}	-0.2995	0.0531	-5.6441	0.000
\multicolumn{5}{c}{Uncompensated price elasticities of input demand}				
ϵ^u_{ww}	-0.8774	0.0651	-13.4798	0.000
ϵ^u_{wl}	0.0459	0.0467	0.9826	0.326
ϵ^u_{wn}	-0.1550	0.0376	-4.1217	0.000
ϵ^u_{lw}	0.1090	0.0648	1.6838	0.092
ϵ^u_{ll}	-0.7838	0.0675	-11.6181	0.000
ϵ^u_{ln}	-0.1738	0.0403	-4.3119	0.000
ϵ^u_{nw}	-0.2149	0.0597	-3.6004	0.000
ϵ^u_{nl}	-0.2221	0.0448	-4.9579	0.000
ϵ^u_{nn}	-0.6992	0.0488	-14.3279	0.000
\multicolumn{5}{c}{Elasticities of expenditures}				
ϵ_{wC}	0.9865	0.0661	14.9305	0.000
ϵ_{lC}	0.8486	0.0618	13.7335	0.000
ϵ_{nC}	1.1362	0.0814	13.9603	0.000
\multicolumn{5}{c}{Elasticities of output}				
ϵ^c_{ww}	0.8790	0.0880	9.9890	0.000
ϵ^c_{ww}	0.7561	0.0888	8.5181	0.000
ϵ^c_{ww}	1.0124	0.1398	7.2436	0.000
\multicolumn{5}{c}{Elasticities of Hicksian technical bias}				
ϵ_{wt}	-0.0098	0.0125	-0.7821	0.434
ϵ_{lt}	-0.0332	0.0134	-2.4802	0.013
ϵ_{nt}	-0.0613	0.0223	-2.7531	0.006
\multicolumn{5}{c}{Allen elasticities of substitution}				
σ_{wl}	1.1481	0.1530	7.5027	0.000
σ_{wn}	0.5459	0.1156	4.7231	0.000
σ_{ln}	0.3545	0.1205	2.9421	0.003

Note: all elasticity estimates are calculated at the sample mean; mean values for expenditure shares for wood, productive labor, and nonproductive labor inputs are 0.36, 0.29, and 0.35, respectively.

5. TECHNICAL CHANGE

Technical change is often treated as the residual component in demand decomposition or, as above, identified as some simple function of time. We begin this section with a definition of technical change and summarize the empirical results regarding technical change.

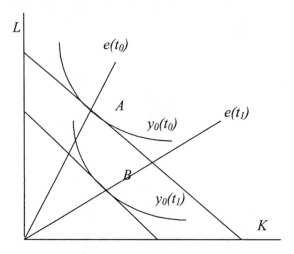

Figure 9.3. Hicks-biased technical change

One definition of technical change is the changes in a production process that come about from the application of scientific knowledge (Antle and Capalbo 1988). Graphically, technical change can be defined as a movement of the isoquants towards the origin. In a multiple-input production case, Hicks (1963) made a distinction between neutral and biased technical changes in terms of the marginal rate of technical substitution (MRTS). Figure 9.3 illustrates Hicks's concept of biased technical change in the simple two-input case, labor and capital, with a homothetic technology. The initial expansion path is $e(t_0)$, and point A is the initial equilibrium with old technology t_0. A new technology (t_1) allows the firm to move to expansion path $e(t_1)$ and point B, where the same output level (y_0) is produced, given the same factor prices. With new technology we use more capital and less labor. Hicks called this capital-using and labor-saving technical change. Technical change is Hicks-neutral if the MRTS and optimal factor proportions are unaffected.

Several methods have been developed to measure productivity. The nonfrontier (or traditional) approach generally assumes that observed production is efficient. Thus, technical change and total factor productivity are used interchangeably in this literature (Hailu and Veeman 2000). Nonfrontier approaches include index number and econometric approaches. Frontier (or efficiency) approaches assume that we rarely observe efficient production and measures inefficiency based on estimated frontier functions (see chapter 7).

Index number and econometric approaches are methods commonly used to measure technical change in the forest industries. Index number approaches relate changes in output to appropriately weighted changes in all inputs to quantify the changes in productivity using index numbers (Abt et al. 1994). Alternatively, econometric approaches use dual cost or profit functions to estimate the components of total factor productivity. These approaches often assume that technical change can be measured as simple shifts in isoquants through time or as isoquant shifts not explained by other included factors. Each approach has its own strengths and limitations. Index number approaches are simpler and more intuitive, but require more maintained hypotheses with respect to the production function.

Stier and Bengston (1992) reviewed studies published from the late 1960s to early 1990s and summarized the results. Studies indicate that technical change in the U.S. forest industry has been capital-using and labor-saving. Our estimates of Hicksian technical change using Holt and Fulcher's methodology indicate that technical change has been labor-saving for both production and nonproduction labor (table 9.3). The insignificant technical bias in the wood input equations implies that technical change has been relatively neutral with respect to wood. Both of these characterizations are consistent with the literature.

The general empirical result that technology has been neutral with respect to the wood input may reflect the physical limitations inherent in the production process, or it may mask problems with product definition and quality measures. While technology may be limited in terms of producing traditional solid wood boards from logs, the emergence of a variety of engineered wood products (oriented strand board (OSB), wood "I" beams, etc.) and the ability to utilize smaller trees of many species means that a result of wood-neutral technical change may reflect dynamic commodity definitions more than static production relationships.

Another development in the production literature particularly relevant to forest products is more complete accounting of externalities in the production process (Repetto et al. 1996). The conventional productivity measurement includes only market outputs and inputs, ignoring the external environmental effects of economic activities. The issues of carbon sequestration, biodiversity, and sustainability of the forest resource imply that internalization of environmental costs and benefits of forest inputs could drive the next wave of forest production analyses.

6. CONCLUSION

Analysis of derived demand for timber is driven by techniques of applied production analysis. The forest product demand literature has followed development of duality theory, flexible functional forms, index number theory, and related work in frontier analysis. Forest applications focus on wood input, spatial heterogeneity of the resource and resource-dependent industries, and the historical assumption of fixed proportions in the production process.

In this chapter, we have shown that the heterogeneity of individual firms may drive the observed aggregate relationships in production and the associated empirical results. Results may also be affected by the use of cross-sectional, time series, or panel data. It is common in the applications cited above to use results based on cross-sectional or panel data without an examination of the applicability of such results to long-term policy analysis, which usually takes the form of projections over time. Spatial heterogeneity of the forest resource means that the same commodity (e.g., pine) may have a location-dependent definition. It is an empirical question whether a cross-section analysis of derived demand for pine, for example, is based on observations of a common production process or perhaps on a less meaningful aggregation of related processes. The emerging techniques of spatial analysis will allow us to exploit and analyze the influence of spatial characteristics on empirical demand analysis.

This is equally true of aggregation across products. At the firm level this is a question of whether traditional product definitions (e.g., sawtimber) are relevant, given current technologies. In the aggregate, it is an empirical issue of whether wood input quality (e.g., size, species) is adequately accounted for in production analyses. To the extent it is not, the empirical results are driven by the distribution of products aggregated rather than by the characterization of a specific production process. These issues are perhaps more relevant to frontier analyses of production where the variation in firms included becomes the efficiency measure. The coevolution of production technologies, forest resource characteristics, and product definitions will undoubtedly be the subject of future analyses.

The recent consolidation of demand into large consumer-oriented retail outlets for wood and office products along with greater consumer activism in areas of sustainability, certification, and green-labeling could indicate the beginning of a new era of demand-driven resource management that goes beyond the traditional price signals. One consequence of this concentration of consumer market power is that public concerns about externalities or sustainability, which have traditionally been expressed through governmental regulation of the supply side (e.g., best management practices), are now being expressed on the demand side. The market power of influential buyers sensitive to consumer concerns is changing the arena for resource policy development. The issue of forest certification and what it means in terms of product definition, market power, and development of new markets should provide a rich data base for future studies of demand.

It may be useful as a conceptual model to posit a meta-analysis of timber demand fully indexed across products, space, and time. The variation in our results to date may be due to the fact that each analysis represents some unique combination of implicit assumptions about aggregating within and across these dimensions. Perhaps our future work will shed more light on whether or how these implicit assumptions predetermine our answers.

7. LITERATURE CITED

ABT, R.C. 1984. Regional production structure and factor demand in the U.S. lumber industry. Unpublished Ph.D. dissertation, University of California, Berkeley, CA. 98 p.

ABT, R.C. 1987. An analysis of regional factor demand in the U.S. lumber industry. For. Sci. 33(1): 164-173.

ABT, R.C., J. BRUNET, B.C. MURRAY, AND D.G. ROBERTS. 1994. Productivity growth and price trends in the North American sawmilling industries: An inter-regional comparison. Can. J. For. Res. 24: 139-148.

ADAMS, D.M., AND R.W. HAYNES. 1980. The 1980 softwood timber assessment market model: Structure, projections and policy simulations. For. Sci. Mono. 22.

ANTLE, J.M., AND S.M. CAPALBO. 1988. An introduction to recent developments in production theory and productivity measurement. *In* Agricultural productivity measurement and explanation, S.M. Capalbo and J.M. Antle (eds). Resources for the Future. Washington DC.

AUST, W.M., R.M. SHAFFER, AND J.A. BURGER. 1996 benefits and cost of forest best management practices in Virginia. So. J. Appl. For. 20(1):23-29.

CHAMBERS, R.G. 1988. Applied production analysis: The dual approach. Cambridge University Press, New York, NY.

deBorger, B., and J. BUONGIORNO. 1985. Productivity growth in the paper and paperboard industries: A variable cost function approach. Can. J. For. Res. 15: 1013-1020.

FISHER, F.M. 1993. Aggregation—Aggregate production functions and related topics: collected papers of Franklin M. Fisher, J. Monz (ed). MIT Press, Cambridge, MA. 280 p.

GREGORY, G.R. 1987. Resource economics for foresters. John Wiley & Sons, Inc.

HAILU, A., AND T.S. VEEMAN. 2000. Output scale, technical change, and productivity in the Canadian pulp and paper industry. Can. J. For. Res. 30: 1041-1050.

HAYNES, R.W. 1977. A derived demand approach to estimating the linkage between stumpage and lumber market. For. Sci. 23: 281-288.HICKS, J.R. 1963. The theory of wages. St. Martins, New York.

HICKS, J.R. 1963. The Theory of Wages. St. Martins, New York. 388 p.

HOLT, M.T., AND C.M. FULCHER. 2001. Hicksian and Marshallian input demand estimation: A differential demands systems approach. Paper presented at Agricultural Economics Workshop. North Carolina State University, Raleigh.

HOUTHAKKER, H.S. 1955-1956. The Pareto distribution and the Cobb-Douglas production function in activity analysis. Review of Economic Studies 23: 27-31.

HUMPHREY, D.B., AND J.R. MORONEY. 1975. Substitution among capital, labor, and natural resource products in American manufacturing. J. Polit. Econ. 83: 57-82.

LANDER, T.B. 1999. The employment impacts of timber harvest constraints: An analysis of the derived demand for labor in the southeastern lumber sector. Unpublished Ph.D. dissertation. North Carolina State University, Raleigh, NC. 59 p.

LICKWAR, P. 1992. Costs of protecting water quality during harvesting on private forestlands in the Southeast. So. J. Appl. For. 16(1):13-2.

MARTINELLO, F. 1985. Factor substitution, technical change, and returns to scale in Canadian forest industries. Can. J. For. Res. 15: 1116-1124.

MARTINELLO, F. 1987. Substitution, technical change, and returns to scale in British Columbian wood products industries. Appl. Econ. 19: 483-496.

MERRIFIELD, D.E., AND R.W. HAYNES. 1983. Production function analysis and market adjustments: An application to the Pacific Northwest forest products industries. For. Sci. 29: 813-822.

MERRIFIELD, D.E., AND W.R. SINGLETON. 1986. A dynamic cost and factor analysis of the Pacific Northwest lumber and plywood industries. For. Sci. 32:220-233.

MORK, K.A. 1978. Cyclical variation in the productivity of labor. MIT-EL78-002WP. MIT Energy Lab.

NAUTIYAL, J.C., AND B.K. SINGH. 1985. Production structure and derived demand for factor inputs in the Canadian lumber industry. Forest Science. 31:871-881.

REPETTO, R., D. ROTHMAN, P. FAETH, AND D. AUSTIN. 1996. Has environmental protection really reduced productivity growth? World Resources Institute, Washington DC.

SINGH, B.K., AND J.C. NAUTIYAL. 1986. A comparison of observed long-run productivity of and demand for inputs in the Canadian lumber industry. Can. J. For. Res.. 16: 443-455.

STIER, J.C. 1980. Estimating the production technology in the U.S. forest products industries. For. Sci. 26: 471-482.

STIER, J.C., AND D.N. BENGSTON. 1992. Technical change in the North American forestry sector: A review. For. Sci. 38(1): 134-159.

WEAR, D.N. 1987. The structure of production in the U.S. solid wood products industries and the derived demand for sawtimber. Unpublished Ph.D. dissertation. University of Montana, Missoula.

---

[1] We begin with well-behaved production and cost functions in our discussion. For properties of well-behaved cost functions, refer to Chambers (1988).

[2] Note that the fixed-proportion technology here refers to the technological relationship between inputs and outputs.

[3] This derivation follows Abt (1984).

[4] Fixed-proportion technology does not require that every mill exhibits the same input-output ratio. Mill proportions vary and maybe either labor intensive or wood intensive.

[5] For the complete derivation and application to U.S. agriculture, refer to Holt and Fulcher (2001).

Chapter 10

Structure And Efficiency Of Timber Markets

Brian C. Murray and Jeffrey P. Prestemon
Research Triangle Institute and USDA Forest Service

Perfect competition has long been the standard by which economists have judged the market's ability to achieve an efficient social outcome. The competitive process, unfettered by the imperfections discussed below, forges an outcome in which goods and services are produced at their lowest possible cost, and market equilibrium is achieved at the point at which the cost of the last unit supplied just equals its value in use to the demander. This point maximizes the amount of utility that consumers obtain and the profit that producers procure through the existence of the market. Therein lies the appeal of perfect competitive markets.

Evidence presented in this chapter suggests that timber markets possess certain structural characteristics that may impede perfectly competitive outcomes and thus warrant further scrutiny. Beyond the general concern about the social costs of any form of imperfect competition, why would a policy maker be particularly concerned about the lack of perfect competition in timber markets? The answer lies in the ability of the market to guide society toward an optimal allocation of resources (labor, capital, land) to forestry and forest-based production. If, for example, lack of competition among buyers were to drive down timber prices, the less profitable timber investments would be, and the less profitable forested land use would be. In principle, a lower price for timber causes less land to be in forest base than would otherwise be the case (Hardie et al. 2000). Not only does this diminish the ability of society's forests to provide the efficient amount of timber and other marketed forest outputs, it also reduces the production of nonmarket goods and services provided by forests, such as wildlife habitat, watershed protection, aesthetics, and biodiversity. Thus private market imperfections can have public goods consequences.

Sills and Abt (eds.), Forests in a Market Economy, 153–176. ©Kluwer Academic Publishers. Printed in The Netherlands.

1. MARKET AND EFFICIENCY CONCEPTS

To evaluate timber market structure and efficiency, we must define what constitutes the relevant market for analysis and describe the characteristics by which markets can be deemed efficient. This section first introduces the general concepts of market definition employed in economic analysis and then defines market efficiency in terms of perfect competition.

1.1 Definition of a Market

Aggregate economic analysis depends critically on the definitions used, either explicitly or implicitly, to define a market. For applications ranging from antitrust analysis to the identification of economic sectors for national income accounting, poorly defined markets can generate misleading results. Nobel Laureate George Stigler wrote extensively on the topic of market definition, and offers with colleague Robert Sherwin this particularly clear description of market definition and its implications (Stigler and Sherwin 1985: 555).

> *The market is the area within which price is determined: the market is that set of suppliers whose trading establishes the price of the good. If one draws demand and supply curves that do not represent the traders in a market, the intersection of the curves is economically meaningless. The infrequency with which one encounters actual market size determination outside the antitrust area is surprising and perhaps disquieting.*

Stigler and Sherwin go on to relate the market for a good to "the area within which the price of the good tends toward uniformity, allowances being made for transportation costs" (Stigler and Sherwin 1985: 555). As the passage above suggests, and the empirical work below will show, price is the critical datum for market definition analysis.

It is commonly asserted that commodities are in the same market if their prices move in a parallel fashion. This phenomenon is referred to as the Law of One Price (LOP). The LOP implies that the price of a commodity is equal across all points of production—i.e., there are no opportunities for spatial arbitrage behavior. However, it should be clarified that prices moving together are a necessary but not sufficient condition for firms to be in the same market (McNew and Fackler 1997, Ravallion 1986, Tirole 1988). Commodity prices may move together due to common macroeconomic forces that have little to do with product similarity. Thus another factor to consider in deciding whether products are in the same market is their degree of substitution with each other. Maize from one farm in Iowa is, for all intents and purposes, perfectly substitutable for maize of the same grade on the farm next door, and thus they are clearly in the same market. We can fully expect their prices to be identical. Whether the maize in Iowa is perfectly substitutable with and in the same market as maize in California is a different matter that requires some empirical analysis to sort out.

Economic theory implies methods for evaluating the extent of the market in the space dimension. Samuelson (1952) described how homogeneous products that are produced in perfectly competitive markets without government intervention should maintain the LOP in that prices in two spatial points of production should differ within the same time period by no more than the cost of product transfer between points. Ravallion (1986) described this as the measure of market integration.

1.2 Perfect Competition and Timber Market Characteristics

The essential structural characteristics of a competitive market are:
1. A large number of buyers and sellers
2. Free entry and exit into and out of the market
3. Products that are identical/homogeneous
4. Buyers and sellers who have perfect information

While economic philosophers since Adam Smith (1776) have argued whether these conditions guarantee perfect competition or, indeed, whether perfect competition guarantees the best possible social outcome, we do not intend to join that debate here. Rather, our emphasis is on whether timber markets possess the procompetitive conditions just referenced and if not, why, and what the implications are. We proceed by briefly evaluating each of the competitive conditions as it relates to the unique characteristics that often pertain in timber markets.

1.2.1 Number of Buyers and Sellers

A large number of buyers and sellers ensures that any one agent cannot deviate from the established market price without either losing their entire market share to the competition or (infeasibly) attracting the entire market from the competition. This fortifies the competitive process to generate a market price that efficiently equates marginal cost on the supply side and marginal benefit on the demand side. Concentrated markets are those with relatively few sellers, buyers, or both. Concentrated sellers (buyers) are more likely to have some power over the price they receive (pay) in the market, because they are large relative to the market and thus can be thought of as less susceptible to losing their entire customer (supplier) base if they raise (lower) the price.

If we judge competition in timber simply by the absolute number of sellers and buyers at the industry level, then the verdict would often fall on behalf of competition rather than concentration. In the United States, for instance, there are millions of private owners of timber. These owners, most states, and the federal government sell timber on the open market to about 4,400 sawmills, 550 pulp and paper mills (U.S. Census Bureau 1999), and thousands of other wood processing outlets. Given these facts, it is unlikely that any single seller or buyer could exert significant market power at the national level. Large numbers of market participants are the norm in other countries as well. Yet despite these numbers, some questions remain as to whether all or even most timber markets have a sufficient number of participants to ensure something close to perfect competition.

Many forms of timber processing, especially pulp and paper manufacture, exhibit scale economies and therefore need relatively large areas of harvestable timber to meet their minimum efficient scale requirements. The existence of these scale economies and the fact that standing timber is land-intensive and immobile suggest a propensity toward a pattern of spatial dispersion of processors. However, timber is a bulky commodity with transport costs that can be high relative to its production value. Consequently, timber transactions between buyers and sellers may be highly localized, with the spatially dispersed processors having strong locational advantages for acquiring the timber closest to their mills (Lofgren 1985, Lowry and Winfrey 1974, Mead 1966, Murray 1992). Such a pattern works to limit the number of buyers with which a seller can effectively contract, which would seem to violate the first requirement of perfect competition. In the context of markets for harvested timber (logs) as a production input, this type of imperfection is referenced as monopsony or oligopsony power. If geography creates these localized zones of potential buyer market power, this can undermine the LOP referenced earlier, as buyers in different locales will have different abilities to deviate from the perfectly competitive price. In this case, price movements in these markets would be nonparallel.

1.2.2   Free Entry and Exit

Free entry and exit from a market is not the same as costless entry and exit. Free entry instead implies that potential market entrants are assumed to exist and are not impeded from entry or exit by technical or institutional barriers. Unimpeded entry allows potential competitors to impose discipline on the incumbents. If a firm that buys a demanded input to production were to try to exert market power by offering a price for the input that is below its marginal value as a production input, a profit incentive would be created that would encourage the entry of new producers into the market. As long as the profits for new entry are greater than the entry cost, new entrants would ultimately bid up the price to the competitive (zero profit) level. Unimpeded exit means that potential entrants would still find it profitable to exert their competitive influence even if their entry into the market were only temporary.

A relevant question is whether free entry and exit conditions exist in most timber markets. The construction of a pulp or paper mill, for example, may cost $1 billion (US), but as alluded to previously, cost itself is not an entry barrier (with respect to the four characteristics of perfect competition enumerated). However, the existence of scale economies (declining marginal and average costs) over a wide range of output can make it unprofitable to enter if the entrant cannot attract enough of the market to operate at the cost-efficient scale. With transportation costs limiting the effective size of the market that the entrant can access, significant scale economies can serve as a technical barrier to entry.

An entry barrier would also exist if an incumbent has a property right that the potential entrant cannot readily obtain. This may be an institutionally granted right, such as a government permit to operate, or a competitive asset such as proprietary technology, name recognition, goodwill, or some other firm-specific asset that is not easily replicable. For timber processors, one asset is timberland that may be used to directly supply the mill. If there are certain economies associated with both supplying and processing timber, then only processors vertically integrated into timber supply (timberland holdings) can enjoy those economies of scope. Whether vertical integration constitutes a significant entry barrier is open to debate. Some forest product firms, indeed, have large timberland holdings that they consider a strategic asset, critical to their competitive stance (e.g., Weyerhauser in the United States). Other firms, however, have succeeded with relatively little of their timber procured from company-owned lands. The strategic role of firm-owned timberland is apparently evolving through time, with some large companies spinning off their timberland divisions into separate entities under the firm's umbrella or divesting of timberland altogether. Georgia-Pacific Corporation has employed both strategies in the late 1990s and early 2000s. Thus the role of timberland holdings as a potential entry barrier remains murky.

1.2.3 Homogeneous Goods

When products are homogeneous, one competitor's product is identical to another's. In these situations, there is no scope for the type of product differentiation that can insulate market participants from direct price competition. From a physical content standpoint, timber—once separated into standard commercial grades—is a fairly homogeneous commodity compared, say, to consumer goods such as pharmaceuticals or wine. However, as discussed above, the location of timber relative to its potential buyers may introduce a form of product heterogeneity that can undermine direct competition in the market. Further, if we can define a product as timber of any quality or simply of a similar species on the stump and recognize that harvest costs and timber quality can vary substantially across space due to site and stand conditions, then product homogeneity diminishes.

1.2.4 Perfect Information

All parties to a transaction should have full information about the relevant terms of the exchange (e.g., price, quantity, and quality) for market outcomes to be efficient. When all information is available, buyers and sellers can refuse to deal when they know that an alternative transaction is more attractive. This knowledge enforces the economic concept of Pareto efficiency by eliminating the possibility that market transactions could be rearranged *ex post* to make some parties better off without making other parties worse off.

Information requirements are not trivial in timber markets. When a tract of timber is up for sale, the quantity of harvestable timber is not known with certainty. As a result, both parties to the exchange have an incentive to obtain quantity information by conducting a timber cruise. Timber cruising involves costs, and both parties may not be willing to incur the costs, leaving one party more informed than the other. This may be exacerbated by the fact that many small timber suppliers are relatively infrequent participants in the timber market and thus may not be as effective in obtaining the information necessary to guarantee the best possible outcome. With asymmetric information, one party is systematically more informed than the other on the terms of the exchange and may have the opportunity to use that information to its advantage, thereby eroding the terms of potential Pareto efficiency.

2. EVALUATION OF MARKET POWER IN THE TIMBER SECTOR

This section describes how the spatial dimension within a timber market might produce an environment conducive to localized buyer market power and then reviews the empirical evidence from the literature on whether such market power is empirically evident in actual market settings.

2.1 Pricing Model for Spatially Differentiated Input Market

Because transport costs are a large component of the delivered cost of wood, the markets for wood inputs might best be described as localized or spatially differentiated in the tradition of Hotelling (1929).[1] Here we discuss how the localized nature of these inputs may provide some degree of local market power for the wood processing mills.[2]

Suppose wood processing mill j offers a price of W_j^W for wood delivered to the mill. Timber growers at a distance of d from mill j receive a price for their stumpage (standing timber) of $p^S = W_j^W - h - t(d)$ when they sell to mill j, where h is a per-unit harvest cost of the timber and $t(d)$ is the transport cost as an increasing function of the distance. An individual timber grower will sell $x^S = f(p^S)$ units of timber $(\partial x^S/\partial p^S > 0)$ to the mill from which they receive the highest price. At some distance from mill j in each direction that they face spatial competitors, timber growers receive the same stumpage price from mill j and a competing mill. Thus, if mill j raises the price offered at the mill, two phenomena may occur: the amount supplied to the mill by its inframarginal sources will increase due to the uniformly higher p^S, and, depending on rival mills' responses, it may expand into those rivals' markets by pushing out the distance at which timber growers are indifferent between mill j and its rivals. Thus, the total amount of wood supplied to mill j from all timber growers in its market, X_j^W, is an increasing function of the mill price it pays, $X_j^W = g(W_j^W)$.

The magnitude of the response to a change in the mill's price depends on two factors, the technological nature of the timber grower's unit supply function and the relative intensity of the border competition with rivals. If transport costs per unit distance are high relative to the full delivered cost of the input, the latter, competitive boundary effect is weak, and the technological response of the inframarginal suppliers dominates, thus enhancing the potential for the exercise of local market power. If transport costs are low and/or mills are densely distributed, spatial differentiation is low, and nearly perfect competition might be expected if other product attributes are undifferentiated.

In addition to the wood input, mill j employs other primary factors of production, denoted by the vector, X_j^W, to produce output, Q. We assume price taking in outputs and all other inputs and that firms maximize mill profits:

$$Max \prod = PQ - W^P X_j^P - W_j^W X_j^W \qquad 10.1$$

Here, P is the price of output and W^P is the price vector of nonwood inputs. $W_j^W = g^{-1}(X_j^W)$ is the inverse of the wood input supply function discussed earlier. First-order conditions are:

$$P(\partial Q / \partial X_j^P) = W^P \qquad 10.2a$$

$$P(\partial Q / \partial X_j^W) = W_j^W + \frac{\partial W_j^W}{\partial X_j^W} X_j^W \qquad 10.2b$$

For the nonwood inputs we get the perfectly competitive result: inputs are employed until their value of marginal product (VMP) equals market price. Optimizing with respect to wood use requires accounting for the effect of the firm's wood consumption level on the price it pays. Manipulating the first-order condition for the wood input equates an input's VMP with its marginal factor cost (MFC).

$$P(\partial Q / \partial X_j^W) = W_j^W \left(1 + \frac{1}{E_j}\right) \qquad 10.3$$

where $E_j = (\partial X_j^W / \partial W_j^W)(W_j^W / X_j^W)$ is the delivered price elasticity facing mill j. Also, $E_j = E/\theta_j$, where $E = (\partial X^W / \partial W^W)(W^W / X^W)$ is the elasticity of aggregate timber supply with respect to a uniform change in delivered price and

$$\theta_j = r_j^I / (r_j^I + r_j^B) \qquad 10.4$$

where r_j^I is the inframarginal supply elasticity and r_j^B is the border supply elasticity. r_j^I is determined by the technology of the unit timber supply function, and r_j^B is determined by the intensity of competition at the spatial border between rivals. If there is no spatial differentiation (e.g., transport costs are zero), then r_j^B is infinite and θ_j equals zero, reflecting perfect competition. At the other extreme, if there is no border competition, then r_j^B equals zero and θ_j is one, reflecting pure spatial monopsony. Thus θ_j, with values in the 0 to 1 interval, can be interpreted as an index of mill j's market power in wood inputs, comparable to Appelbaum's (1982) conjectural elasticity (CE) term.

Under imperfect competition, the firm faces a finite elastic input supply function, presumably positive with respect to price (i.e., $0 < E_j < 1$). Consequently, VMP$_j$ exceeds the input price, W_j^W. Alternatively, under perfect competition ($q_j = 0$), the firm is a price taker ($E_j = 4$) and price equals VMP, as in the case of the nonwood inputs. If the correspondence between price and VMP can be determined empirically, then we can infer how competitive the market is.

2.2 Empirical Evidence on the Degree of Market Power in Timber Markets

The estimation of market power has been a prominent component of empirical industrial organization for years, stemming from the classic work of Lerner (1934) and continuing with the fusion of game theory, producer theory, and econometrics to provide structural measures of market power (e.g., Appelbaum 1982, Bresnahan 1987, Iwata 1974).[3]

Historically, most of the attention in empirical studies of market power has been directed toward the analysis of output markets, but more emphasis has recently been placed on imperfect competition in inputs. Our focus is on timber as a production input. Livestock commodity markets, though, which have some of the same structural features as timber markets (high transport costs, scale economies in processing), have received a fair amount of recent attention in NEIO-based empirical studies of input market power (see the 14 studies referenced in Azzam [1998, table 2] and Muth and Wohlgenant [1999]).

Mead's (1966) study of market power by lumber producers in the U.S. Pacific Northwest marked the earliest comprehensive analysis of timber market structure. Mead employed a combination of an industry case study and structure-conduct-performance methods to examine the competitiveness of both lumber output markets and sawlog input markets. Mead concluded that the lumber markets he studied were competitive, while sawlog markets were moderately oligopsonistic.

Lowry and Winfrey (1974) examined oligopsony in the pulpwood markets of the U.S. South. Using informal methods, they argued that the (assumed) oligopsonistic structure of pulpwood markets both dissuades forest investment by nonindustrial private forest landowners and encourages vertical integration by processing firms. But Lowry and Winfrey do not formally test the hypothesis that oligopsony exists in the market.

Scandinavian economists have taken up much of the remaining empirical work in timber oligopsony power. Sweden, in particular, has a historically unique institutional structure of roundwood markets—something akin to a bilateral monopoly of buyer and seller cooperatives—that has made the issue particularly relevant there. Johansson and Lofgren (1983, 1985) employed theoretical arguments to explain seemingly irrational responses, such as excess demand for roundwood, via a model of monopsony price discrimination and capacity constraints. Lofgren (1985) uses a spatial oligopsony model and an empirical characterization of roundwood supply functions to explain the outcomes of a roundwood price negotiation process between the buyer and seller cooperatives. Brannlund (1989) measures the social welfare costs of Swedish pulpwood market power under the assumption that the markets are purely monopsonistic and sawlog markets are perfectly competitive. The welfare costs he estimates are large, but monopsony power is assumed rather than tested.

The first studies to use a formal theoretical structure and econometric methods to test hypotheses about timber market oligopsony power were produced concurrently by Murray (1995a) and Bergman and Brannlund (1995). Murray employs a dual profit function approach to determine the degree of oligopsony power within the estimation of a system of output supply and factor demand equations. Using aggregate data for the United States during the period 1958 to 1988, the econometric results suggest that sawlog markets are perfectly competitive throughout the period, while pulpwood markets exhibit some episodes of oligopsony power. The Murray study is examined in more detail in section 2.2.1.

Bergman and Brannlund (1995) estimate the degree of oligopsony in the Swedish pulpwood market. Their methods and findings are similar to Murray's for the United States, suggesting that Swedish pulpwood markets are largely oligopsonistic with varying degrees of market power over time. Bergman and Nilsson (1999) extend the earlier work of Bergman and Brannlund by adding detail on pulpwood inputs from Nordic countries. They find, in contrast, that perfect competition cannot be rejected in Swedish pulpwood markets. Ronnila and Toppinen (2000) test for pulpwood market oligopsony in Finland and, assuming constant market power over time (1965 to 1994), suggest Finnish pulpwood markets have, on average, been competitive over time, with some evidence of market power in the wood chip market.

2.2.1 Empirical Example: Estimation of Oligopsony Power in U.S. Timber Markets

This section briefly describes in more detail Murray's (1995a) study of oligopsony power in U.S. sawlog and pulpwood markets. Murray's study combines modern production theory (duality) with econometric methods to test hypotheses in their structural form. The basis of Murray's model is a timber processor's restricted profit function:

$$\Pi_t = \Pi(P_t, W_t^L, W_t^M, Z_{Kt}, Z_{Wt}, t) \qquad 10.5$$

P_t is the price of the processed output (e.g., lumber, paper) in period t, W_t^L is the labor wage, W_t^M is the price of a composite material input, Z_{Kt} is the quantity of capital used as a production input, and Z_{Wt} is the quantity of wood processed. With an unrestricted profit function, only prices would occur as arguments in the function, indicating that input quantities can freely move to their optimal level in response to changes in prices. A restricted function, though, imposes constraints by replacing the input price with an input quantity, the Z variables in equation 10.5, which are referred to as quasi-fixed factors (QFFs) of production.

The QFFs typically are often imposed on time series models, such as the one employed by Murray, in recognition of rigidities in the ability of capital to optimally adjust in the short run. That is the nature of Z_{Kt}'s role as a QFF in Murray's model. However, wood is employed as a QFF by Murray (1995a) for a different reason. One of the problems hindering the estimation of input market power in the studies predating Murray was the inherent difficulty in measuring the VMP of the input in question. Under perfect competition, the market price of the input is assumed to equal VMP. However, as indicated above, the market price does not equal VMP under imperfect competition. Thus, the analyst must use some other information to estimate VMP. Studies prior to Murray (Azzam and Pagoulatos 1990, Schroeter 1988) either used production functions, with the attendant econometric problems, or impose symmetry between input and output market power, which may not be appropriate for evaluating the forest products sector, to estimate VMP. Murray's solution was to specify wood, the oligopsonized input, as a QFF in the restricted profit function, Z_{Wt}. This specification enables the computation of a shadow price for wood by taking the derivative of the restricted profit function with respect to the wood QFF.

$$\lambda^W = \frac{\partial \Pi}{\partial Z_W} \qquad 10.6$$

Using t to index the time period, the shadow price, λ_t^W, provides an estimate of wood's VMP in the production process, thereby providing a means around the VMP estimation described earlier. Taking together equations 10.6 and 10.3, Murray imposes an oligopsony condition to be estimated econometrically:

$$W_t^W\left(1+\frac{\theta_t}{E}\right) = \lambda_t^W \qquad 10.7$$

where W_t^W is the market price of wood, E is the wood supply elasticity, and θ_t is the CE parameter which, as described earlier, serves as the market power index bounded by zero and one. The CE parameter is specified as a function of several exogenous variables, X_t:

$$\theta_t = \theta(X_t) \qquad 10.8$$

The market power index can be written in terms of the markdown between the shadow price and the industry wood price:

$$\theta_t = E\left(\frac{\lambda_t^W - W_t^W}{W_t^W}\right) \qquad 10.9$$

Estimating the gap between the shadow price and the industry price lies at the heart of empirically revealing the structural parameter, θ.

The bracketed term above is the input analog to the well-known Lerner's index of the magnitude of the monopoly price distortion, [(P–MC)/P], so that the relationship between the CE parameter and the Lerner input index is:

$$L = \frac{\theta}{E} \qquad 10.10$$

Equations 10.6 through 10.10 are combined by substitution and estimated jointly with an output supply function and input demand functions for labor and materials. The econometric system is estimated to determine the aggregate degree of oligopsony power in wood markets for the two largest wood processing sectors in the United States: the sawmilling sector, which processes sawlogs, and the combined paper and paperboard mill sectors, which process pulpwood. The system of equations is estimated using an iterative nonlinear variant of a seemingly unrelated regression system.

Among the parameter estimates presented by Murray, the CE parameters are of most interest to this study. The values for the sawlog and pulpwood markets are presented separately in table 10.1. CE values are computed for 5-year intervals throughout the sample period. The parameter covariance matrix is used to compute standard errors and t-statistics for each CE value. Computing t-statistics provides a test of the price-taking assumption by suggesting acceptance or rejection of price taking ($\theta = 0$) at a specified level of significance. For the entire period, the average degree of oligopsony power in sawlog markets is relatively low, as indicated by a mean θ estimate of 0.042. The highest mean value of approximately 0.10 is found in the earliest years of the sample. The value of θ declines throughout the period and falls below typical levels of significance after 1978, as indicated by the low t-statistics. The t-statistic at the sample mean is approximately 2.2, indicating rejection of the price-taking hypothesis, $\theta = 0$, at the 5% level for the sample period as a whole.

Table 10.1. Market power indicators 1958 through 1988 (from Murray 1995a)

Industry or Period	Conjectural Elasticity (θ)	T-statistic	Lerner Index (L)
U.S. sawmilling (sawlogs)			
1958–62	0.0977	3.3463	0.5229
1963–67	0.0696	3.2442	0.3713
1968–72	0.0432	2.6896	0.2669
1973–77	0.0252	1.9052	0.1662
1978–82	0.0149	1.3985	0.1061
1983–88	0.0102	0.9193	0.0637
1958–88	0.0424	2.2076	0.2435
U.S. paper/paperboard (pulpwood)			
1958–62	0.4146	3.6957	0.7083
1963–67	0.2621	3.3382	0.4322
1968–72	0.1169	2.1975	0.1806
1973–77	0.0686	1.4423	0.0892
1978–82	0.0553	1.3264	0.0857
1983–88	0.1342	2.2990	0.2296
1958–88	0.1740	2.3805	0.2857

The degree of market power is higher for pulpwood than for sawlogs. The sample mean value of θ is 0.174, with a t-statistic of 2.4, indicating statistical significance at the 5% level. Statistical significance at 5% is maintained throughout most of the period, as implied by the periodic mean values for the t-statistic. The exception is the two periods spanning 1973 through 1982, where t-statistics are well below 2. The mean value of θ is highest in 1958 through 1962 at 0.415 and declines from 1978 through 1982 to a statistically insignificant level of 0.055, at which point the trend reverses to a statistically significant value of 0.134 for 1983 through 1988.

The estimates of higher oligopsony power in pulpwood markets than in sawlog markets are not surprising, given the presence of larger, relatively isolated mills in the pulp and paper sector and smaller, more densely distributed sawmills. These findings are consistent with other commentaries on the structure of timber input markets referenced in the preceding section.

The prevalent decline in market power over time indicated by the results is intuitive as new entrants and capacity expansion eat away at the bargaining strength of incumbent processors. This is consistent with the erosion in output market power over time found by Appelbaum (1982). However, the late period reversal of the declining trend in pulpwood market power is a curious phenomenon, which Murray speculates might be explained by the time pattern in the use of wood chips by pulp mills.

The low but statistically significant values of θ for sawlogs in the earlier years of the sample indicate near price-taking behavior. However, even a small amount of market power corresponds with nontrivial price distortions, as evidenced by L values in excess of 0.25 for the early years. This occurs because of the relatively limited short-run price responsiveness of sawtimber growers, which is reflected in low price elasticity for sawlogs. Consequently, even the intense border competition implicit in the low values of θ afforded the capture of some oligopsony rents by sawlog processors in the earlier time periods of the study. Those rents more or less dissipated over time.

In a related paper, Murray (1995b) uses the results from table 10.1 to estimate the magnitude of social welfare costs from the pulpwood market distortions in the later years of the sample (when sawlog markets are held to be competitive). Applied to data from the southern United States, the study finds that the primary welfare effects of the oligopsony distortion are a sizeable transfer of wealth from open market timber suppliers to the pulp and paper industry, but relatively small absolute welfare efficiency costs. Murray (1995b) also examines the implications of assuming *a priori* that pulpwood markets are purely monopsonistic (i.e., setting $\theta = 1$) rather than using the econometric estimates of θ in table 10.1. Under that assumption, absolute welfare estimates are of similar magnitude to Brannlund's (1989) estimate for Sweden, using the same monopsonistic assumption.

3. EMPIRICAL EVIDENCE ON TIMBER MARKET EFFICIENCY

Economic theory of market efficiency (Fama 1970, 1991; Muth 1961) implies methods for evaluating the degree of market perfection in the time and space dimensions. Recall from section 1, that products produced in different places or by different firms are identical, and that buyers and sellers have perfect information regarding factors that shift supply and demand. Together these factors imply that time series of market prices for a good should not contain predictable changes between periods (LeRoy 1989) and that time series of prices of the same good produced in different places should move together (the LOP). Otherwise, economic agents would be able to exploit opportunities for arbitrage across locations and time, and thus the market would not be in equilibrium.

Several empirical analyses have tested the statistical and economic importance of spatial separation and product homogeneity on the functioning of wood product markets. These studies start with the contention that if spatial separation or product distinctions generate conditions for imperfect markets, then statistical analyses of prices should reveal the effects of these factors. Notable examples span a range of forest products and include studies by Buongiorno and Uusivuori (1992), Jung and Doroodian (1994), Hänninen et al. (1997), Murray and Wear (1998), and Nagubadi et al. (2001). These studies broadly conclude that LOP holds across the spatial dimension, albeit over broadly aggregated areas and time periods.

As in studies of spatial market price behavior, empirical studies of prices in the time dimension are partly motivated by a desire to better understand the effects of policies and shocks on markets. Analyses of timber markets in North America and Europe showed mixed results with regard to price behavior. Washburn and Binkley (1990, 1993), Hultkrantz (1993), and Prestemon and Holmes (2000) tested the price behavior of timber in the southern United States. Most of these studies were simply tests of the theory of efficient markets (that all available information is reflected in the current price, thus no intertemporal arbitrage opportunities exist), not perfect markets (see section 1.2 of this chapter). Williams and Wright (1991) and Deaton and Laroque (1992, 1996) offer one set of critiques in the context of storable products about whether efficiency can even be evaluated using univariate time series tests when a commodity is storable.

Market efficiency research also addresses the applicability of alternative harvest timing rules. An active area of research in resource economics since the mid-1980s has been the development of harvest timing rules in the presence of stochastic prices (Abildtrup et al. 1997, Brazee and Bulte 2000, Brazee and Mendelsohn 1988, Clarke and Reed 1989, Forboseh et al. 1996, Gong 1999, Haight and Holmes 1991, Lohmander 1988, Norstrøm 1975, Plantinga 1998, Thomson 1992, Yin and Newman 1996). A key question arising from the harvest timing literature is how observed price behavior, harvest timing rules, and market efficiency are related. For instance, if prices are predictable, then harvests can be timed to take advantage of high price periods. This would imply unexploited opportunities for intertemporal arbitrage and thus inefficient markets.

The harvest timing research has implications for optimal resource allocation. If prices are mean-reverting (stationary), then the applicable harvest timing rule, such as that developed by Brazee and Mendelsohn (1988), implies that timber can yield much higher profits than previously supposed (e.g., compared to profits from a static Faustmann harvest timing rule), because suppliers can capitalize by harvesting more (less) when prices are above (below) the stationary mean. This finding, derived from simulations, implies that many timber producers may not be using all available information to time timber sales and harvests, leading to lower than possible profits. The result is an aggregate misallocation of resources toward alternative uses of land and away from timber production. Harvest timing rules based on integrated (e.g., random walk) price behavior (e.g., Haight and Holmes 1991), on the other hand, imply no significant misallocation of resources. If prices are indeed nonstationary and producers time harvests according to such price expectations, then all the available investment strategies and options for land use are being appropriately evaluated, indicating an efficient market. Haight and Holmes (1991) quantify through simulation that timing with nonstationary prices offers small or insignificant extra advantage over the profits and hence equilibrium land values yielded by using a Faustmann approach. In the section that follows, we evaluate whether all useful information is being used to determine the current market price, a measure of how well characteristic 4 of the perfect market is being met. In this analysis, if prices are found to be stationary processes, then substantial misallocation of resources may be occurring in land-intensive production. If prices are found to be nonstationary, then perhaps no substantial misallocation is occurring. The example is applied to the U.S. South, one of the world's primary timber producing regions, where private production dominates and where markets are largely left to themselves to determine going prices and production levels.

3.1.1 Time Series Behavior of U.S. Southern Pine Stumpage Markets

To illustrate the empirical work in understanding market efficiency in the time dimension, we describe results of tests of whether prices are stationary or nonstationary. In previous work on the topic, tests of timber price behavior have been mainly limited to procedures with nonstationarity as the null hypothesis, which have been shown to be weak for near-unit root processes (e.g., Schwert 1989), which timber prices might be.[4] These include Dickey-Fuller type tests (Dickey and Fuller 1979, Said and Dickey 1984). But there are tests that take the opposite null, for example, that the series follows a stationary AR(p) process. Those developed by Kwiatkowski et al. (1992) and Leybourne and McCabe (1994) are examples of these. Use of these might add greater confidence to conclusions or dampen our certainty about the true nature of timber price processes. In fact, the principal contribution of the price behavior research presented below is the assessment of price behavior using alternative testing procedures.

In our empirical example, we apply augmented Dickey-Fuller (ADF) and Leybourne-McCabe (Leybourne and McCabe 1994) approaches. The timber price series evaluated are for southern pine sawtimber and pulpwood stumpage from the U.S. South. These include 18 series that emanate from substate regions. Although species may differ in characteristics across the region, the southern pines (especially *Pinus taeda, P. elliottii,* and *P. echinata*) are quite uniform in appearance and application. The lumber deriving from them is classified using a common set of grading rules (Forest Products Laboratory 1987:1-17), and few differences exist among these primary species in the quality of their fiber for pulp. Together, pulp and lumber comprise over 80% of their demands (Prestemon and Abt 2002). Southern pine timber prices have been the subject of a large part of the timber price research occurring in the United States, given that this market is so active and dominated by private producers and thus sheds light on the market efficiency question.

The ADF test is conducted by regressing the current change in price (dy_t) on lagged changes in prices and a single lag (y_{t-1}) of the current price:

$$dy_t = \alpha y_{t-1} + \sum_{i=1}^{I} \beta_i dy_{t-i} + \gamma + e_t \qquad 10.11$$

where $dy_t = y_t - y_{t-1}$, the size of I may be determined by a model selection procedure (e.g., Hall 1994), and γ is a constant. The existence of a unit root can be determined by whether an estimate of α in OLS estimation of equation 10.11 differs significantly from one.

The Leybourne-McCabe method evaluates the null hypothesis that a series is a stationary autoregressive of order p [AR(p)] process against an ARIMA (p,1,1) alternative. In terms of market efficiency testing, the Leybourne-McCabe test seeks sufficient empirical evidence to support an alternative hypothesis that market prices are consistent with an efficient market (prices are *not* stationary). This is in contrast to the ADF, whose alternative is that prices may be consistent with market inefficiency (prices *are* stationary). In other words, the Leybourne-McCabe test is powerful against the hypothesis for which the ADF is weak. This score-based stationarity test begins with the maximum-likelihood estimate of the parameters $\phi = (\phi_1, \phi_2, ..., \phi_J)$ obtained by fitting the ARIMA model:

$$dy_t = \beta + \sum_{j=1}^{J} \phi_j dy_{t-j} + \varsigma_t + \theta \varsigma_{t-1} \qquad 10.12$$

and then constructing the series,

$$y_t^* = y_t - \sum_{j=1}^{J} \phi_j^* y_{t-j} \qquad 10.13$$

where the ϕ_j^* are estimates of ϕ_j from equation 10.12, and the size of J in equations 10.12 and 10.13 are set beforehand. Two possible least-squares regressions can be estimated: regressing y_t^* on an intercept (no-trend case), or regressing y_t^* on an intercept and a trend (deterministic time trend case). The test is conducted on the residuals, $\{\varepsilon^*\}$, from either of these two possible regressions, in the following manner:

$$s^* = \sigma_\varepsilon^{*-2} T^{-2} \varepsilon^{*\prime} V \varepsilon^* \qquad 10.14$$

where $\sigma_\varepsilon^{*2} = \varepsilon^{*\prime} \varepsilon^* / T$ and V is the covariance matrix of a nonstationary series, where the elements v_{jk} of V are the min(j,k). Critical values for the statistic s^* are tabulated for both the no-trend and the deterministic time trend case by Kwiatkowski et al. (1992) and are applicable to the Leybourne-McCabe test. Leybourne and McCabe showed in simulations how the number of lags (J) of differenced prices in equation 10.12 does not appreciably affect the outcome of the test. In our analysis, we use the deterministic time trend case, since a time trend, consistent with the stationary null, can exist in an inefficient market. Thus, if we find that the value obtained from equation 10.14 is greater than that expected, then sufficient evidence exists that prices are not stationary and support nonstationarity. That result would be consistent with an efficient market.

Price data used in the analysis are the quarterly price data available from Timber Mart-South (Norris Foundation 1977 to 2002). Timber Mart-South is a report for timber submarkets throughout the U.S. South including prices for up to 22 price regions, two regions per state. The periods of monthly reported prices (1977 to 1987) were converted to quarterly by mid-month sampling. Our results rely on a kind of temporal aggregation of price series likely to be present in the Timber Mart-South data, which could lead to power reduction in tests for a unit root (e.g., Haight and Holmes 1991, Taylor 2001). This should be kept in mind when evaluating the following results. In states where region redefinitions occurred in 1992, the weighting correction approach recommended by Prestemon and Pye (2000) was applied to the pre-1992 series. A few series had missing data, so the ADF and Leybourne-McCabe tests were conducted only for 18 of the 22 series. The southern pine sawtimber and pulpwood stumpage prices were deflated by the consumer price index for all urban consumers. The ADF was conducted using the Hall (1994) general-to-specific procedure, beginning with $I = 16$ lags, finding the specification with the minimum of the Schwarz Information Criterion, holding the number of usable observations constant at 85. The Leybourne-McCabe included $J = 4$ lagged difference terms.

Results (table 10.2) show that both the ADF and the Leybourne-McCabe tests concur that southern pine sawtimber stumpage prices are nonstationary.

Table 10.2. Results of Leybourne-McCabe and ADF tests for southern pine sawtimber stumpage prices deflated by the consumer price index (1977:1-2002:2)

Submarket	Leybourne-McCabe Test		ADF Test			
			Statistic		Obs.	Lags
Alabama-1	1.74	***	-2.43		85	0
Alabama-2	1.45	***	-2.44		85	0
Arkansas-1	1.83	***	-2.65	*	85	0
Florida-1	1.51	***	-2.19		85	1
Florida-2	0.82	***	-1.57		85	5
Georgia-1	0.84	***	-1.54		85	11
Georgia-2	0.12		-1.07		85	5
Louisiana-1	1.78	***	-1.40		85	6
Louisiana-2	1.85	***	-2.44		85	0
Mississippi-1	1.38	***	-2.27		85	0
Mississippi-2	1.63	***	-1.59		85	2
North Carolina-1	1.73	***	-1.29		85	4
North Carolina-2	1.89	***	-0.86		85	5
South Carolina-1	1.45	***	-1.47		85	11
South Carolina-2	1.69	***	-2.30		85	0
Texas-1	1.78	***	-2.39		85	0
Texas-2	1.78	***	-1.35		85	5
Virginia-2	0.08		-3.07	**	85	0

Asterisks = rejection of tests' respective nulls, at 10% (*), 5% (**), and 1% (***) significance.

Two exceptions, at 5% significance, are those reported for Georgia (region 2), where the Leybourne-McCabe test could not reject the null of a stationary AR(p) process, and Virginia (region 2), where the Leybourne-McCabe and the ADF (at 5% significance) favored the AR(p) process over a nonstationary process.

For pulpwood (table 10.3), both the Leybourne-McCabe and the ADF tests agreed that series are stationary in two cases, with the former not rejecting the null of an AR(p) process (at 1% significance) and the latter rejecting the null of a nonstationary price (at 10% significance) in favor of an AR(p) process for Louisiana's two price regions. The tests appear at odds for North Carolina (region 2) and Texas (region 1). We can conclude, however, based on these results, that intertemporal arbitrage has worked successfully to eliminate much predictability of both southern pine pulpwood as well as sawtimber stumpage prices, with perhaps slightly more evidence of unpredictability and thus efficiency in the sawtimber market.

Table 10.3. Results of Leybourne-McCabe and augmented ADF tests for southern pine pulpwood stumpage prices deflated by the consumer price index (1977:1 to 2002:2)

Submarket	Leybourne-McCabe Test		ADF Test			
			Statistic		Obs.	Lags
Alabama-1	0.70	***	-2.24		85	12
Alabama-2	0.76	***	-1.35		85	0
Arkansas-1	0.45	***	-1.45		85	1
Florida-1	0.59	***	-1.89		85	0
Florida-2	1.02	***	-0.68		85	8
Georgia-1	0.92	***	-2.39		85	15
Georgia-2	0.76	***	-0.96		85	12
Louisiana-1	0.11		-2.65	*	85	0
Louisiana-2	0.10		-3.90	***	85	0
Mississippi-1	0.70	***	-2.42		85	13
Mississippi-2	0.87	***	-1.43		85	6
North Carolina-1	0.72	***	-1.47		85	11
North Carolina-2	0.35	***	-3.02	**	85	11
South Carolina-1	0.97	***	-1.56		85	4
South Carolina-2	0.76	***	-1.50		85	7
Texas-1	0.34	***	-3.00	**	85	4
Texas-2	0.38	***	-2.52		85	0
Virginia-2	0.07		-2.30		85	12

Asterisks = rejection of the tests' respective nulls, at 10% (*), 5% (**), and 1% (***) significance.

This analysis carries several conclusions. First, southern pine sawtimber and pulpwood stumpage prices appear to be broadly nonstationary, suggesting intertemporal efficiency in the U.S. South timber sector. An implication of this finding is that southern timber prices tend to retain the effects of market shocks, so that a nonstationary price harvest timing rule (e.g., Thomson 1992) would also seem most applicable to landowners managing southern pine.

4. SUMMARY

This chapter examines several key issues related to the structure and performance of private timber markets. It starts by introducing key aspects of market definition, perfect competition, and market efficiency, then describes how these conditions are met in typical private timber market settings, and then presents empirical evidence on timber market structure and efficiency. Two key features of timber markets are that (1) high transportation costs for timber and processing scale economies mean that the distance between suppliers and demanders matters, and (2) timber's role as both capital and product implies complex interaction between market price expectations and harvesting behavior. These technical and spatial factors play an important role in determining the efficiency of and movement of prices in timber markets.

The chapter presents empirical evidence on the extent of buyer market power (monopsony) in timber market settings in the United States and Scandinavia, which largely addresses the first two components of a competitive (efficient) market defined in section 1 (numerous buyers and sellers, free entry and exit). The findings broadly suggest that these timber markets function closer to perfect competition than to pure monopsony, especially sawtimber markets. There is some evidence of oligopsony power in pulpwood markets, though the degree has varied over time.

The chapter also presents some empirical evidence on the spatial and temporal efficiency of timber markets, using data from the U.S. South, which largely addresses the third and fourth efficiency conditions (product homogeneity and perfect information). The evidence, while mixed, broadly supports the notion of temporal efficiency of timber prices. Page constraints preclude a full assessment of spatial market integration here, though much of the literature suggests a higher degree of integration at the final product (i.e., lumber and paper) than at the timber stage, as the latter are more spatially constrained than the former.

Despite the substantial research advances outlined here, much remains unknown about the structure, function, and efficiency of timber markets in North America, Scandinavia, and other major producing regions. Better understanding of these markets could enable better government policies, more efficient investment strategies, and greater aggregate benefits for producers and consumers. Fortunately, enhancements in economic theories of the firm and the market, improvements in econometric methods, and increased availability of more disaggregated and longer time series of price and production data all bode well for the research required to advance this understanding.

5. LITERATURE CITED

ABILDTRUP, J., J. RIIS, AND B.J. THORSEN. 1997. The reservation price approach and informationally efficient markets. J. For. Econ. 3:229-245.

APPELBAUM, E. 1982. The estimation of the degree of oligopoly power. J. Economet. 19: 287-299.

AZZAM, A. 1998. Competition in the U.S. meatpacking industry: Is it history? Agr. Econ.: Int. J. 18:107-126.

AZZAM, A.M., AND E. PAGOULATOS. 1990. Testing oligopolistic and oligopsonistic behavior: An application to the U.S. meat-packing industry. J. Agr. Econ. 41(3):362-370.

BERGMAN, M.A., AND M. NILSSON. 1999. Imports of pulpwood and price discrimination: A test of buying power in the Swedish pulpwood market. J. For. Econ. 5(3): 365-387.

BERGMAN, M.A., AND R. BRANNLUND. 1995. Measuring oligopsony power: An application to the Swedish pulp and paper industry. Rev. Ind. Org. 10:307-321.

BRANNLUND, R. 1989. The social loss from imperfect competition: The case of the Swedish pulpwood market. Scan. J. Econ. 91(4):689-704.

BRAZEE, R.J., AND R. MENDELSOHN. 1988. Timber harvesting with fluctuating prices. For. Sci. 34(2):359-372.

BRAZEE, R.J., AND E. BULTE. 2000. Optimal harvesting and thinning with stochastic prices. For. Sci. 46(1):23-31.

BRESNAHAN, T.F. 1987. Competition and collusion in the American automobile industry: The 1955 price war. J. Ind. Econ. 35(4):457-482.

BUONGIORNO, J., AND J. UUSIVUORI. 1992. The law of one price in the trade of forest products: Cointegration tests for U.S. exports of pulp and paper. For. Sci. 38(3):539-553.

CLARKE, H.R., AND W.J. REED. 1989. The tree-cutting problem in a stochastic environment: The case of age-dependent growth. J. Econ. Dyn. Control. 13:565-595.

DEATON, A., AND G. LAROQUE. 1992. On the behaviour of commodity prices. Rev. Econ. Stud. 59:1-23.

DEATON, A., AND G. LAROQUE. 1996. Competitive storage and commodity price dynamics. J. Polit. Econ. 104(5):896-923.

DICKEY, D.A., AND W.A. FULLER. 1979. Distribution of the estimators for autoregressive time series with a unit root. J. Am. Stat. Assoc. 74(366):427-431.

FAMA, E.F. 1970. Efficient capital markets: A review of theory and empirical work. J. Financ. 25:383-417.

FAMA, E.F. 1991. Efficient capital markets: II. J. Financ. 46:1575-1617.

FORBOSEH, P.F., R.J. BRAZEE, AND J.B. PICKENS. 1996. A strategy for multiproduct stand management with uncertain future prices. For. Sci. 42(1):58-66.

FOREST PRODUCTS LABORATORY. 1987. Wood handbook: Wood as an engineering material. Ag. Handbook 72. USDA. 466 p.

GONG, P. 1999. Optimal harvest policy with first-order autoregressive price process. J. For. Econ. 5:413-439.

HAIGHT, R.G., AND T.P. HOLMES. 1991. Stochastic price models and optimal tree cutting: Results for loblolly pine. Nat. Res. Mod. 5:423-443.

HALL, A. 1994. Testing for a unit root in time series with pretest data-based model selection. J. Bus. Econ. Stat. 12:461-470.

HARDIE, I., P. PARKS, P. GOTTLIEB, AND D. WEAR. 2000. Responsiveness of rural and urban land use to land rent determinants in the U.S. South. Land Econ. 76(4): 659-673.

HÄNNINEN, R., A. TOPPINEN, AND P. RUUSKA. 1997. Testing arbitrage in newsprint imports to United Kingdom and Germany. Can. J. For. Res. 27(12):1946-1952.

HOTELLING, H. 1929. Stability in competition. Economic J. 39:41-57.

HULTKRANTZ, L. 1993. Informational efficiency of markets for stumpage: Comment. Am. J. Agr. Econ. 75(1):234-238.

IWATA, G. 1974. Measurement of conjectural variations in oligopoly. Econometrica 42:947-966.

JOHANSSON, P.O., AND K.G. LOFGREN. 1985. The economics of forestry and natural resources. Basil Blackwell, Oxford. 304 p.

JOHANSSON, P.O., AND K.G. LOFGREN. 1983. Monopsony, conjectural equilibria, and the Swedish roundwood market. For. Sci. 29(3):439-449.

JUNG, C., AND K. DOROODIAN. 1994. The Law of One Price for U.S. softwood lumber: A multivariate cointegration test. For. Sci. 40(4):595-600.

KWIATKOWSKI, D., P.C.B. PHILLIPS, P. SCHMIDT, AND Y. SHIN. 1992. Testing the null of stationarity against the alternative of a unit root: How sure are we that economic time series have a unit root? J. Economet. 54(1-3):159-178.

LERNER, A.P. 1934. The concept of monopoly and the measurement of monopoly power. Rev. Econ. Stud. 1:157-175.

LEROY, S. F. 1989. Efficient capital markets and martingales. J. Econ. Lit. 27(4):1583-1621.

LEYBOURNE, S.J., AND B.P.M. MCCABE. 1994. A consistent test for a unit root. J. Bus. Econ. Stat. 12(2):157-166.

LOFGREN, K.G. 1985. The pricing of pulpwood and spatial price discrimination: Theory and practice. Eur. Rev. Agr. Econ. 12: 283-293.

LOHMANDER, P. 1988. Pulse extraction under risk and a numerical forestry application. Syst. Anal. Model Simul. 5(4):339-354.

LOWRY, S.T., AND J.C. WINFREY. 1974. The kinked cost curve and the dual resource base under oligopsony in the pulp and paper industry. Land Econ. 50:185-192.

MCNEW, K. AND P.L. FACKLER. 1997. Testing market equilibrium: Is cointegration informative? J. Agr. Res. Econ. 22(2):191-207.

MEAD, W.J. 1966. Competition and oligopsony in the Douglas-fir lumber industry. Berkeley, CA: University of California Press.

MURRAY, B.C. 1995a. Measuring oligopsony power with shadow prices: U.S. markets for pulpwood and sawlogs. Rev. Econ. Stat. 77(3):486-498.

MURRAY, B.C. 1995b. Oligopsony, vertical integration, and output substitution: Welfare effects in U.S. pulpwood markets. Land Econ. 71(2):193-206.

MURRAY, B.C., AND D.N. WEAR. 1998. Federal timber restrictions and interregional arbitrage in U.S. lumber. Land Econ. 74(1):76-91.

MURRAY, B.C. 1992. Testing for imperfect competition in spatially differentiated input markets: The case of U.S. markets for pulpwood and sawlogs. Ph.D. dissertation, Duke University, Durham, NC.

MUTH, J.F. 1961. Rational expectations and the theory of price movements. Econometrica 29(3):315-335.

MUTH, M.K., AND M.K. WOHLGENANT. 1999. A test for market power using marginal input and output prices with application to the U.S. beef processing industry. Am. J. Agr. Econ. 81(3):638-643.

NAGUBADI, V., I.A. MUNN, AND A. TAHAI. 2001. Integration of hardwood stumpage markets in the southcentral United States. J. For. Econ. 7(1):69-98.

NORRIS FOUNDATION. 1977-2002. Timber Mart-South. The Daniel B. Warnell School of Forest Resources, University of Georgia, Athens.

NORSTRØM, C.J. 1975. A stochastic model for the growth period decision in forestry. Swed. J. Econ. 77:329-337.

PLANTINGA, A.J. 1998. The optimal timber rotation: An option value approach. For. Sci. 44(2):192-202.

PRESTEMON, J.P., AND R.C. ABT. 2002. Timber products supply and demand, P. 299-325 in The southern forest resource assessment, D.N. Wear and J.G. Greis (eds.). USDA For. Serv. Gen. Tech. Rpt. SRS-53, Asheville, North Carolina.

PRESTEMON, J.P., AND T.P. HOLMES. 2000. Timber price dynamics following a natural catastrophe. Am. J. Agr. Econ. 82(1):145-160.

PRESTEMON, J.P., AND J.M. PYE. 2000. Merging areas in Timber Mart-South data. South. J. App. For. 24(4):219-229.

RAVALLION, M. 1986. Testing market integration. Am. J. Agr. Econ. 68(1):102-109.

RONNILA, M., AND A. TOPPINEN. 2000. Testing for oligopsony power in the Finnish wood market. J. For. Econ. 6(1): 7-22.

SAID, S.E., AND D.A. DICKEY. 1984. Testing for unit roots in autoregressive moving average models of unknown order. Biometrika 71(3):599-607.
SAMUELSON, P. A. 1952. Spatial price equilibrium and linear programming. Am. Econ. Rev. 42(3):283-303.
SCHROETER, J.R. 1988. Estimating the degree of market power in the beef packing industry. Rev. Econ. Stat. 70:158-162.
SCHWERT, G.W. 1989. Tests for unit roots: A Monte Carlo investigation. J. Bus. Econ. Stat. 7:147-160.
STIGLER, G. J., AND R. A. SHERWIN. 1985. The extent of the market. J. Law Econ. 28:555-585.
TAYLOR, A.M. 2001. Potential pitfalls for the purchasing power parity puzzle? Sampling and specification biases in mean-reversion tests of the law of one price. Econometrica 69(2):473-498.
THOMSON, T.A. 1992. Optimal forest rotation when stumpage prices follow a diffusion process. Land Econ. 68(3):329-342.
TIROLE, J. 1988. The theory of industrial organization. MIT Press, Cambridge, MA.
U.S. CENSUS BUREAU. 1999. Manufacturing industry series (EC97M-3211A: sawmills; -3221A: pulp mills; -3222B: paper mills; -322C: newsprint mills; -322D: paperboard mills). U.S. Department of Commerce, Economics and Statistics Administration.
WASHBURN, C.L., AND C.S. BINKLEY. 1990. Informational efficiency of markets for stumpage. Am. J. Agr. Econ. 72(2):394-405.
WASHBURN, C.L., AND C.S. BINKLEY. 1993. Informational efficiency of markets for stumpage: Reply. Am. J. Agr. Econ. 75(1):239-242.
WILLIAMS, J.C., AND WRIGHT, B.D. 1991. Storage and commodity markets. Cambridge University Press, New York. 502 p.
YIN, R., AND D.H. NEWMAN. 1996. The effect of catastrophic risk on forest investment decisions. J. Environ. Econ. Manage. 31:186-197.

---

[1] The data used in this study indicate that the stumpage price paid to timber growers, which is net of transport and harvest costs, is on average roughly one-fourth the delivered (mill) price for pulpwood and two-thirds the delivered price for sawlogs.

[2] A more formal presentation of the spatial input market and local market power determinants can be found in Murray (1992).

[3] The more modern empirical methods have been referred to as the new empirical industrial organization (NEIO) methods (Bresnahan 1989).

[4] An ARIMA(p,d,q) is an autoregressive integrated moving average time series process, where p is the lag order of the autoregression, d is the order of integration, and q is the lag order of the moving average. "Near unit-root processes," such as a pure first-order autoregressive process, ARIMA(1,0,0), with an autoregressive parameter approaching 1.0, are mean-reverting. In these, a portion of each period's price change is predictable, tending toward the long-run mean price level. In contrast, certain unit-root processes, such as the random walk, an ARIMA(0,1,0) process, have changes from one period to the next that are not predictable using previous price levels or innovations.

Chapter 11

International Trade In Forest Products

Jeffrey P. Prestemon, Joseph Buongiorno, David N. Wear, and Jacek P. Siry
USDA Forest Service, University of Wisconsin-Madison, USDA Forest Service, and University of Georgia

The 21st century continues a trend of rapid growth in both international trade of forest products and a concern for forests. These two trends are connected. Forces causing trade growth are linked to the loss of native forest resources in some countries and the accumulation of nonnative forest resources in other countries. Factors increasing trade include relaxation of trade barriers, income growth, and improvements in wood growing, harvest, and manufacturing technologies. But environmental concerns are increasing as consumer preferences change, and as native forests recede and plantation forests become more prominent.

Efforts to address environmental concerns often involve changes in trade policies. The effects of these changes can be evaluated with trade models. In this chapter, we describe the context of forest products trade and major policy issues, explain three approaches to policy modeling, and present three applications of partial equilibrium modeling of trade policies.

1. FOREST PRODUCTS TRADE TODAY

World forest products trade has grown rapidly in volume and value, fuelled by both world economic growth and falling trade barriers. Tariffs on forest products have been decreasing as a result of consecutive rounds of the General Agreement on Tariffs and Trade (GATT) and World Trade Organization (WTO). The last round of GATT resulted in agreements to significantly reduce average worldwide tariffs on forest products (Barbier

Sills and Abt (eds.), Forests in a Market Economy, 177–199. © Kluwer Academic Publishers. Printed in The Netherlands.

1996), although these reductions may be matched by an increase in nontariff barriers.

Tariffs are declining mainly because of the creation of new multilateral trade liberalization accords and through modification of existing agreements, including the WTO and other multilateral regional accords. Principal among the regional accords are the North American Free Trade Agreement (NAFTA), the European Union, and the European Free Trade Association (Frankel et al. 1997). As of June 2000, there were 134 regional trade agreements (World Trade Organization 2001).

Trade liberalization, growing trade volumes, and the rise in concern for the environment have created ample subject matter for trade policy analysis. Major issues arising from the growing importance of trade include effects on import-competing domestic industries, biodiversity protection, global warming, forest sustainability, and the hardening of trade blocs. When governments negotiate freer trade, certain sectors are harmed and others benefit. Freer trade's effects on long-protected domestic industries can create political turmoil and lead to efforts to erect trade barriers to limit negative effects on them. Depending on the mix of ownership, tenure enforcement, and whether a country is a net importer or exporter of forest products, trade policies can encourage or discourage forest loss (Prestemon 2000) and have uncertain influences on biodiversity and carbon balances. Active forestry and efforts to preserve forests may be partial solutions to global warming, as some of the tropical forest loss is matched by temperate forest gains (e.g., Reddy and Price 1999, Sohngen and Sedjo 2000). In many countries, efforts to slow deforestation through trade instruments could even have some perverse effects on forests.

A country's environmental protection policies and related laws can have effects on trade and on global welfare as well. One set of policies and laws concerns timber or forest certification. Certification is the formal affirmation by an objective party that forest management and harvesting on a forested tract are done in environmentally, economically, and socially sustainable ways. Certification may have the effect of increasing timber production costs, leading to lower outputs and, presumably, less trade. The potential economic costs of reduced trade resulting from realignment and output decreases merit quantification (Sedjo et al. 1998).

Many of the multilateral and bilateral trade arrangements negotiated in the last 50 years conflict with the WTO and threaten to segment world markets for forest products. Such segmentation—resulting in groups of regions with freer trade among members than with nonmember nations—could ultimately be harmful from a global perspective (Frankel 1998, Frankel et al. 1997, Ito and Krueger 1997, Shiells 1995). These trade blocs

2. INTERNATIONAL TRADE THEORY AND TRADE MODELING

2.1 Comparative Advantage

The magnitude and direction of forest product flows are determined by geography (distance), size of economies (demand), character of forest endowments (supply), and government policies (interventions and historical relationships). Classical trade theory prescribes that trade occurs because there are differences among trading partners in their relative costs of production. In forest products, empirical research based on theoretical and applied work done by Vanek (1959) and Leamer (1984) suggests that the principal factor determining the direction and magnitude of net trade is the size of a country's forest endowment relative to the size of its economy (Bonnefoi and Buongiorno 1990, Prestemon and Buongiorno 1997). But people can change a country's forest endowment by planting and creating faster-growing forests, improve transport networks to lower the cost of product movement, invent and acquire better machines and technologies to more efficiently use raw materials, and train workers to facilitate technology development and implementation. Examples of human efforts to affect comparative advantage are large-scale softwood plantation establishment in New Zealand and Chile, countries without substantial marketable native softwood endowments but which today have a comparative advantage in softwood forest products.

Governments, through unilateral and cooperative actions, affect how completely comparative advantages are expressed. Government interventions and cooperative actions designed to encourage certain outcomes therefore have effects important to understand and quantify. Economic analyses can quantify tradeoffs with far-reaching implications for welfare, industry structure and competitiveness, tariff revenues, and the environment.

2.2 Trade Modeling

A focus of trade policy analysis has been on the economic consequences of restricting trade—i.e., quantifying the costs of government interventions in markets of tradable products. The literature developed from this describes how governments can set tariffs to maximize domestic welfare and to promote infant industries. Much of the earlier theoretical and applied trade policy analyses assumed perfect industry competition (e.g., Corden 1974, Dixit and Kyle 1985, Krugman and Obstfeld 1988). In some commodity markets and small economies, however, imperfect competition may be common, hence the development of trade theory describing best and second-best policies in that context. The perfect competition model has also been enriched by considering negative and positive externalities of production and consumption, strategic trade policy (Helpman and Krugman 1989), and trading blocs (Frankel 1998, Frankel et al. 1997, Ito and Krueger 1997, Shiells 1995).

Trade modeling approaches for evaluating effects of policies fall into three categories: computable general equilibrium, spatial partial equilibrium, and nonspatial partial equilibrium. Each approach has its own particular uses, advantages, and disadvantages.

Computable general equilibrium (CGE) models provide answers to questions about policies at the level of whole sectors and economies. These models are systems of many equations relating key components of an economy to each other and to other economies through commodity trade, currency exchange rates, and capital flows. CGE models are built from functional relationships between supply and demand for all products and aggregate supplies and demands for inputs to production. A strength of CGE models is that, because all sectors are allowed to be affected by all other sectors, effects of a policy on prices and quantities of outputs and inputs of all affected markets can be measured most completely. Second, the effects of trade policy changes can be measured in terms of aggregate national economic welfare. The primary weakness of CGE models is their lack of model resolution. Because of tradeoffs between model accuracy and size on the one hand and model resolution on the other, the effects of trade policies at the subsector levels are usually not revealed. This means that it is difficult to evaluate the effects of policies on specific products. Simpler models, which treat a particular sector in detail and other sectors in more aggregate terms, are possible (e.g., Yúnez-Naude 1992) and perhaps a direction for further development in forest product sector trade modeling.

Partial equilibrium models assume that the feedbacks from sectoral changes to aggregate macro variables (gross domestic product [GDP] growth, national investment, wages, etc.) are negligible, and that assumption allows modelers to quantify the detailed effects of a policy on specific commodity markets. In partial equilibrium models, macro variables affect the sector, but are not affected significantly by it. One example is the spatial partial equilibrium model (Samuelson 1952, Takayama and Judge 1964), which exploits concepts of market spatial distribution and transportation and transaction costs to observe spatial reallocations of production and consumption given a policy change. Spatial partial equilibrium models have been used to study the effects of trade policies on forest product trade levels (e.g., Frankel 1998, Boyd and Krutilla 1992, Boyd et al. 1993, Buongiorno and Gilless 1984, Gilless and Buongiorno 1987).

Nonspatial partial equilibrium models of trade (e.g., Olechowski 1987) remain the most widely used means of studying the effects of changes in trade policies on specific commodities. These models rely on estimates of the elasticity of import demand and export supply with respect to prices and with respect to other variables in order to measure the effects of an exogenous economic shock (e.g., a policy change) on markets. Therefore, the key element of partial equilibrium analysis is the estimate of the import demand equation for the product(s) of interest. The transportation and transaction costs of trade are left out of nonspatial partial equilibrium analysis. Presumably, empirical elasticity estimates contain the effects of these costs, but a consequence of ignoring spatial distribution is the inability of such models to anticipate the appearance of wholly new trade flows between two points of production and consumption.

3. EMPIRICAL STUDIES OF THE EFFECTS OF GOVERNMENT POLICIES ON FOREST PRODUCT TRADE

3.1 Case 1: The U.S.-Canada Memorandum of Understanding on Softwood Lumber Trade

The longest running trade dispute in the U.S. forest products sector concerns softwood lumber production and imports from Canada. Here we address one episode in this debate. The crux of the dispute is that U.S. producers view Canadian producers' long-term forest management contracts with Provincial governments as a subsidy. The U.S. firms successfully argued for relief in the form of a countervailing duty on imports of softwood lumber from Canada. Subsequent negotiations resulted in a Memorandum of Understanding (MOU) that allowed Canadian authorities to collect an export tax equivalent to the countervailing duty from the beginning of 1987 to October of 1991.

This study (details are contained in Wear and Lee 1993) estimated the influence of the MOU on softwood lumber trade and lumber markets in the United States. The study faced an issue common in applied policy analysis: evaluation of a recent policy-relevant event with few data.

3.1.1 An Impact Model

The first step was to model the share of Canadian lumber in the softwood lumber market in the United States as a function of supply and demand factors for the U.S. market and a set of variables that recognized imports from Canada as an excess supply (i.e., Canada's home market variables). We hypothesized that these variables would predict variation in market share before and after the imposition of the MOU and that the MOU would shift this relationship.

The impact model was defined as follows:

$$m^c = g(HS, GNP, t, CLUM, XLUM, XLOG, XCH, M) \qquad 11.1$$

where m^c is the ratio of Canadian imports in the United States to total U.S. consumption. U.S. demand factors include housing starts (HS), gross national product (GNP, a proxy for nonhousing uses of lumber), and a time index (t), which allows for a changing relationship between demand factors and lumber consumption. U.S. supply factors include domestic export of comparable lumber ($XLUM$) and logs ($XLOG$). Additional variables are the Canadian-U.S. exchange rate (XCH), Canadian lumber consumption ($CLUM$), and a vector of impact variables (M), used to test for changes in market share coincident with the period of the MOU. All righthand-side variables were assumed to be exogenous, and equation 11.1 was estimated by ordinary least squares.

The vector of impact variables (M) consisted of six separate dummy variables (M_{85}, M_{86}, M_{87}, M_{88}, M_{89}, M_{90}), to allow for different impacts in each year. Coefficients for the dummy variables for 1985 and 1986 capture unexplained variation in import share in years prior to the policy.

Equation 11.1 was estimated with annual data from 1960 to 1990. The model explained most of the variation in import share (adjusted $R^2 = 0.96$). All variables affected the share significantly, except M_{85}. The significant effect of M_{86} reflected a strong anticipatory response by Canadian producers in the fourth quarter, after a countervailing duty had been announced and would have been retroactively applied to fourth-quarter exports. The effects of M_{86} to M_{90} were not statistically different, so they were replaced by a single dummy variable. The value of the coefficient for this dummy variable was –0.048, indicating that the MOU reduced market share by about 4.8%.

3.1.2 Price-Quantity Analysis

The next step was to translate the results of the impact model into effects on lumber quantities and prices. The method for estimating market impacts is summarized in figure 11.1, which describes the U.S. softwood lumber market with an aggregate demand function for lumber (D) and a two-part lumber supply with total supply (Q) equal to the sum of Canadian supply (S^C) and other supply (almost exclusively domestic, S^O). The subscripts 0 and 1 refer to supply before and after the policy, respectively. P is price.

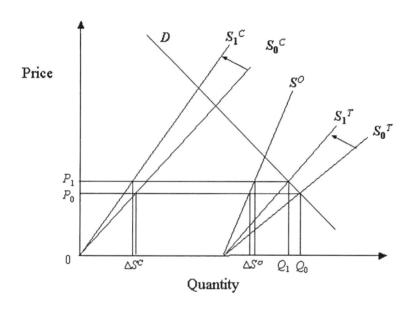

Figure 11.1. Structure of the U.S. softwood lumber market

The export tax is implemented as a proportional assessment on lumber price, so the policy results in a pivotal inward shift in Canadian supply (ΔS^C) analogous to the implementation of an ad valorem tax, and a consequent shift in total supply from S_0^T to S_1^T. The shift in total supply lowers the equilibrium quantity, $\Delta Q = Q_1 - Q_0$, and raises the equilibrium price, $\Delta P = P_1 - P_0$. These shifts cause a movement along the other supply curve (S^O). The policy can be evaluated through its impact on four variables: ΔS^O, ΔS^C, ΔQ, and ΔP as follows. First, movement along the lumber demand curve (ΔQ) and the 'other lumber supply' curve (ΔS^O) is defined by their own price elasticities of η and δ, resulting in two linear equations.

$$\Delta Q = \eta \frac{Q_0}{P_0} \Delta P \qquad\qquad 11.2$$

$$\Delta S^O = \delta \frac{S_0^O}{P_0} \Delta P \qquad 11.3$$

The policy impact on market share is modeled by the differential of Canadian market share:

$$dm^C = d\left[\frac{S^C}{Q}\right] = \frac{Q_0 \Delta S^C - S_0^C \Delta Q}{Q_0^2} \qquad 11.4$$

Because the shares must sum to one, a symmetrical result holds for the other market share:

$$-dm^C = dm^O = d\left[\frac{S^O}{Q}\right] = \frac{Q_0\, dS^O - S_0^O\, dQ}{Q_0^2} \qquad 11.5$$

By setting dm^C equal to the impact on share estimated by the impact model (-4.8 percent), and imposing the market identity, we have four linear equations and four unknowns which can be solved with estimates of the two elasticities (δ and η), determined from the literature to be $\eta = -0.17$, $\delta^O = 0.4$, and $\delta^C = 1$.

Applying the four-equation model (11.2-11.5) to production, consumption, and price data for the years 1987-1990, the MOU resulted in a 2.6 billion board-feet (bbf) annual reduction of imports from Canada to the United States (table 11.1).

Table 11.1. Estimated total policy impacts for the years 1987 to 1990

	Total Change	Annual Change
Market impacts		
U.S. lumber consumption (bbf)	-3.06	-0.76
U.S. lumber production (bbf)	+7.32	+1.83
Lumber imports from Canada (bbf)	-10.43	-2.61
Lumber price (1982 US$/mbf)	—	+19.91
Welfare impacts		
U.S. producer surplus (millions of 1982 US$)	2632.4	658.1
U.S. consumer surplus (millions of 1982 US$)	-3789.5	-947.4
U.S. total impact (millions of 1982 US$)	-1157.0	-289.3

In response, U.S. lumber production increased by 1.8 bbf, and price increased by roughly $20 per thousand board-feet (mbf). U.S. lumber consumption was reduced by about 0.8% over this period.

3.1.3 Welfare Analysis

To calculate consumer surplus, we describe total supply by the import and domestic supply elasticities and base year observations:

$$P = b_T + \beta_T Q; \quad \beta_T = \frac{P}{\delta^T Q}; \quad b_T = P_0 - \beta_T S_0^T \qquad 11.6$$

where δ^T is the own price elasticity of total supply, equal to

$$\delta^T = \frac{\delta^C S^C + \delta^O S^O}{S^T} \qquad 11.7$$

Consumer surplus, the area under the demand curve and above the price line, defines consumer benefits and is altered by shifts in the price-quantity equilibrium. Change in consumer surplus is defined by the total quantity and price changes as shown in figure 11.1:

$$\Delta CS = -\left[(P_1 - P_0)Q_1 + \frac{1}{2}(Q_0 - Q_1)(P_1 - P_0)\right] \qquad 11.8$$

Producer surplus, the area above the supply curve and under the price line, defines producer profits. Change in domestic producer surplus is:

$$\Delta PS^O = (P_1 - P_0)S_0^o + \frac{1}{2}(P_1 - P_0)(S_1^o - S_0^o) \qquad 11.9$$

The computation of change in producer surplus for Canadian firms following the policy is based on treating the shifted supply curve as an effective supply, and defining producer costs using the prepolicy supply curve.

$$\Delta PS^C = (1-\alpha)P_1 S_1^C - \frac{1}{2}P_1' S_1^C - \frac{1}{2}P_0 S_0^C \qquad 11.10$$

$$P_1' = P_0 S_1^C$$

where α is the export tax rate and P_1' is the price defined by the prepolicy supply curve at S_1^C. The tax revenue is defined as $R = \alpha P_1 S_1^C$.

We estimated total quantity and price changes from the 4.8% reduction in Canada's market share for the years 1988 to 1990 by simulating removal of the policy impact (S^C = 4.8%) from market results (P_o, Q_o) in each of these years, using the equations and elasticities given above. The total as well as the average annual impacts of the policy are listed in table 11.1. Over the full course of the MOU, U.S. producers gained about US$2.6 billion (in 1982 dollars), while consumers lost about $3.8 billion. Thus, the net U.S. cost is approximately $1.2 billion over this 4-year period.

3.1.4 Discussion

The estimated welfare impacts reflect effective rent seeking by domestic lumber producers. Large positive returns accrued to domestic firms but also to the exporting country, and they are accounted for by increased consumer and efficiency costs. However, Canada's termination of the MOU indicates that these benefits, complicated by the MOU's prohibition on the redistribution of export tax revenue to wood products firms, were not adequate compensation for infringing on its resource sovereignty.

In the quantity terms of the original countervailing duty complaint, the MOU succeeded. Our analysis indicates that imports fell by about 2.6 bbf per year, while U.S. production increased by about 1.8 bbf per year. This, coupled with the consequent change in lumber price, led to considerable improvement in the competitive position of domestic producers in home markets. This improved competitiveness provided significant economic benefits to U.S. firms, even as domestic environmental issues caused available softwood timber inventories to contract.

3.2 Case 2: The Long-Run Effects of NAFTA

NAFTA took effect on January 1, 1994, after much debate and critical analysis of its potential impacts on North American economies. NAFTA created the world's largest free-trading zone: Canada, the United States, and Mexico. After a 15-year phase-in period, most commodity tariffs (including forest products) and nontariff barriers will be eliminated, affecting trade in a region of 400 million consumers and aggregate output of over 7 trillion dollars. United States-Canada-Mexico trade in forest products accounts for most forest products trade in all three, so NAFTA may have important implications for the forest sectors of each. In order for producers and consumers of traded forest products in each country to make informed decisions on investments in new plants and equipment, projections of the effects of NAFTA on trade in individual product categories will be useful.

Much modeling of NAFTA prior to its implementation was with CGE models (e.g., Francois et al. 1992). These models provided estimates of the net effects of NAFTA on exchange rates, price levels, aggregate economic output, wages, and interest rates or capital flows. However, CGE models provided results for large sectors, not commodities. Prestemon and Buongiorno (1996) described a method for calculating the disaggregated effects of NAFTA on U.S. and Canadian exports to Mexico using these CGE models and a system of partial equilibrium models. Their approach was to (1) estimate partial equilibrium demand equations for Mexico's imports of forest products by product class, equations that made imports a function of import prices, Mexican forest product-consuming sector outputs, and Mexican production input prices; (2) estimate reduced-form price equations for each product class, which were functions of Mexican, U.S., and Canadian input prices and output levels; and (3) predict changes in Mexico's forest product import prices and quantities using CGE models' estimates of the effects of NAFTA on tariffs, output levels, and input prices.

The Mexican import demand function for each product is:

$$M_d = M_d(Y, P_M, \mathbf{w}) \qquad 11.11$$

where M_d is the quantity imported by Mexico, Y is domestic output of the forest product-consuming industries in Mexico, P_M is the import price of the product, $\mathbf{w}$ is a vector of input prices relevant to wood product producers and consumers. The total derivative of equation 11.11 shows the effects of changes in each variable on imports of each product, other things being equal:

$$dM_d(Y, P_M, \mathbf{w}) = \frac{\partial M_d}{\partial P_M} dP_M + \frac{\partial M_d}{\partial Y} dY + \sum_{i=1}^{I} \frac{\partial M_d}{\partial w_i} dw_i \qquad 11.12$$

In relative terms,

$$\frac{dM_d(Y, P_M, \mathbf{w})}{M_d} = \beta_{P_M} \frac{dP_M}{P_M} + \beta_Y \frac{dY}{Y} + \sum_{i=1}^{I} \beta_i \frac{dw_i}{w_i} \qquad 11.13$$

where β is the elasticity of import demand for the product with respect to the subscripted variable. Hence, counterfactual estimates of the trade agreement on individual products can be made by inserting estimates of import demand equations and price equations, which are functions of driving sector- and economy-wide variables, whose estimates can derive from CGE studies.

Estimation of output price changes deriving from NAFTA start by specifying the price function for each product, in pesos, as:

$$P_M = \theta E P(\mathbf{z}) \qquad 11.14$$

where $\theta = 1 + \tau$, τ is the ad valorem tariff applied by Mexico to U.S. or Canadian forest product imports, and **z** are exogenous factors affecting the price. Predictions of the effects of NAFTA on the final price of forest products imports are obtained by taking the total derivative of equation 11.14 with respect to each variable and then converting the resulting equation into elasticities and proportional changes in the righthand-side variables:

$$\frac{dP_M}{P_M} = \alpha_E \frac{dE}{E} + \alpha_\theta \frac{d\theta}{\theta} + \sum_{j=1}^{J} \alpha_j \frac{dz_j}{z_j} \qquad 11.15$$

where α is the elasticity of import price with respect to the subscripted variable.

The effect of NAFTA on trade quantity comes from substituting equation 11.15 into the first term on the righthand side of equation 11.13. The result is a model of the effect of NAFTA on the product import quantity:

$$\frac{dM_d(Y,P_M,w)}{M_d} = \beta_{P_M}\left[\alpha_E \frac{dE}{E} + \alpha_\theta \frac{d\theta}{\theta} + \sum_{j=1}^{J} \alpha_j \frac{dz_j}{z_j}\right] + \\ \beta_Y \frac{dY}{Y} + \sum_{i=1}^{I} \beta_i \frac{dw_i}{w_i} \qquad 11.16$$

Prestemon and Buongiorno (1996) estimated import demand equations (11.11) and price equations (11.14) for six categories of lumber, four categories of wood panels, newsprint, scrap plus waste paper, and seven kinds of wood pulp imported by Mexico from the United States. Similar equations were estimated for newsprint and three kinds of wood pulp imported by Mexico from Canada.

Three alternative scenarios of NAFTA's impacts on sector- and economy-wide variables were evaluated. These scenarios made alternative assumptions regarding the effects of the agreement on intraindustry competition, capital markets, and labor markets. Together, they provided a range of possible long-run effects of NAFTA on macroeconomic indicators in the United States, Canada, and Mexico.

Results are summarized in table 11.2. Minimum and maximum change estimates of the three scenarios are reported. Results show that impacts differ widely by product. NAFTA's net effect is to increase lumber, plywood, and scrap and waste paper imports from the United States and newsprint imports from the United States and Canada by large percentages relative to those observed in 1992. Effects on other products (other panels and wood pulp) would have been small, reflecting their inelastic import responsiveness imports to prices.

Table 11.2. The long-run percentage change in value of NAFTA on U.S. and Canadian exports to Mexico, by selected forest product

Product	1992 Import quantity [a]	1992 Import value (1994 US$ m)	Minimum change (%)	Maximum change (%)
U.S. forest products				
Douglas fir lumber	101 Mm^3	15	73	207
Ponderosa pine lumber	312 Mm^3	95	17	47
Southern pine lumber	93 Mm^3	16	10	77
Other softwood lumber	457 Mm^3	83	10	77
Oak lumber	72 Mm^3	26	19	58
Other hardwood lumber	56 Mm^3	17	74	247
Hardwood veneer	3,576 Mm^2	5.0	4	25
Softwood plywood	181 Mm^3	37	89	266
Hardwood plywood	76 Mm^3	15	26	72
Particle board	81 Mm^3	22	0	0
Bleached sulfate pulp	285 Mmt	126	1	43
Semibleached sulphate pulp	25.8 Mmt	10	0	3
Unbleached sulphate pulp	6.6 Mmt	22	0	2
Bleached + semibleached sulphite pulp	8.5 Mmt	2.8	0	15
Unbleached sulphite pulp	4.6 Mmt	1.1	0	0
Dissolving grades of pulp	79.9 Mmt	38	0	0
Mechanical + semichemical pulp	11.5 Mmt	4.2	9	50
Scrap + Waste paper	828 Mmt	110	1	20
Newsprint	119 Mmt	63	47	136
Canadian forest products				
Bleached + semibleached. sulphate pulp	7.3 Mmt	0.4	2	41
Unbleached sulphate pulp	1.0 Mmt	13	2	14
Mechanical + semichemical pulp	36 Mmt	16	3	20
Newsprint	46 Mmt	23	12	44
Total, US [b]		686	21	83
Total, Canada [c]		39	21	117

[a] Mm^3 = 1000 m^3, Mm^2 = 1000 m^2, Mmt = 1000 metric tonnes
[b] Excluding 1992 imports of paper and paperboard worth US$283 million and logs and panels worth US$40 million
[c] Excluding 1992 imports of paper and paperboard worth US$3 million and pulp worth less than US$1 million in 1992

3.2.1 Discussion

This model of trade analyzed what would have been the effect of NAFTA had the agreement been fully implemented in 1994. The partial equilibrium modeling was a simple way of obtaining disaggregated, product-specific assessments of the net effects of the trade agreement. The approach also took into account the general equilibrium effects of the agreement on industry-relevant cost factors and output levels. Although these estimates of trade effects were not projections of the future (i.e., they were counterfactual), the empirical models of import demands and prices could have been used to project. To project trade volumes by product into the future, the analyst would insert projected values of the macroeconomic variables for selected future years and then calculate the quantities and prices of forest product imports that would result in those years.

The policy implications of the modeling on the forest sectors complement the predicted net effects of NAFTA on economy-wide variables. First, NAFTA's effects on exports from the U.S. will be concentrated in solidwood products, especially lumber. U.S. lumber exporters to Mexico could use this finding to ready themselves for greater exports there. Second, U.S. and Canadian newsprint exporters should find NAFTA expanding their export opportunities to Mexico; exporters of paper products that were not analyzed here might expect a similar effect from NAFTA. Third, NAFTA will have little effect on pulp exports to Mexico, indicating that the agreement should have little effect on their pulp prices, implying ultimate benefits for the North American paper industry.

3.3 Case 3: Freer Global Trade and the WTO

The 1994 Uruguay Round of GATT, incorporated into the 1994 Marrakech agreement establishing the WTO, was intended to further the goal of worldwide free trade. Part of this WTO round involved significant reductions in forest products tariffs. Because U.S. tariffs on most forest products were already low before the agreement, the substantial reduction in tariffs by other countries benefited the U.S. forest industry (Wisdom 1995).

Nevertheless, in spite of the free trade advances of the Uruguay Round, tariffs remain a significant barrier to trade for forest products worldwide (Bourke and Leitch 1998). At the urging of the United States, Canada, Indonesia, and New Zealand, in 1998 the Asia Pacific Economic Cooperation (APEC) countries proposed the forest sector for accelerated tariff liberalization (ATL). The effects of this agreement were analyzed with a variant of the global forest products model (GFPM) developed at the University of Wisconsin-Madison in collaboration with the Food and

Agriculture Organization (FAO) of the United Nations and the USDA Forest Service (Zhu et al. 2001). After reviewing the GATT and WTO agreements, this section describes the application of the GFPM to predict the impacts of accelerated tariff liberalization on the global forest sector.

3.3.1 The Uruguay Round of GATT and Accelerated Tariff Liberalization

Barbier (1996) summarized the 1994 Uruguay Round agreements for forest products, which served as the base scenario for this application of the GFPM. The tariffs on most forest products would be reduced 33% on a trade-weighted basis. Most major importers agreed to tariff elimination on pulp and paper by 2004. The major developed countries were also committed to reducing tariffs by 50% on solid wood products over 5 years starting in 1995. For developed countries, the average tariff of forest products (wood, pulp, paper, and furniture) would be reduced from 3.5% to 1.1%. In the United States, average tariffs would be cut from about 3.1% to 1.8%. The tariff escalation for processed products in developed country markets would be reduced significantly. For wood, the reduction ranged from 30% to 67%; for paper and paperboard, tariff escalation would be eliminated. The Uruguay Round agreement committed all major developed countries and a high proportion of developing countries to bind forest product tariff rates, thus reducing the market risk greatly.

The implications of the agreement for nontariff barriers are less clear, although the Agreement on the Application of Sanitary and Phytosanitary Measures and the Agreement on Technical Barriers to Trade would improve the market access (Barbier 1999). Barbier (1996, 1997) and Brown (1997) found that the impact of these provisions resulting from the Uruguay Round on trade in forest products would be small. This is partly because tariffs in Organization for Economic Cooperation and Development (OECD) countries would remain high for some products, such as wood-based panels. In other countries, tariffs between 10% and 20% are common, and they can reach 40%. The ATL proposal covered all forest products, such as logs and wood products, pulp, paper, and paper products. Parties to the Uruguay Round of GATT "zero-for-zero" agreement (i.e., a market access agreement where all the participating countries eliminate the same tariff and [sometimes] nontariff barriers on the same products) meant to move up the elimination of tariffs on pulp, paper, and paper products from January 1, 2004, to January 1, 2000. Others would attempt to remove tariffs by the same date, but could delay removal until January 1, 2002. The proposal called for the elimination of tariffs on all other products by January 1, 2002.

This is the alternative scenario modeled with the GFPM, with all tariffs set to zero beginning in 2000.

3.3.2 The Global Forest Products Model

The GFPM uses price-endogenous linear programming (Zhang et al. 1993), which is based on the technique of solving for a spatial equilibrium in competitive markets (Samuelson 1952). The GFPM integrates the classical four major components of forest sector models (Kallio et al. 1987): timber supply, processing industries, product demand, and trade. The GFPM shows how production, consumption, imports, exports, and prices are likely to change in response to changes such as economic growth, tariffs, or technology.

The GFPM works by optimizing the short-run allocation of resources in global product markets. Long-run resource allocation is partly governed by market forces (e.g., capacity expansion and trade) and political forces. The latter include wood supply shifts determined by forest policy, waste paper recovery rates mandated by environmental policy, tariffs, and technology. The GFPM solves a sequence of annual global equilibria by maximizing the value of the products minus the cost of their production, subject to material balance and capacity constraints in all countries. In each projection year, for each country and commodity, net supply (domestic production plus imports) is equal to net demand (final consumption, plus input in other processes, plus exports). Final demand is price responsive; demand for wood or intermediate products derives from the demand for final products through input-output coefficients that describe technologies in each country. The supply of raw wood and nonwood fiber in each country is price responsive. The supply of recycled paper is constrained by the waste paper supply, which itself depends on past paper consumption and the recycling rate. Each country exports to the world market and imports from it. Imports and exports for each country are constrained to simulate inertia in yearly changes in trade (Buongiorno and Gilless 1984, Kallio et al. 1987). The shadow prices of the material balance constraints are the market-clearing prices at which demand equals supply for all countries and commodities (Hazell and Norton 1986).

From one year to the next, demand changes in each country due to changes in GDP. Wood supply shifts exogenously according to a chosen scenario. The rate of fiber recycling depends on technology and recycling policy. Capacity changes depend on past production and the shadow price of capacity in different countries. Tariff changes affect the cost of imports, ad valorem. Then, a new equilibrium is computed subject to the new demand and supply conditions, new technology, new capacity, and new tariff.

The GFPM models 180 countries separately. Disaggregation to the national level is useful because political power on matters of international trade rests largely at national levels. It also facilitates data verification and enhances modeling flexibility. Lastly, it facilitates review and evaluation of results, because expert knowledge is more available at the country level than at more aggregate levels. Fourteen commodities are considered in the model: from fuelwood to paper and paperboard, covering all the forest products in the FAO statistical yearbook (FAO 1999).

The GFPM uses econometric demand equations for end products, summarized by elasticities with respect to national income (measured by real GDP) and real product price, in constant U.S. dollars (Baudin and Lundberg 1987, Buongiorno 1978). The assumptions on the GDP growth rate of each country were those used in the 1999 FAO Global Forest Products Outlook study (Zhu et al. 1998).

In any given year and country, the supply of industrial roundwood is a function of its price. The supply also shifts over time, reflecting estimates of change at constant prices. The shift rates vary by country and are based on past production, forest area and stock, growth rates, extent of plantations, and policy. The shift rates used here were the annual percentage changes in the commercially available wood supply projected by the Global Fiber Supply Model (Bull et al. 1998). The price elasticities were also estimated by the participants of the GFPM study, based on their knowledge of countries and on existing literature. The waste paper supply curves were horizontal, with upper bounds determined by the previous year's paper consumption. Waste paper recovery rates were assumed to increase over time (Ince 1994, Mabee 1998).

Manufacturing is represented in the GFPM by input/output coefficients and associated manufacturing costs. The estimation procedure was such that the implied consumption, production, and trade was as close as possible to the national statistics, while the input-output coefficients were within a plausible range, given prior knowledge of the technology. The manufacturing cost (capital, wages, energy, etc.), excluding the cost of raw materials explicit in the model, was estimated as the unit value of output minus the cost of all inputs, all expressed at world prices. So, net profits were constrained to equal zero, consistent with a competitive equilibrium. The technology was held constant at the 1997 level, except for the paper industry. There, it was assumed that the amount of recycled paper input would increase gradually between 1997 and 2010, to reach levels predicted by other studies (Ince 1994, Mabee 1998).

In the GFPM, each country imports from and exports to the world market. The world price was calculated as the average of import (cost, insurance, freight) and export (free on board) unit values. In the base year, all the supply curves for raw materials and the demand curves for end products were calibrated to supply or demand the amounts observed in each country, at world price. In this way, the solution of the model in the base year was almost identical to the observed quantities and prices.

The model was calibrated for 1997 and used to project prices and quantities to 2010. The parameters were estimated econometrically when possible and taken from other studies when not, and some had to be determined by judgment. The import tariffs were simulated as decreases in transportation costs. The tariff data for 1997 were obtained from the U.S. Department of Commerce, International Trade Administration. Tariffs were ad valorem, so the effect of a change in tariff changed with the price level. The tariff changes for each commodity and each country were based on Barbier (1996).

3.3.3 Effects of Tariff Liberalization to the Year 2010

With or without tariff elimination, the model projected changes between 1998 and 2010. Results show that wood consumption would continue to rise, especially in Asia. Worldwide, it would reach 4.4 billion m^3 by 2010, 29% more than in 1997. Still, the real price of products would increase only slightly. North and Central America, a net exporter in 1997, would become a net importer by the year 2010, with or without tariff elimination. This would be due, in part, to timber harvest restrictions in the United States and Canada. Europe and Asia would remain major importers, while Oceania, South America, and South America would increase their exports.

Table 11.3 summarizes the effects of eliminating the tariffs, other things being equal, from 1998 to 2010, for selected products and regions and for the United States. The predicted effects of freeing trade on world production were quite small, totalling less than 0.5% average increases in yearly production. In fact, tariff elimination would cause industrial roundwood production to decrease slightly in Asia and Europe. In the United States, the main effects on production would be an average 1.2% increase in annual sawnwood output, and it would be negligible for other products.

Accelerated liberalization would have a larger relative effect on trade than on production. World imports of industrial roundwood would decrease by about 2% under free trade. The main decrease (4%) would take place in Asia, mostly in Japan. The U.S. imports of industrial roundwood would be unaffected. World imports of manufactured products would increase with free trade, by 2% for sawnwood and by 1% for pulp and paper. U.S. imports of sawnwood would decrease by about 3%, and those of pulp would decrease by 2%. Instead, U.S. imports of wood-based panels and papers would increase.

International Trade in Forest Products

Table 11.3. Predicted effects of tariff reductions on production and trade quantities, 1998 to 2010, for selected regions and products

Region	Industrial roundwood			Sawnwood		
	Production %yr^{-1}	Import %yr^{-1}	Export %yr^{-1}	Production %yr^{-1}	Import %yr^{-1}	Export %yr^{-1}
Africa	0.6	3.1	1.0	1.0	-0.4	7.6
North and Central America	0.0	3.2	-15.5	0.8	-2.9	0.3
United States	0.4	0.3	6.3	1.2	-3.4	1.4
South America	1.1	0.9	0.0	1.0	5.9	11.5
Asia	-0.3	-3.7	-0.3	-1.3	6.4	3.9
Oceania	0.4	1.0	1.0	0.0	0.5	0.4
Europe	-0.1	-0.1	-2.8	-0.3	4.8	5.0
Former USSR	0.5	-0.3	2.0	0.0	1.1	0.2
World	0.1	-1.7	-1.9	0.0	2.0	2.0
	Wood pulp			Paper and paperboard		
Africa	1.4	-1.0	3.5	0.4	0.0	2.5
North and Central America	0.1	-1.5	1.7	-0.3	2.0	-0.1
United States	0.0	-1.8	2.6	-0.4	3.3	0.5
South America	0.4	-0.5	2.6	-0.4	-1.1	-1.3
Asia	-1.3	3.4	-4.3	0.5	0.7	0.7
Oceania	-1.1	-3.7	0.3	-1.4	2.5	-0.7
Europe	0.3	0.2	-0.7	0.2	0.8	1.6
Former USSR	-1.0	2.2	0.8	-1.5	3.0	1.4
World	-0.1	0.7	0.7	0.1	0.9	0.9

While U.S. imports of industrial roundwood were almost unchanged by tariff elimination, exports decreased by 6%. Instead, U.S. exports of sawnwood and panels would increase by 1% to 1.5%. U.S. exports of wood pulp would increase by 3%, while exports of paper and paperboard would remain nearly unchanged. The main gains in exports, in relative terms, would occur in South America for sawnwood and panels.

3.3.4 Discussion

With or without free global trade, world consumption of forest products projected with the GFPM continued to grow along the historical trends, and real world prices increased moderately. Eliminating tariffs caused only small changes in world production and consumption, but the trade shifted toward more processed products. Exports increased from northern Europe, Oceania, South America (Chile), and Asia (Indonesia and Malaysia). For the United States, the production and consumption was barely affected, but the trade quantity changed more. The U.S. exports of logs decreased while exports of most processed products increased.

Based on GFPM projections, global timber harvest would increase little due to tariff elimination. Projected timber harvesting would increase most in South America (by about 1% per year) but decrease slightly in Asia and Europe. The increase in harvest, where it occurs, is likely to come in large part from plantations (Tomberlin and Buongiorno 2001). Thus, at a broad scale, further tariff liberalization should have little effect on harvesting in primary forests.

We caution, finally, that these results depend on arguable assumptions and model parameters, which are estimates, so improvements in baseline data and methods would reduce uncertainties. Furthermore, tariff elimination in forest products is only a part of a broader set of reductions in other trade barriers. Those other reductions may contribute to increasing income and rising standards of living in poor countries, accompanied by decreases in fuelwood use and increases in demand for forest amenities (Raunikar and Buongiorno 1999).

4. CONCLUSION

Rapid growth in international trade of forest products has been observed at least partly because countries have reduced trade barriers, especially tariffs and quotas. In spite of this, significant additional barriers to trade remain, and there is growing evidence of largely negative economic and environmental outcomes from such trade constraints. Policy makers and economists have held out competitive pricing and free trade policies as the best means of promoting the efficient allocation of scarce forest resources. Barriers to trade distort market signals and may result in inefficient utilization of timber and nontimber forest resources. This is a common outcome in countries where competitive markets do not exist, but this also happens in countries with reasonably competitive forest products markets. Remaining barriers range from persistent tariffs and export quotas to trade-limiting new and existing technical, environmental, and sanitary standards. Timber trade restrictions are usually used as tools to (1) protect (subsidize) or develop domestic wood processing industries so that they generate employment, value added, and export revenues; and (or) (2) better fund forest management and protect forest resources. These policies have mixed results. While they may succeed in generating employment or value added, they also may be associated with high overall economic and environmental costs. These shortcomings are clearly recognized in the Forest Principles (1992 Earth Summit in Rio de Janeiro), which call for trade barrier reductions to achieve sustainable forest management.

Forest product trade liberalization and trade agreements have several documented economic advantages, including providing a framework for more effective institutions to protect, manage, and profit from forests and strengthening the economy. Trade models can help to quantify the effects of such changes. Nevertheless, trade policies and trade barriers can be surprisingly immovable. The persistence of forest products trade barriers may arise out of reluctance by governments to relinquish control over trade, or it may arise because these governments see trade restrictions as effective means of achieving important economic development goals. Modeling the effects of the status quo in comparison to what could be, however, can help policy makers better assess whether to chart a new economic course.

5. LITERATURE CITED

BARBIER, E.B. 1996. Impact of the Uruguay Round on international trade in forest products. Rome: Food and Agriculture Org. of the United Nations. 51 p.

BARBIER, E. B. 1997. The effects of the Uruguay Round tariff reductions on the forest products trade: A partial equilibrium analysis. Paper prepared for the Economic and Social Research Council conference on international dimensions of tax and environmental policies. Coventry, UK: University of Warwick.

BARBIER, E.B. 1999. Timber trade and environment. P. 106-117 *in* World forests, society and environment, Palo, M., and J. Uusivuori (eds.). Dordrecht: Kluwer Academic Publishers.

BAUDIN, A., AND L. LUNDBERG. 1987. A world model of the demand for paper and paperboard. For. Sci. 33(1): 185-196.

BONNEFOI, B., AND J. BUONGIORNO. 1990. Comparative advantage of countries in forest-products trade. For. Ecol. Manage. 36:1-17.

BOURKE, I.J., AND J. LEITCH. 1998. Trade restrictions and their impact on international trade in forest products. Rome: Food and Agriculture Org. of the United Nations. 42 p.

BOYD, R., K. DOROODIAN, AND S. ABDUL-LATIF. 1993. The effects of tariff removals on the North American lumber trade. Can. J. Agr. Econ. 41:311-328.

BOYD, R., AND K. KRUTILLA. 1992. The impact of the free trade agreement on the U.S. forestry sector: A general equilibrium analysis. P. 236-253 *in* Emerging issues in forest policy, Nemetz, P. (ed.). Vancouver: University of British Columbia Press.

BROWN, C. 1997. The implications of the GATT Uruguay Round and other trade arrangements for the Asia-Pacific forest products trade. Rome: Food and Agriculture Org. of the United Nations.

BULL, G.Q., W. MABEE, AND R. SCHARPENBERG. 1998. The FAO Global Fibre Supply study: Assumptions, methods, models and definitions. Global Fibre Supply Study working paper series. GFSS/WP/01. Rome: Food and Agriculture Org. of the United Nations. 48 p.

BUONGIORNO, J. 1978. Income and price elasticities in the world demand for paper and paperboard. For. Sci. 24(2):231-246.

BUONGIORNO, J., AND J.K. GILLESS. 1984. A model of international trade of forest products, with an application to newsprint. J. World For. Res. Manage. 1:65-80.

CORDEN, W.M. 1974. Trade policy and economic welfare. Oxford: Clarendon Press. 318 p.

DIXIT, A., AND A.S. KYLE. 1985. The use of protection and subsidies for entry promotion and deterrence. Am. Econ. Rev. 75:139-152.

FAO. 1999. Online FAO Yearbook of Forest Products, FAOSTAT statistics database. (http://apps.fao.org/) Rome: Food and Agriculture Org. of the United Nations.

FRANCOIS, J.F., C.R. SHIELLS, H.M. ARCE, K. JOHNSON, K.A. REINERT, AND S.P. TOKARICK. 1992. Economy-wide modeling of the economic implications of an FTA with Mexico and a NAFTA with Canada and Mexico (report on investigation No. 332-317 under Section 332 of the Tariff Act of 1930). Washington DC: U.S. Intern. Trade Comm. 711 p.

FRANKEL, J.A. 1998. The regionalization of the world economy. National Bureau of Economic Research Project Report Series. Chicago: University of Chicago Press. 285 p.

FRANKEL, J.A., E. STEIN, AND S.-J. WEI. 1997. Regional trading blocs in the world economic system. Washington, DC: Institute for International Economics. 364 p.

GILLESS, J.K., AND J. BUONGIORNO. 1987. PAPYRUS: A model of the North American pulp and paper industry. For. Sci. Monogr. 28.

HAZELL, P.B.R., AND R.D. NORTON. 1986. Mathematical programming for economic analysis in agriculture. New York: MacMillan. 400 p.

HELPMAN, E., AND P.R. KRUGMAN. 1989. Trade Policy and Market Structure. Cambridge, Massachusetts: MIT Press. 191 p.

INCE, P.J. 1994. Recycling and long-range timber outlook. USDA For. Serv. Gen. Tech. Rpt. RM-242. 23 p.

ITO, T., AND A.O. KRUEGER. 1997. Regionalism versus multilateral trade arrangements. National Bureau of Economic Research-East Asia seminar on economics, volume 6. Chicago: University of Chicago Press. 419 p.

KALLIO, M., D.P. DYKSTRA, AND C.S. BINKLEY. 1987. The global forest sector: An analytical perspective. New York: John Wiley and Sons. 718 p.

KRUGMAN, P.R., AND M. OBSTFELD. 1988. International economics: Theory and policy. Chicago: Addison Wesley Longman, Inc. 800 p.

LEAMER, E.E. 1984. Sources of international comparative advantage. MIT Press, Boston, MA. 53 p.

MABEE, W.E. 1998. The importance of recovered fibers in global fiber supply. Unasylva 49(193):31-36.

OLECHOWSKI, A. 1987. Barriers to trade in wood and wood products. P. 371-390 *in* The global forest sector: An analytical perspective, Kallio, M., D.P. Dykstra, and C.S. Binkley (eds.). New York: John Wiley and Sons. 718 p.

PRESTEMON, J.P. 2000. Public open access and private timber harvests: Theory and application to the effects of trade liberalization in Mexico. Env. Res. Econ. 17(4):311-334.

PRESTEMON, J.P., AND J. BUONGIORNO. 1996. The impacts of NAFTA on U.S. and Canadian forest product exports to Mexico. Can J. For. Res. 26(5):794-809.

PRESTEMON, J.P., AND J. BUONGIORNO. 1997. Comparative advantage in U.S. interstate forest product trade. J. For. Econ. 3(3):207-228.

RAUNIKAR, R., AND J. BUONGIORNO. 1999. Le rôle joué par l'économie dans la gestion forestière: de Faustmann à la courbe environnementale de Kuznets. Revue Forestière Française. No spécial 1999: 102-116.

REDDY, S.R.C., AND C. PRICE. 1999. Carbon sequestration and conservation of tropical forests under uncertainty. J. Agr. Econ. 50(1):17-35.

SAMUELSON, P. 1952. Spatial price equilibrium and linear programming. Am. Econ. Rev. 42(3):283-303.

SEDJO, R., A. GOETZL, AND S. MOFFAT. 1998. Sustainability of temperate forests. Washington DC: Resources for the Future. 102 p.

SHIELLS, C. 1995. Regional trade blocs: Trade creating or diverting? Finance Develop. 32(1):30-32.

SOHNGEN, B., AND R. SEDJO. 2000. Potential carbon flux from timber harvests and management in the context of a global timber market. Clim. Change 44(1-2):151-172.

TAKAYAMA, T., AND G.G. JUDGE. 1964. Equilibrium among spatially separated markets: a reformulation. Econometrica 32(4):510-523.

TOMBERLIN, D., AND J. BUONGIORNO. 2001. Timber plantations, timber supply, and forest conservation. Pp. 85-94 *in* World forests, markets, and policies. M. Palo, et al. (eds.) Amsterdam: Kluwer Academic.

VANEK, J. 1959. The natural resource content of foreign trade, 1870-1955, and the relative abundance of natural resources in the United States. Rev. Econ. Stat. 41:146-153.

WEAR, D.N., AND K.J. LEE. 1993. U.S. policy and Canadian lumber: Effects of the 1986 Memorandum of Understanding. For. Sci. 39(4):799-815.

WISDOM, H. 1995. NAFTA and GATT: What do they mean for forestry? J. For 93(7):11-14.

WORLD TRADE ORGANIZATION. 2001. Regionalism facts and figures. http://www.wto.org/english/tratop_e/region_e/regfac_e.htm, obtained March 19, 2001.

YUNEZ-NAUDE, A. 1992. Hacia un tratado de libre comercio Norteamericano; efectos en los sectores agropecuarios y alimenticios de México. Invest. Agrar. Econ. 7: 209-230.

ZHANG, D., BUONGIORNO, J., AND INCE, P.J., 1993. PELPS III: A microcomputer price endogenous linear programming system for economic modeling. Version 1.0. USDA For. Serv. Res. Pap. FPL-RP-526. 43 p.

ZHU, S., J. BUONGIORNO, AND D.J. BROOKS. 2001. Effects of accelerated tariff liberalization on the forest products sector: A global modeling approach. For. Pol. Econ. 2(1):57-78.

ZHU, S., D. TOMBERLIN, AND J. BUONGIORNO. 1998. Global forest products consumption, production, trade, and prices: GFPM model projections to 2010. Working Paper GFPOS/WP/01. Rome: Food and Agriculture Org. of the United Nations. 333 p.

Section Two

MULTIPLE PRODUCTS FROM FORESTS

Subhrendu K. Pattanayak
Research Triangle Institute

Forest management is characterized by joint production, with multiple outputs obtained from a common forest resource base. As illustrated in the previous section, classical forest economics focuses on timber production. However, forest owners and managers are concerned with a wide range of outputs from their forests, either because of the direct benefits or in response to government regulation and incentives. In this section, we examine the behavior of various forest owners and users regarding multiple outputs, including economic stability, carbon sequestration, amenities (e.g., wildlife), non-timber forest products (e.g., fuelwood), and soil quality. Methods for quantifying the value of these outputs are considered in section three.

As expected given the diversity of owners and outputs of concern, the chapters in this section apply a range of empirical methods to data at various scales and from various sources. Chapters 12 and 13 start with the optimal rotation framework introduced in chapter 4 and use simulation methods to predict how policy prescriptions for managing public forests and for carbon sequestration change stand-level and market outcomes. The remaining chapters use the household production framework to specify econometric models, which are estimated using plot and household level data. The economic agents modeled in these chapters are simultaneously producers and consumers of forest outputs. Chapter 14 considers non-industrial private forest landowners who jointly produce timber and amenities. Chapter 15 addresses farming households who collect and consume non-timber forest products. Chapter 16 considers farmers who adopt agroforestry, the combination of trees and crops. In each case, the empirical models identify socioeconomic and biophysical determinants of the behavior underlying joint production of multiple forest outputs. These chapters cover a wide geographic range, from the US to the Philippines, Mexico, India, and Brazil.

Sills and Abt (eds.), Forests in a Market Economy, 201. ©Kluwer Academic Publishers. Printed in The Netherlands.

Chapter 12

Public Timber Supply under Multiple-Use Management

David N. Wear
USDA Forest Service

In many parts of the world, substantial shares of timber inventories are managed by government agencies. The objective of this chapter is to examine the potential influence of public timber production on market structure as well as on prices, harvest quantities, and economic welfare. National forest management in the United States is used as a tractable case study, but findings provide general insights into the potential market effects of interactions between public and private producers in timber markets.

The United States has held a substantial portion of its forested lands in public ownership for more than 100 years. Currently the vast majority of public forests are managed by two different agencies, the USDA Forest Service and the Bureau of Land Management, with the largest share held by the Forest Service. The national forests (originally Forest Reserves) were controversial at their inception, and their management remains controversial today. While contemporary conflict has focused on various aspects of management and planning, the essence of the debate is discord regarding the ultimate ends for which public forests should be managed. In this regard, today's debates over public forests are little different from debates at the turn of the 20th century between Gifford Pinchot, the first chief of the Forest Service, and John Muir, the founder of the U.S. wilderness movement (Frome 1962). The issues are less about how these forests are managed and more about whether they should be managed at all.

Many of the debates regarding national forest management have revolved around economic issues, especially the harvesting of timber. The management of public forests in a market economy raises a number of interesting economic questions. Producing timber from public lands defines

Sills and Abt (eds.), Forests in a Market Economy, 203–220. ©*Kluwer Academic Publishers. Printed in The Netherlands.*

an unusual situation where the government takes part in a private market (Wear 1989a). How might activities of a public producer, with a potentially large share of production, shape the market? How might abrupt changes in timber production caused by policy shifts allow for windfalls or impose costs on other market players? The provision of nonmarket benefits, including public goods, is another important aspect of public land management. How are values formed regarding national forests and their sometimes unique benefits? National forests are highly concentrated in parts of the United States, where they represent dominant landowners especially in rural regions. What impacts do national forests have on the structure of these rural economies? Because they are spatially concentrated, public lands can also act as a vehicle for transferring wealth to or from these small areas (Wear and Hyde 1992). What are the distributional implications of national forest management?

In this chapter, we focus on the first set of questions—that is, on how the activities of a large public timber producer might influence the structure of timber markets and how these impacts might be addressed within a forest planning approach. We explore a theoretical structure of timber markets with a dominant public player, address the impacts that changes in timber management regimes might have on timber markets and the distribution of economic activity, and discuss how market effects could be incorporated within national forest planning.

1. WHY PUBLIC FORESTS?

An investigation of the role of national forests in private timber markets requires some clarity on the general rationale behind government participation in markets. Producing timber can be viewed as an intervention in the market, in some ways little different from more traditional means of market intervention—e.g., taxation or subsidy. There are three traditional reasons for government intervention in a market: stabilization, efficiency, and equity (Atkinson and Stiglitz 1980, Boyd and Hyde 1989). All three have been invoked to justify national forest management.

The initial motivation for national forests focused on stability. During the frontier epic of timber harvesting in the United States, forest regeneration and management was absent as the product was mined from successive regions of the country. The forestry movement of the day sought mechanisms to guard against eventual timber famine. Careful husbandry of forests by the federal government was seen as a logical instrument for stabilizing future timber markets.

The efficiency rationale can be seen in Forest Service concerns for nonmarket benefits, and these also date from the inception of the agency. Dueling with timber stability as justification for national forests were concerns regarding the protection of watersheds, in particular protection against flooding that could result from deforestation. This is classic market failure, where markets fail to endogenize offsite costs (floods) while procuring market goods (timber). The number of nonmarket services formally addressed by national forest management—e.g., recreation, aesthetics, and wildlife habitat—has expanded, but they all are conceptually similar to the flooding case in that they address services derived from *in situ* forests. Recent concerns regarding the role of public lands in well-functioning ecosystems shifts the debate somewhat in that management now addresses the role of national forests in broader dynamic landscapes that are shaped by both private and public land management. Still, this is an efficiency rationale.

Distributional motivations have also played a role in the management of national forests. Concerns for the stability of forest-dependent communities date to the Great Depression, when the general notion of redistribution via government intervention was more widely accepted. The community-stability rationale for public forest management suggests that careful federal management of forests can insulate communities in the hinterlands from wide swings in lumber markets—in this sense it can also be viewed as a stability rationale—and implies a redistribution of income to these areas from the economy at large—an equity rationale. Community stability persisted as an important rationale for public forest management until the 1990s, especially as it was tied into broader government initiatives to encourage rural development (see generally Schallou and Alston 1987).

The breadth of expectations and objectives for national forest management has expanded considerably since the turn of the 20th century as the public interest in public lands has expanded both in terms of the range of goods and services and the spatial scale of the interested publics. The scope of national forest planning has increased dramatically in response.

Understanding the breadth and complexity of motivations for public forestry is important, because it helps illuminate the controversy surrounding public and increasingly private forestry and illustrates the difficulties of isolating the role of timber production from the broader scheme of public forest management. Knowing the evolution of rationales also allows for insights into the long-run flexibility of public forestry institutions. Changes in motivation and management reflect the evolution of institutions over time in response to changing public perspectives and preferences as well as changing resource scarcities.

2. NATIONAL FOREST PLANNING

Translating the rationale of public forest management into operations and forest management activities is a complex undertaking. The technologies of planning have evolved considerably on several fronts. As objectives have become more complex, so have regulations that govern the planning of national forests. Here we look back on a planning system that has been in place for 20 years. At the time of this writing, the regulations that prescribe planning procedures are being revised.

The planning of national forests is possibly the most complex application of natural resource economics to date. Several factors which contribute to its difficulty are significant: (1) forests are used by society to a variety of ends, and many uses are incompatible with others; (2) because most of the national forest system is concentrated in rural western states, public planning decisions can have strong effects on local economies; (3) resource tradeoffs must be evaluated with imperfect information on the values of many resources; (4) the actual resource tradeoffs are unclear because the production responses of forests to management are uncertain; and (5) the complexity of the problem creates an enormous number of plausible management alternatives. The national forest planning process has been an attempt to structure this complex and sometimes ambiguous resource management problem in a way that leads to well-informed decisions.

The primary tool for forest planning analysis has been an optimization model used to bring together data to describe a national forest, the production relationships which describe how the forest will develop and respond to different management activities, the values of different resource outputs, and the costs of management. The model is solved using an optimization approach that defines an economically efficient management plan for the forest subject to a set of policy and management constraints. This reduces the decision space by eliminating from consideration the many suboptimal management plans that could achieve the same level of outputs. Of course, the degree to which a solution actually reflects an optimal plan depends on the construction of the model, especially the constraints.

Cost/benefit analysis of this type addresses the relative efficiency of forest management alternatives and attempts to define that management plan which gives rise to the highest net discounted benefits. In addition, forest planning clearly addresses distributive or equity questions as well. These distributive issues, often encapsulated as a community stability policy, are largely concerned with a redistribution of resource wealth from the public at large to the rural areas, which are dependent on public forests for input to their wood products industries. These concerns for local production levels and their derivative employment and income are often used to justify departures from the efficient solutions defined by cost/benefit analysis. Impact analysis utilizes an input-output model (IMPLAN) that describes the historical impacts of forest outputs on local economies and projects the economic impacts of various production alternatives.

3. THE MARKET VIEW IMPLIED BY NATIONAL FOREST PLANNING

Forest planning addresses production decisions at a national forest level using two economic analyses: a cost/benefit analysis and an impact analysis. While each national forest plans separately without explicit consideration of its market interactions, a market view is implied by the structure of these analyses. The cost/benefit analysis views each national forest as a typical producer in natural resource markets. That is, each forest is assumed to not be able to influence total production in these markets, and, therefore, its production decisions have no bearing on resource prices or on input quantities. The impact or distributional analysis using IMPLAN, on the other hand, assumes considerable market power for individual national forests[1]. When each forest compares the derivative jobs for its alternative production levels, the implication is that the forest is omnipotent in resource markets, defining its production as an increment to total production and employment levels within the local area. The extent of the market power actually held by most national forests likely falls somewhere between these two extremes.

Alternative market structures for use in forest planning have been discussed to a limited degree. The discussion has centered on whether a downward-sloping demand curve should be applied in management models instead of the horizontal demand curves that have been used (e.g., Chappelle 1977, Walker 1971). There are theoretical as well as computational implications of this approach. Using a downward-sloping demand curve would imply that each national forest would control total production in its market area. If this were the case, the approach would be appropriate, but only with some important modifications. Simply replacing the horizontal demand curve with a downward sloping curve and solving the linear planning model would lead to a monopoly solution with reduced production and higher prices relative to a competitive solution. This solution would increase public timber revenues at the expense of timber consumers, thereby redistributing wealth from the local area to the federal government. In order to account for total consumer and producer benefits, and to simulate the fair market solution, the objective function would need to be redefined as the maximum of the sum of producer and consumer welfare (Samuelson 1952).

This provides a useful approach for places where the public sector is the sole or strongly dominant timber producer.

In most cases, however, the actual structure of the market problem lies between the extremes of a powerless and an omnipotent resource supplier. Because the agency does not ordinarily control the entire resource base in a region, it cannot completely control production and therefore total price. What's more, harvest levels are not set by administrative edict. Rather, sale quantities are set by plans, and harvest rights are sold at auction. The actual quantity sold is very much influenced by market conditions (Adams et al. 1991).

Where the government does control a large share of the resource base, its actions have some potentially strong influence on the production and investment decisions of private producers so that total production may be influenced indirectly. If this is the case, then an examination of direct effects alone is inadequate. Because changes in public production may lead to changes in private production, the total influence of forest planning decisions may not be captured in individual forest plans.

The role of the Forest Service in certain timber markets has been extraordinary. It is unique because only in the case of forestry has the federal government taken such a large and active role as a resource producer in an otherwise private market. The national forests contain the largest share of the nation's softwood sawtimber (51% in 1997; Smith et al. 2001), and the share is much higher in western regions (e.g., 74% in the Rocky Mountain region). Because of this dominance and because of a multiple-use agenda that allows the agency to manage timber at a financial loss, the Forest Service cannot be considered a typical timber producer. Indeed, in many cases it can be argued that planning decisions largely shape markets for timber. Therefore, production plans and actions should hold at least some influence over timber production from private and other public producers and consequently determine market prices for timber. [2]

4. THE MECHANICS OF REGIONAL TIMBER MARKETS

A discussion of the role of the Forest Service in timber markets is best built as a departure from the perfectly competitive case. A market is the place where producers and consumers interact and production levels and prices are established. In the case of timber, the outcome is timber harvest quantities and timber prices. Putting aside the public role in these markets for the moment, consider first the actions of private timber producers and consumers that would define timber supply and demand, respectively, in a perfectly competitive market.

For the purpose of this discussion, consider timber demand at the regional level as derived from the national market for wood products (see also chapter 9). For a small production region, the demand for delivered logs is dependent on prices for products, such as lumber and plywood, determined in national and perhaps international markets, and on local manufacturing costs. The value of the logs in production is the value of the lumber and other outputs produced minus all the relevant production costs. Logs are consumed by wood products manufacturers who adjust their demands for logs and other inputs such as labor and machines, based on the prices of products, logs, and these other inputs. A similar demand relationship holds for labor as well as any other input; employment in local wood products industries is therefore dependent not only on national product prices but also on local wages and timber prices.

The demand for standing timber or stumpage is derived through the next step in production, logging and hauling. The value of stumpage to a logging firm is defined by the prevailing price of delivered logs and the costs of transporting standing trees to the mill yard (logging plus hauling costs). Timber consumers adjust their demands for timber and other inputs based on these prices and costs.

The supply of timber from forest owners is derived in a somewhat different way. The dominant question facing owners of merchantable timber is whether to harvest today or to hold onto standing timber (chapter 8). This decision is based on owner preferences, market expectations, and the costs of bringing timber to market. The forest owner must decide whether revenues exceed costs and then whether the returns to harvesting today are better than the future opportunities for timber revenues. This conceptual model of supply is complicated beyond the usual analysis of commodity supply because a typical producer cannot be defined (see chapter 8). This is because the complement of forests is not homogenous but varies in terms of site productivity, operability, and accessibility. The result—and this is important for this analysis—is a segmented supply relationship: as prices rise, discrete bundles of timber may enter the market.

This relationship defines how profit-seeking forest owners would respond to any set of timber prices. For illustration, consider that forests, especially in an area like the Rocky Mountains, are composed of stands of trees of various qualities and accessibility. These attributes define the costs of bringing each type of stand to market, which include the cost of accessing the timber and preparing a sale. Assuming for the moment that future prices or opportunity costs are constant across these timber stands, the timber supply response function is defined as the quantity of timber placed on the market for any given price of timber. An example is shown in figure 12.1. At some minimum price (p_1) forest owners will produce timber from a certain quality/accessibility class (a1). Production begins when the market price is equal to p_1, which is the opportunity costs plus the cost of bringing timber of class a1 to the market. As price increases (to p_2, p_3, and so on), more costly timber classes are brought to market.

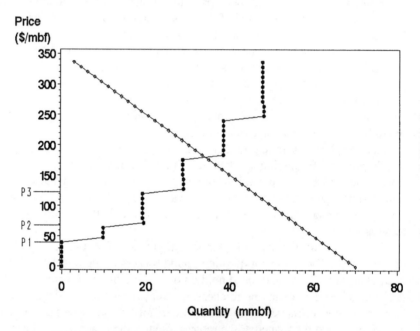

Figure 12.1. A hypothetical timber market described by a supply curve (stair-stepped) and demand curve (linear)

If we apply the demand curve in figure 12.1, then the market is completely defined. The quantity of timber produced is defined where the demand price for timber—the value of the derived products minus the costs of logging, hauling, and manufacturing—is just equal to the supply price—the opportunity costs of harvesting plus the marketing costs. Our key assumption is that the market is perfectly competitive, so that all producers and consumers seek to maximize their profits and utility, respectively, from timber production and that no producer or consumer is large enough to influence the market. If such a case holds and externalities do not exist, then social welfare is well served by this solution. That is, the net discounted benefits arising from timber production are at a maximum.

5. THE FOREST SERVICE ROLE IN TIMBER MARKETS

Extending this analysis to distinguish between private and public production requires adjusting some of the basic assumptions. The Forest Service is not bound to profit-maximizing behavior because of important externalities involved in the production of timber and because of other objectives guiding public land management, such as the distributive goals discussed earlier. At an operational level, federal forestry agencies are not funded by timber receipts, so do not face efficiency incentives. Because of differences in motives and the large share of timber stocks controlled by the agency in some regions, production decisions on the national forests can change the shape of a supply response function. This, in turn, will feed back to the production decisions of other producers. In this section, we explore these feedbacks—that is, the potential interactions between private and public producers in a market for timber.

When the assumption of a profit motive for a major timber producer is dropped, the behavioral basis for the supply response curve shown in figure 12.1 is lost. That is, because the agency makes decisions based on nonmarket as well as market concerns, the amount of timber brought to market is no longer based strictly on marketing and opportunity costs. For example, the Forest Service may decide to depart from efficient production levels to address distributive goals. This redefines the shape of the timber supply response function because some timber is brought to market at a price that does not completely cover marketing and opportunity costs. Therefore, and quite regardless of motive, the agency may alter the aggregate supply response in a region.

In a sense, this is analogous to a Stackelberg model of oligopsony (Shapiro 1989), where a dominant producer leads the market and all other producers—who comprise what is referred to as the competitive fringe—react to the dominant producer's plans. The departure from the Stackelberg model is that, in this case, the leader (public lands) is not driven by profit maximization but is, in effect, a profit-indifferent leader. This does not necessarily imply that the public producers are inefficient, but that their objectives are not characterized by a strict financial optimization. As a result, their behavior appears less than optimal from the perspective of a timber profit function. The next section examines the potential implications of this type of market structure.

5.1 Simulating Timber Supply

Our analysis uses an engineering approach to simulate long-run timber supply relationships for hypothetical private and public timber owners within a region (e.g., Hyde 1980). Assume that timber grows according to the following logistic biological production function (Swallow et al. 1990):

$$f(T) = \frac{K}{1 + e^{(\alpha + \theta T)}} \qquad 12.1$$

where $f(T)$ defines volume as a function of stand age, T; K defines a carrying capacity of 15.055 mbf/acre; and the coefficients are defined as $\alpha=6.1824$ and $\theta=-0.0801$.

If we assume that this production function applies to both private and public landowners, that both ownerships apply a discount rate of 4%, and that both maximize the net present value of timber production, then the standard Faustmann solution for the rotation age (see chapter 4) occurs where the present net value of the infinite series of rotations is at a maximum

$$V = \max_T \left[(p - c_1) f(T) e^{-rT} - c_2 \right] \left(1 - e^{-rT} \right)^{-1} \qquad 12.2$$

where p is stumpage price, c_1 is a harvesting cost, c_2 is a forest regeneration cost, and r is the discount rate. Long-run supply is derived by solving 12.2 for a range of price scenarios. The contribution of a stand to timber supply is defined by multiplying volume at age T ($f[T]$) by the area of the stand and dividing by T (i.e., this is the average annual contribution of the forest type to supply). This method implies that the forest has reached a long-run equilibrium where the forest has a uniform age distribution between zero and age T. While this simplification limits insights into short-run dynamics, the equilibrium engineering approach provides a useful mechanism for evaluating long-run implications of various management approaches (chapter 8).

With only one stand type, however, the resulting timber supply curve is trivial because the solution to the Faustmann problem is invariant to changes in long-run price levels. To generate a realistic supply curve, we need to account for a distribution of stand types consistent with what we observe in nature. For this analysis, we generate a hypothetical forest with six stand types and hold the biological production function constant across these types. We recognize differences in stand types by varying the cost terms ($c1$ and $c2$) by type so that forest type 1 is the least costly and forest type 6 is the most costly to harvest. This is consistent, for example, with a forest in a mountainous landscape where costs can vary substantially between sites. Area and cost factors for the hypothetical forest are displayed in table 12.1.

Table 12.1. Distribution of area (acres) and costs ($/acre) for six land classes and two owners

Variable	Owner	LC 1	LC 2	LC 3	LC 4	LC 5	LC 6
Area	Private	100.	100.	100.	100.	100.	100.
Area	Public	100.	100.	100.	100.	100.	100.
Cost (c2)	Private	$10.	$20.	$35.	$50.	$70.	$100.
Cost (c2)	Public	$5.	$10.	$18.	$25.	$35.	$50.
Cost (c1)	Private	$10.	$10.	$20.	$30.	$40.	$50.
Cost (c1)	Public	$10.	$10.	$10.	$10.	$10.	$10.

LC= Land class

We next apply these long-run models to explore the implications of various public management strategies for the aggregate supply from a region. As shown in table 12.1, we assume that the public and private sectors have an identical distribution of forest types. This is consistent with the checkerboard ownership pattern observed in the western United States where alternating sections of land are arbitrarily assigned to the two different owner groups. It is very different from ownership patterns in other parts of the country where the public sector occupies the lands nobody wanted, thereby skewing the distribution toward low-quality land (see Shands and Healy 1977). However, assuming equal land-quality distributions serves our purpose here by allowing us to focus exclusively on the impacts of differences in management approaches.

We examine three scenarios for public timber management while treating private timber management as if it were driven strictly by timber profit maximization. (1) As a benchmark, we start by examining the case where public lands are managed identically to private lands and with the costs listed for private lands—labelled the market scenario. (2) The subsidized scenario applies the Faustmann criterion to public lands but allows the costs to be subsidized as outlined in table 12.1. (3) The CMAI scenario forces timber harvest to occur at the age of maximum timber volume production (the culmination of mean annual increment or CMAI) as long as the net present value (equation 12.2) is positive. The third scenario also uses the subsidized costs for timber management shown in table 12.1. Use of the subsidized cost figures is consistent with observed management and with using timber harvesting to accomplish other ecosystem goals, thereby incurring both joint benefits and joint costs (a rationale for so-called below-cost timber sales). The CMAI rule has long been applied to national forest management.

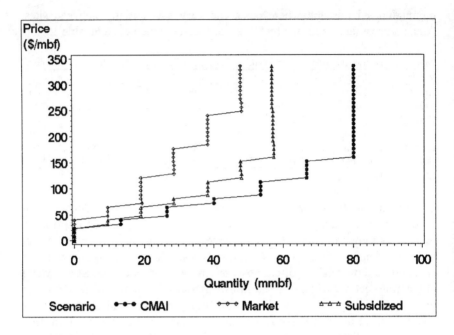

Figure 12.2. Supply curves estimated for three public forest scenarios: (1) profit maximization (market), (2) profit maximization with some costs subsidized (subsidized), and (3) profit maximization constrained by a culmination of mean annual increment rule (CMAI)

Public intervention of the two varieties listed above yield flatter supply curves (figure 12.2), as anticipated from the theoretical development. The benchmark market scenario is the steepest of the supply curves. Maximum annual production of 48 mmbf is reached at a price of about $270/mbf. Subsidized timber production on the public lands results in a flatter supply curve with more timber produced throughout the price range. Maximum annual production of 57 mmbf is reached at a price of about $160/mbf. The CMAI scenario is flatter yet, with maximum annual production of 80 mmbf reached at $160/mbf. As modelled here, public intervention has the general effect of rotating total supply outward.

5.2 Market-Level Implications

To fully develop the market implications of public intervention, we specified a demand function and conducted simulations of market clearing behavior with the three specifications of supply, compared to the case of no public supply. To smooth out the mechanistic definition of supply described above, we fit OLS regression equations for the supply schedules shown in figure 12.2. Supply equations were given the general form:

$$p = aQ + bQ^2 \qquad 12.3$$

and are specified for private and public owners. The demand equation has the form:

$$p = c + dQ \qquad 12.4$$

Coefficients are shown in table 12.2. Simulations are used to address how these interventions could influence (1) total timber harvests and the price of timber in the region, (2) private production and private producer surplus, (3) consumer surplus, and (4) stability. The first three issues can be addressed using a deterministic solution to the markets defined by equations 12.3 and 12.4. To examine effects on stability, we specified a stochastic demand equation by defining distributions for the parameters c and d. We solve for market clearing price and quantities by finding where producer plus consumer surplus is maximized (chapter 8).

Table 12.2. Estimated coefficients for supply and demand equations (equations 12.3 and 12.4)

Equations	Parameters	
	a	b
Supply – market	4.23	0.039
Supply – subsidy	0.133	0.072
Supply – CMAI	0.044	0.037
	c	d
Demand equation	400	-3

The simulation model was implemented using a FORTRAN program. Draws from a pseudo-random number generator were used to specify the demand equation. A simple search algorithm was used to define the optimal solution for each iteration. Producer and consumer surpluses are calculated for each scenario and supply and demand equations (12.3 and 12.4). The simulations were repeated 1000 times for each scenario and results summarized using means and coefficients of variation for all variables listed above. Simulation results are summarized in table 12.3.

Table 12.3. Harvest quantities and coefficients of variation resulting from Monte Carlo market simulations for four scenarios where public timber is managed: (1) None: without any harvests from public inventories, (2) Market: to maximize timber profit, (3) Subsidy: to maximize net present value but with costs subsidized as shown in table 12.1, and (4) CMAI: to maximize volume but only if subsidized net present value is positive

Variables	Scenarios			
	Quantities			
	None	Market	Subsidy	CMAI
Price ($/mbf)	162.98	120.1	97.27	80.64
Quantity (mmbf)	30.45	46.8	55.18	61.87
Q-private (mmbf)	30.45	23.2	19.24	16.28
Q-public (mmbf)	0	23.6	35.9	45.59
Consumer surplus ($1000s)	1300.16	2951.4	4080.6	5069.98
Private producer surplus ($1000s)	2715.31	1574.3	1106.7	804.34
	Coefficients of variation			
	None	Market	Subsidy	CMAI
Quantity (mmbf)	0.16	0.23	0.23	0.26
Q-private (mmbf)	0.16	0.23	0.32	0.38
Q-public (mmbf)	—	0.23	0.18	0.21
Consumer surplus ($1000s)	0.27	0.22	0.20	0.20
Private producer surplus ($1000s)	0.37	0.52	0.70	0.82

5.2.1 Price and Quantity Effects

Simulation results show that both the subsidy and the CMAI scenarios lead to reduced prices and increased total harvest levels. For the CMAI scenario, price falls by about 25% while harvest quantity expands by about the same percentage. Proportional response is also found for the subsidy scenario with price decreasing and quantity increasing by about 12%.

5.2.2 Private Production

Compared to the case where public timber is managed for profit, the subsidy and CMAI scenarios result in reductions in timber production from private lands. Private output is reduced by 10% and 21% under subsidy and CMAI scenarios, respectively. However, the impact on private producer surplus is nearly double, with a reduction of about 20% for the subsidy scenario and about 38% for the CMAI scenario. This reflects the simultaneous reduction in output from private forests and in output prices.

5.2.3 Consumer Surplus

Consumers of timber are unambiguously better off with the interventions modelled here—that is, without regard to other forest goods and services that may be substitutes for public timber production. Consumer surplus is 25% higher for the subsidy scenario and 57% higher for the CMAI scenario.

5.2.4 Stability

We evaluate the impact of intervention on stability by examining the coefficients of variation for the variables described above. For timber prices, the coefficient of variation increases under the intervention scenarios (as much as 40% for the CMAI scenario). The effect on quantity is much less emphatic—there is little impact on the coefficient of variation for total harvest. However, the coefficient of variation is increased for private production with the intervention scenarios (+24% for subsidy and +40% for CMAI). Likewise, the coefficient of variation for the producer surplus for private landowners increases by similar percentages. The variability of the consumer surplus measure decreases but only slightly.

Public intervention in timber markets therefore harms private timber producers—private output and timber prices are reduced. As a result, producer surplus for the private timber-producing sector falls substantially. This has the unambiguous result of discouraging timber harvest and investment in timber management on private lands. What's more, the intervention increases the variability in private returns. For private owners, intervention leads to a less stable market.

However, total harvests do increase with the subsidy, as expected, and the variability of total harvest does not increase with level. Consumers of timber benefit from increased output and reduced prices but face increased price uncertainty as a result. To the extent that timber production is correlated with labor used in manufacture of wood products, the increased harvests would have a positive impact on employment in this sector, without increased variability, though this impact would be tempered somewhat by the degree of substitutability between labor and timber in the wood products sector (see Wear 1989b and chapter 9).

6. CONCLUSIONS

Simulations developed here provide useful insights into the direction of certain impacts but little in the way of the actual magnitude of these impacts. Therefore the findings presented here should be viewed as working hypotheses regarding the long-run implications of public intervention in timber markets.

The simulation experiments presented here suggest that public timber harvests can have direct as well as indirect impacts on the operation of timber markets. The interventions described here cause output to expand and prices to fall consistent with economic intuition. There are two types of indirect impacts. One is the decrease in private production and the substantial reduction in rent accruing to private timber owners. While investment was not directly modelled here, this result suggests that public harvesting reduces the incentive to invest on private lands. Additionally, the flattened total supply curve results in prices that are relatively more volatile (measured in terms of the coefficient of variation). This suggests that a harvest policy at least partly justified by a stability rationale may in fact expand and smooth harvests but at the cost of making prices more variable.

Depending on the share of public ownership and the extent of public participation, these findings suggest that market feedbacks should be considered in forest planning. Recent developments in forest planning have focused on public forests as components of broader landscapes and ecosystems rather than as isolated systems. The same logic can be extended to the economic perspective regarding public forest management. The analysis presented here suggests that where the public sector is a dominant producer, public forest management should also be evaluated in a broader market context.

Findings also highlight a need to coordinate forest policies that address private and public forest management. Public harvest policy may counteract policies and programs intended to encourage management on private forests.

The market model defined here was simplified to focus on the interactions among producers and abstracted from the very important intertemporal aspects of forest plans and timber supply. Extending the time horizon of the market analysis would add considerable complexity but would not change the overall result; private timber production decisions are influenced by public decisions. Accordingly, the total production of timber in a region and the price of timber will be influenced by public production decisions, not only directly through the availability of public timber but also indirectly through the effects of these decisions on regional timber prices and therefore private production levels. The extent of these influences is proportional to the share of timber held by the public sector and the condition of public and private inventories.

In the United States, timber harvest from public land has dropped precipitously over the past decade, resulting in a redistribution of welfare between consumers and private producers (see Murray and Wear 1998). This illustrates another mechanism through which public management—in this case the sudden reduction of public harvests—may influence private decisions and timber markets as a whole.

These issues have increasing currency in many parts of the world, especially where countries have undergone transitions to market economies. In such places, governments face questions regarding how to organize forest ownership and how to market public timber resources in a new market economy. The salient findings are that feedbacks between public and private sectors can have compounding impacts on timber markets and that these interactions can distort markets and social benefits.

7. LITERATURE CITED

ADAMS, D.M., C.S. BINKLEY, and P.A. CARDELLICHIO. 1991. Is the level of national forest timber harvest sensitive to price? Land Econ. 67:74-84.

ATKINSON. A.B. and J.E. STIGLITZ. 1980. Lectures on public economics. McGraw-Hill, New York. 619 p.

BOYD, R.G. and W.F. HYDE. 1989. Forestry sector intervention: The impacts of public regulation on social welfare. Iowa State University Press, Ames IA. 295 p.

CHAPPELLE, D. 1977. Linear programming for forestry planning in forestry and long range planning.P. 129-163 *in* Forestry and Long Range Planning, F.J. Convery and C.W. Ralson (eds.). Duke University School of Forestry and Environmental Studies, Durham, NC.

FROME, M. 1962. Whose woods these are: The story of the national forests. Doubleday and Co. Inc., New York, 360 p.

HYDE, W.F. 1980. Timber supply, land allocation and economic efficiency. Johns Hopkins Press, Baltimore, MD. 224 p.

MURRAY, B.C. AND D.N. WEAR. 1998. Federal timber restrictions and interregional arbitrage in U.S. lumber. Land Econ. 74(1):76-91.

SAMUELSON, P.A. 1952. Spatial price equilibrium and linear programming. Am. Econ. Rev. 42:283-303.

SCHALLOU, C.H. and R.M. ALSTON. 1987. The commitment to community stability: A policy or shibboleth? Environ. Law 17(3):429-482.

SHANDS, W.E. and R.G. HEALY. 1977. The lands nobody wanted. The Conservation Foundation, Washington DC. 282 p.

SHAPIRO, C. 1989. Theories of oligopoly behavior. P. 329-414 *in* Handbook of Industrial Organization, Volume I, R. Schmalensee and R.D. Willig (eds.). Elsevier Science Publishers, New York, NY.

SWALLOW, S.K, P.J. PARKS, and D.N. WEAR. 1990. Policy relevant nonconvexities in the production of multiple forest benefits. J. Environ. Econ. Manage. 19(2): 264-280.

SMITH, W.B., J.S. VISSAGE, D.R. DARR, AND R.M. SHEFFIELD. 2001. Forest Resources of the United States, 1997. Gen. Tech. Rpt. NC-219. USDA Forest Service, North Central Station, St. Paul, MN 198 p.

WALKER, J.L. 1971. An economic model for optimizing the rate of timber harvesting. Unpublished Ph.D. Dissertation, University of Washington, Seattle, WA.

WEAR, D.N. 1989a. The market context of national forest planning. The Public Land Law Rev. 10:92-104.

WEAR D. N. 1989b. Structural change and factor demand in Montana's wood products industries. Can. J. For. Res. 19(5):645-650.

WEAR, D.N. AND W.F. HYDE. 1992. Distributive issues in forest policy. J. Bus. Admin. 21:297-314.

---

[1] IMPLAN does not assume market power—i.e., prices are fixed—but the tacit assumption behind the analysis of employment impacts of individual timber sale programs is that increased national forest production will influence total market output and therefore total employment in the wood products sector.

[2] Shares of timber production had approached shares of inventory in some western states until the early 1990s. Since then, timber production has fallen precipitously (see Murray and Wear 1998).

Chapter 13

Economics of Forest Carbon Sequestration

Brian C. Murray
Research Triangle Institute

Concern over rising greenhouse gas (GHG) concentrations in the atmosphere and the corresponding potential for harmful global climate change has spawned international effort to mitigate the build-up of GHGs. In addition to targeting reductions of GHG emissions from fossil fuel combustion, global mitigation efforts include the sequestration of atmospheric CO_2, the most critical GHG, into terrestrial carbon (C) stocks, often referred to as the creation of a C "sink."

Terrestrial ecosystems play an important role in regulating the atmospheric abundance of CO_2. Roughly one-half of the C in the Earth's terrestrial ecosystems is found in forest vegetation and soils (Watson et al. 2000). Carbon exchange between forests and the atmosphere occurs when a forest ecosystem transforms atmospheric CO_2 into forest C stock components (trees, roots, other vegetation, litter, and soils) through photosynthesis. Intact forests store C, but may ultimately release it through natural or anthropogenic disturbances.[1] About one-third of all CO_2 emissions since 1850 are a result of land use activities, predominately forest clearing, and about one-fourth of CO_2 emissions have been absorbed back into terrestrial ecosystems, such as forests (Watson et al. 2000). Thus, it is evident that policy measures to reduce atmospheric concentrations of CO_2 and other GHGs must take into account the role that forests play as a prime component of terrestrial ecosystems. As a result, forest C sequestration has been identified as one of an array of potential strategies to reduce the build-up of atmospheric CO_2 (Clinton and Gore 1993, Dixon et al. 1993, Sampson and Sedjo 1997).

This chapter demonstrates the effects of C sequestration incentives on the optimal management of an individual forest stand. In general, the incentive

Sills and Abt (eds.), Forests in a Market Economy, 221–241. ©Kluwer Academic Publishers. Printed in The Netherlands.

to sequester C will modify the optimal timing of forest harvests and the monetary return to forested land use. An analytical model of a timber-C rotation is presented and applied to data from different forest settings to demonstrate empirically the effect of different C prices on the optimal rotation, the land value of forests, and the amount of C sequestered at the stand level.

The stand-level view is limiting because it does not capture the interaction of all timber and C suppliers in the marketplace. Aggregate (i.e., national or large geographic regions) analysis is addressed in a separate section. Because C prices can alter the value of land in forests, they can also affect the way that land is allocated among uses. Many researchers have examined the incentives to change land use in response to C values and thereby estimate an aggregate cost function. The chapter reviews and categorizes the numerous aggregate empirical analyses that have been conducted within the last decade. These studies vary in their treatment of market processes and therefore in their estimates of the marginal cost of sequestering a ton of C.

The chapter concludes with a summary of the key issues related to C sequestration facing researchers and those charged with designing policy instruments to induce higher levels of sequestration from forests.

1. A CARBON SUPPLY FUNCTION

The purpose of this chapter is to conceptually develop a supply function for C sequestration from forests and to present empirical information on key parameters of such a function. For our purposes, a C supply function relates variation in the quantity of C supplied by forests to variations in price or other monetary value assigned to C.

The amount of C sequestered in forests is a subset of the amount sequestered in all terrestrial ecosystems, and the C supply function must be specified accordingly. The level of C sequestration from terrestrial ecosystems is a function of the distribution of land uses across the landscape and the rate of sequestration from each use. Consequently, an aggregate supply function can be specified:

$$S_A = \int_i^N c_i^*(v,p,Z)\, \phi_i(v,p,Z)\, L\, di \qquad 13.1$$

where i indexes one of N different possible land uses; $c_i^*(v,p,Z)$ is the C density of land use i as a function of the C price (v), a vector of non-C output and input prices (p) and land quality (Z); $\phi_i(v,p,Z)$ is the share of land allocated to use i; and L is the total area of the land base. The C price represents the monetary compensation a party receives (pays) for sequestering (releasing) a unit of C. The price could be determined in a market for tradable C permits or through a tax/subsidy set administratively by the government. The socially optimal level for the C price equals the monetized value of the marginal environmental damages caused by another unit of C's contribution to atmospheric CO_2. However, in a tradable permit market, the price will be determined by the marginal cost of abatement. This will equal the marginal environmental damages only if the institution setting the permit level selects the optimal level; i.e., if they properly anticipate the level at which the marginal social cost of reducing C equals the marginal social benefit. Likewise, an administratively set C price will equal the optimal price if the administering body precisely selects a tax/subsidy equal to the monetized marginal damages.

Market compensation for C will change the amount of C sequestered if it alters the distribution of land use and/or modifies the C density of land uses. Equation 13.1 indicates that variations in forest C can come from the *intensive* margin—variation in the C density of forested (F) land use, C_F—and from the *extensive* margin—variation in the amount of land allocated to forest, ϕ_F. Each of these factors is analyzed separately following.

Across any given landscape, land will vary by physical characteristics, such as soil composition, altitude, slope, and other determinants of fertility, as well as by economic characteristics determined by the proximity to processors of harvested outputs and human populations. These factors affect the relative profitability of land use alternatives. It is this inherent variability in land attributes, as well as the feedback between land markets and commodity markets discussed below, that produces a landscape with multiple land uses rather than a single dominant use (Hardie and Parks 1997, Stavins and Jaffe 1990).

Heterogeneity of land quality implies that different units of land will engender different C responses. As discussed above, a price for C will set the conditions for marginal changes in land use to higher C-density uses. Suppose we move from a world in which C is not accorded a price ($v = 0$) to one in which it is accorded a positive price. This is depicted in figure 13.1 With a positive price for C, some land now in agriculture (A), ostensibly because it is more profitable, may in fact be more profitable as forest (F), thereby establishing incentives for afforestation. But there is other land that, because of its physical characteristics and location, is much more profitable in crops than in forests. It would take a very high C price to make this land more profitable as forest, all else being equal. Thus, we envision a continuum in which successively higher C prices are necessary to induce higher levels of C on the extensive margin of land use change.

In a similar vein, there is a continuum of practices that can be employed to sequester more C on a given unit of land. A relatively small number of these practices may be economic at low C prices, e.g., small modifications of the harvest rotation length and low-level fertilization, but higher prices would yield more management response and higher C sequestration rates.

Taking together the responses on the extensive and intensive margins, one can construct an upward-sloping function for the marginal cost of C sequestration at the aggregate level. At progressively higher C prices, more land is devoted to higher C-density uses, more C-intensive practices are employed, and more C is sequestered in terrestrial ecosystems.

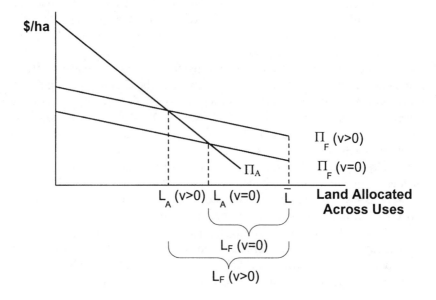

Figure 13.1. Land rent, land allocation, and C pricing

2. CARBON SUPPLY ON THE INTENSIVE MARGIN

The intensive margin involves changes in C on a per unit area basis. In forest management, the purest form of analysis for a given area of land is the stand-level perspective.[2] Davis and Johnson (1987: 29) define a forest stand as a "homogeneous, geographically contiguous parcel of land, all of the same [forest] type and larger than some defined minimum size." Stand-level analysis is the kernel of the forest supply function, as all other forms of analysis (i.e., forest-level, regional, national, global) are simply aggregations of the stand-level problem. The optimal timber rotation (see chapter 4) is a stand-level formulation of the harvest decision. We extend that framework here to evaluate the optimal time to harvest a stand when both timber and sequestered C generate monetary value for landowner.

The example below focuses entirely on the rotation length as one intensive response to a C price. Other options for modifying management intensity, including vegetation control, fertilization, and thinning, are potentially important but are not explicitly modeled in this section.

2.1   Optimal Rotation Model with Carbon Compensation

Consider both timber and the amount of C sequestered in biomass as explicitly valued forest outputs. Modeling C and timber together is a natural extension of the optimal rotation framework for timber-only management regimes, first introduced by Faustmann (1849) and is essentially a variant of the amenity value inclusion in the rotation problem by Hartman (1976). Extensions in this direction can be found in Murray (2000), Hoen and Solberg (1997), Van Kooten et al. (1995), and Hoen (1994).

In this specification, a forest generates credits as C is stored and incurs debits when C is released through harvest activity and the subsequent decay of harvested products. Modifying Hoen (1994), the bare land value of a forest when both C and timber are priced is:

$$BLV_{TC} = [p^T Q(T)e^{-rT} - R + \int_0^T vC'(t)e^{-rt}dt - [vC(T)\int_0^D d(s)e^{-rs}ds]e^{-rT}]$$
$$[1-e^{-rT}]^{-1} \qquad 13.2$$

where BLV_{TC} is bare land value of a timber and C forest management regime; T is the rotation age; $Q(T)$ is timber volume at the time of harvest; R is the cost of forest establishment; p^T is the real price of timber (assumed constant over time); $C(t)$ is the amount of C sequestered as a function of stand age (t); v is the price per unit of C sequestered or released; r is the real discount rate; $d(s)$ is the amount of C released s years after harvest on site or from wood products; and D is the length of time after harvest that C releases occur.

The optimal rotation is determined by the first-order condition of the bare land value function with respect to T, which is derived from equation 13.2 to express the optimal harvest-timing rule:

$$p^T Q'(T) + vC'(T)[1 - \int_0^D d(s)e^{-rs}ds] + r[vC(T)\int_0^D d(s)e^{-rs}ds]$$
$$= r[p^T Q(T)] + r[BLV_{TC}] \qquad 13.3$$

At the optimum, the marginal revenue of extending the rotation another year just equals the marginal cost of rotation delay. The marginal revenue of rotation delay includes three successive terms on the lefthand side representing the following components: (1) additional value from timber growth, (2) the net value of additional C credits, and (3) interest on the forestalled payment of C debits from harvest. On the marginal cost (righthand) side of the equation, an increase in the rotation age includes interest on the value of the growing stock and the land.

Consider the situation in which C is not valued ($v = 0$) but timber is ($p^T > 0$), then equation 13.2 collapses to the familiar Faustmann optimal rotation. The first-order condition of the Faustmann equation differs from equation 13.3 only by the omission of the C terms and substitution of a land value term in a timber-only regime (BLV_T) for the land value term in a timber–C regime:

$$p^T Q'(T) = r[p^T Q(T)] + r[BLV_T] \qquad 13.4$$

The optimal time to harvest timber, in the absence of C compensation, is when the marginal value of stand growth equals the opportunity cost of tying up the harvestable inventory and land for another period.

Comparing the optimal harvest rules in equations 13.3 and 13.4 indicates that rotations will typically be delayed when C is priced, because the benefits of additional net C receipts outweigh the costs of holding land and growing stock for another period (see, empirically, Hoen and Solberg 1997, Murray 2000, Van Kooten et al. 1995). Landowners, then, would be willing to accept suboptimal timber regimes when timber profit losses are more than compensated by C revenue gains. Numerical simulations demonstrate this point below.

The modifications in rotation age brought about by C pricing affect the time profile of C storage. Discounting is necessary to make C flows in the future equivalent in value to C flows today. Here the time-discounted C stock effects are given by the sum of C quantity terms given in equation 13.5:

$$S_0 = \int_0^T C'(t)e^{-rt}dt - [C(T)\int_0^D d(s)e^{-rs}ds]e^{-rT}][1 - e^{-rT}]^{-1} \qquad 13.5$$

Longer rotations extend the period for which C is accumulating in the forest and delay the reintroduction of C to the atmosphere through harvesting and product emissions. Thus, S_0 represents the present value of the C stock benefits of bare land allocated to rotational forestry.

2.2 Empirical Example: Optimal Rotation, Bare Land Values, and Carbon Quantity under Carbon Pricing in the United States

The analytical model described in section 2.1 is now used to numerically compute bare land values for timber–C rotations across three different forest settings in the United States: (1) pine plantations in the southeastern United States, (2) Douglas fir plantations in the Pacific Northwest, and (3) natural hardwoods of maple and beech in the Lake States region. Data sources are described in Murray (2000).[3] Table 13.1 presents optimal rotation, bare land value, and C stock results for each forest type. Within each forest type, results are evaluated with respect to their sensitivity to variations in C price.

2.2.1 Bare Land Value and Optimal Rotation Effects

When only timber is valued ($v = 0$), the optimal rotation of planted pine is 20 years, and the bare land value is $1,424/ha. Compared to the forest types discussed below, the optimal rotation for pine is relatively insensitive to the C price. At a C price of $10/megagram (Mg; metric ton), the optimal rotation stays at 20 years, and no extra C is sequestered. At a price of $30/Mg, the optimal rotation lengthens to 24 years, and bare land value equals $2,035/ha. At $v = 40, the rotation lengthens slightly to 25 years, but an additional $10/Mg does not extend the rotation beyond that.

The second panel in table 13.1 presents results for Douglas fir. Baseline optimal rotation ages are about twice as long for this forest type as for pine. Yet C pricing has roughly the same relative effect on Douglas fir rotations as it does on pine. C prices do have a more pronounced effect on the bare land values of Douglas fir than on pine. When $v = 50, the bare land value of Douglas fir is almost $1,900/ha (173%) above its baseline value, while this same price raises the bare land value of pine by about $1,600 (115%).

Results for the final forest type, maple-beech, are shown in figure 13.2. The optimal rotation length for each scenario is the peak of the bare land value function. This forest type has the longest baseline rotation age (47 years) and the largest absolute effect of C prices on rotation age. A C price of $30 raises the optimal rotation 18 years to age 65, and a price of $50 raises the optimal rotation age to 85 years. The baseline bare land value is lowest among this forest type ($185/ha) but nearly quadruples as the C price rises to $50/Mg. Note that for the higher C prices, the bare land value function is very flat at the higher rotation ages, indicating that the landowner would be largely indifferent across a wide range of older rotations, including, possibly, an infinite rotation with no timber harvesting at all.

Table 13.1. Optimal rotation length, bare land value, and C stock effects under different C prices for three U.S. forest types

Forest type	C price ($/Mg)					
	0	10	20	30	40	50
Planted pine						
Bare land value ($/ha)	1,424	1,724	2,035	2,362	2,705	3,048
Optimal rotation length	20	20	22	24	25	25
C stock effect (Mg/ha)[a]	30.01	30.01	31.83	33.52	34.32	34.32
Douglas fir						
Bare land value ($/ha)	1,089	1,429	1,787	2,166	2,558	2,977
Optimal rotation length	39	41	44	46	49	54
C stock effect (Mg/ha)[a]	32.95	34.64	36.98	38.55	40.52	43.48
Maple-beech						
Bare land value ($/ha)	185	266	357	460	571	691
Optimal rotation length	47	52	57	65	75	85
C stock effect (Mg/ha)[a]	7.65	8.55	9.43	10.59	11.63	12.35

See Murray (2000), table 1, for data details.
Mg = megagram, or metric ton.
[a] Time-discounted sum of C flows, evaluated at year 0.

2.2.2 Carbon Supply Effects

Table 13.1 presents the time-discounted C stock consequences of different C prices (refer to equation 13.4). In this regard, it provides a measure of (per hectare) C supply as a function of the C price across different forest types. For planted pine, raising the C price from $10 to $20/Mg raises the rotation age by 2 years and increases the present value of future C flows by 1.82 Mg/ha (= 31.83 − 30.01). Raising the C price from $20 to $30/Mg extends the pine rotation 2 more years to 24, and enhances C storage 1.69 units to 33.52 Mg/ha. Note that incremental changes in the C price from $0 to $10/Mg and from $40 to $50/Mg induce no incremental C storage on pine plantations. Thus, incremental payments to landowners in these price ranges would be pure transfers to the landowners, with no corresponding opportunity cost or C benefit.

Economics of Forest Carbon Sequestration

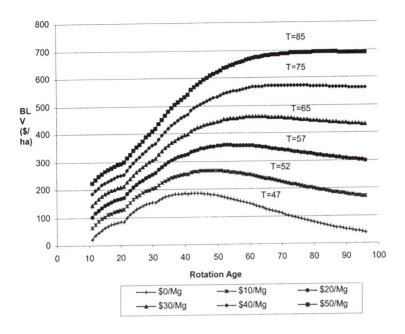

Figure 13.2. Optimal rotation and bare land value (BLV) by C price level: maple-beech. Maple-beech optimal rotations are relatively sensitive to the C price. At higher C prices, optimal rotations approach an infinite solution (preservation).

Douglas fir rotations sequester more C than the other forest types, both in the absence of C prices and in response to incremental changes in the C price. A $50 C price would generate 10.5 additional Mg/ha of C stock. The maple-beech forest type has the lowest absolute C storage potential of the forest types analyzed. However, this type is the most responsive in terms of rotation age responses to C price increments. Those responses mean that maple-beech have generally larger incremental C responses than pine (e.g., up to 4.7 additional Mg/ha compared to 4.3 Mg/ha, respectively) but smaller than the Douglas fir C effects.

3. CARBON SUPPLY ON THE EXTENSIVE MARGIN

This section formalizes the land allocation part of the discussion in section 1 and presents some empirical evidence from the literature on land use change in response to C monetary incentives.

Land use theory, dating from the days of the classical economists Ricardo and von Thunen, stipulates that the use of a specific unit of land is determined by its highest economic return among alternatives. First, we can express the returns (rents) to land use i as follows:

$$\pi_i = \pi_i(v, p, Z) = px_i^*(v, p, Z) + vc_i^*(v, p, Z) \quad 13.6$$

where v is the C price, p is a vector of non-C commodity and input prices, Z indexes exogenous land quality, x_i^* is the profit-maximizing vector of commodity outputs and input demands, and c_i^* is the profit-maximizing level of net C sequestration. Following standard Ricardian land use theory, land is allocated to its most profitable use. Given that land quality varies, the proportion of the land base in use i is expressed as $\phi_i = \phi_i(v,p,Z)$.

Since the total land base is fixed at L, land allocation can be expressed in terms of ϕ_i. The total change in land allocated to use i following the introduction of or change in C price is:

$$d\phi_i = (\partial \phi_i / \partial v)dv + \sum_{j=1}^{M}(\partial \phi_i / \partial p_j)dp_j \quad 13.7$$

The C price affects ϕ_i directly ($\partial \phi_i/\partial v$) by altering use i's rent (π_i) relative to alternative uses. A change in land allocation can affect the terms of trade in other markets (dp_j), such as those for timber and agricultural commodities. Thus we must consider the potential feedback effects from other markets on land allocation. This can be considered a more general equilibrium approach to modeling the land use effects of a C price, whereas ignoring external market feedback is best described as a pure partial equilibrium approach. Examples of the alternative approaches follow.

Consider the effects of an increase in the C price on forest area. As discussed above, since forest is a relatively C-dense land use, a positive C price will tend to raise forest rents relative to alternative uses. Thus, we should generally expect the direct effect of a C price on forest area, i.e., the first term in equation 13.7, to be positive. If the geographic region of interest is just a small supplier in the relevant commodity markets, this land use shift may have no appreciable effect on commodity prices (i.e., regional producers are price takers in the market). If that is the case, we can ignore the second term in equation 13.7, capturing price feedback from the commodity markets. However, when the affected region is too large to be considered a price taker in the broader market—for instance, when the land use movements are at the national scale for a large timber-producing country—then the interactions of the commodity and land markets should be taken into account.

Commodity market feedback is illustrated in figure 13.3. Here, the imposition of a C price shifts out the demand for forested land use relative to agriculture, as indicated by the outward shift in the land rent profile from $\pi_F(v = 0, p_0)$ to $\pi_F(v > 0, p_0)$, where p_0 indicates the baseline M-dimension commodity price vector. The direct response is for land to shift from agriculture to forest to satisfy that demand [from $L_A(v = 0, p_0)$ to $L_A(v > 0, p_0)$]. As the graphs for the commodity markets show, the land transfer shifts out the supply of timber and shifts in the supply of agricultural commodities.[4]

Given downward-sloping demand functions for each commodity—again, invoking the assumption that the affected region faces a price-dependent demand function rather than a fixed demand price—timber prices will fall, and agricultural prices will rise. These price changes affect forest rents (negatively) and agricultural rents (positively) and thus the relative returns that affect land allocation in the first place. This is illustrated by the offsetting shifts in land demands caused by the price changes. Thus, price feedback from the commodity markets dampens the incentives for land use change initially brought about by the C price, i.e., $\Sigma(\partial \phi_F / \partial p_j) dp_j < 0$. In this situation, the first and second terms in equation 13.7 work in opposite directions. Thus, the net effect of C price on forest area is, in theory, ambiguous. Researchers and policy analysts typically view the second (market feedback) term as weaker than the first (direct incentive) term—indeed, the second term is often omitted from analysis. But ultimately this becomes an empirical question. If empirically important, ignoring these market interactions runs the risk of overstating the expected sequestration benefits of direct C sequestration incentive programs.

3.1 Empirical Studies of Carbon Supply from Changes in Land Use and Forestry

This section presents several empirical studies that have addressed the issue of C sequestration supply from an aggregate rather than stand-level perspective. All of the studies are from the United States, which is where most of the work in aggregate C supply modeling has been performed since the inception of this type of work in the early 1990s.[5]

Empirical studies of aggregate C sequestration can be categorized along the following dimensions: (1) whether landowner response functions are characterized by a normative or positive behavioral model, and (2) the extent to which the model captures market price feedback resulting from the C policy.

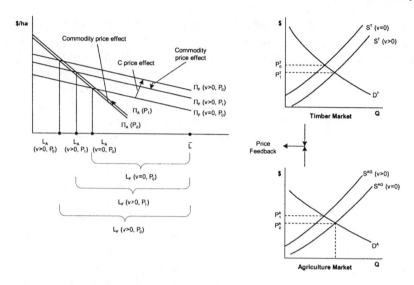

Figure 13.3. Supply curve for terrestrial C sequestration. A positive C price shifts out the demand for forest land relative to agricultural land. As this land moves from agriculture to forest, agricultural prices rise, and timber prices fall. This weakens the incentives for continued land movements.

Alternative models of landowner behavior include the normative model, in which the analyst asserts an explicit objective function of land rent maximization. Landowners are assumed to optimally select the land use that achieves that objective, and the model solves for the optimal allocation, subject to the parameters and data underlying the objective function. The problem is that landowners may not always select the land use that generates the highest land rent specified by the analyst, either because of errors in (or constraints on) optimization by the landowner or errors by the analyst in specifying the landowner's objectives. Alternatively, the positive or revealed behavior model is typically based on econometric analysis of observed land use and management decisions. Econometric models still rest on the assumption that landowners are economically rational, but they hold open the possibility that landowner decisions may be determined by factors not directly measurable by the analyst. The challenge in econometric modeling is to properly account for the unobserved factors underlying the decisions, so that the resulting empirical model is not biased.

Regarding the coverage of markets, consider a government-sponsored afforestation program. This will generate two kinds of costs, direct and indirect. Direct costs include government subsidy payments to landowners as compensation for the opportunity costs of afforestation, as well as any costs of administration and enforcement. Indirect costs and benefits include the costs and benefits to producers and consumers of the commodities affected by the movement of land from nonforested use to forested use (general equilibrium effects) in response to the subsidies. The total social cost of a policy includes the net sum of direct and indirect costs. Some of the studies reviewed next confine the analysis to the size of the payment necessary to induce the land and C response (direct costs), while other studies examine the welfare consequences in the related commodity markets (indirect costs).

3.1.1 Normative Land Use Change Studies

The first significant effort to estimate an aggregate C supply function from forests was conducted by Moulton and Richards (1990). In that study, the authors employ detailed data on the characteristics of marginal agricultural land in the United States, the potential physical yield of timber and C from those lands, their historical land rental rates, and their cost of conversion to forest. These data are used to construct an engineering supply function of the amount of C that could be sequestered by a tree planting program and forest management on marginal agricultural lands in the United States. Their results suggested a rather significant potential for such a program—over 700 Mg per year, or roughly one-half of U.S. CO_2 emissions—at a cost of $19.5 billion. The cost per ton sequestered varied widely by geographic region and type of forest management activity undertaken. No economic studies since Moulton and Richards (1990) have suggested such a large annual sequestration potential from U.S. lands.

Parks and Hardie's 1995 paper uses data from the 1987 Natural Resources Inventory to identify 116 million acres of crop and pasture in the contiguous United States with the potential to establish a stand of trees. The program they were simulating was assumed to run for a 10-year period. They aggregate the data to 433 geographic regions within the United States and estimate (1) forest C yields, (2) forest establishment costs, (3) agricultural opportunity costs, and (4) hypothetical government payments required to change land use. The hypothetical bid rate was calculated as the net difference in agricultural and forest rents plus the landowner's share of forest establishment costs: bid rate = $\pi^{*A} - \pi^{*F} + c$ where π^{*A} = value of agricultural rents for the 10-year contract period, π^{*F} = value of forest rents for the 10-year contract period, and c = landowner's share of forest establishment costs.

Parks and Hardie sorted the 433 geographic regions in ascending order of hypothetical bid rates to construct a least-cost supply function (and acreage function) for C sequestration. By their estimates, a program of roughly the same magnitude of the Conservation Reserve Program (CRP)—net present value of payments over 10 years = 3.7 billion dollars (1987 dollars)—would yield about 43 million Mg of sequestered C per year and would cover approximately 22.2 million acres.[6]

3.1.2 Positive Land Use Change Studies

One common shortcoming of the Moulton and Richards (1990) and Parks and Hardie (1995) studies is the implicit assumption that all land would be converted to forest at the hypothetical bid rate offered. Since the bid rate covers the difference in rents between agriculture and forestry and forest establishment costs, this implies that all landowners would switch to forest as soon as the conversion subsidy made forest marginally more profitable than the current agricultural use. This is consistent with the simplest version of Ricardian land rent theory, but does not seem to accurately reflect how all landowners behave in actual market settings.

In a follow-up paper to the 1995 article, Parks and Hardie (1996) address the issue of hypothesized versus revealed behavior. The authors reference the fact that only a small portion of landowners eligible for the U.S. CRP—and for whom switching to trees and receiving the CRP rental would be more profitable than keeping land in agriculture—actually participated in the program. Parks and Hardie estimate a probit model of the decision to participate in the CRP tree planting program as a function of the bid rate offered and other variables.[7] They then use predicted participation rates from the probit model as a replacement for the 100% participation assumption implicit in the least-cost analysis in their 1995 paper. The results are dramatic. Sequestration levels at the predicted participation rates are only about 10% of the levels at the full participation rate. If we use the predicted 10% rate to scale down the C sequestration estimates derived using the Parks and Hardie (1995) model, then perhaps only about 4 million Mg of C would be generated on about 1.7 to 1.8 MM acres, at a one-time cost of $380 million from an entirely voluntary program like CRP.

Articles by Stavins (1999) and Newell and Stavins (2000) also pivot off of the notion that landowners' revealed land use behavior is the key to more accurately estimating the marginal cost of a C sequestration program. Using an econometric land use model (Stavins and Jaffe 1990) applied to panel data for 36 counties in the Delta region of Arkansas, Louisiana, and Mississippi, Stavins (1999) estimates the effect of a two-part policy instrument, wherein land moving from agriculture to forest receives a subsidy of Z per acre, and land moving from forest to agriculture (deforestation) is taxed at a rate of Z per acre. Stavins uses the econometric model to estimate baseline forested acres without a subsidy/tax ($Z = 0$) and then estimates forested acres at different levels of Z. It is implicitly assumed that landowners will respond to the subsidy and tax in essentially the same way they have responded to all other components of land rent differentials over the time period covered in the data used to estimate the model (1935 to 1984). While the Stavins marginal cost (MC) function lies to the right of the Parks and Hardie function from their latter (1996) study—i.e., higher estimated sequestration levels for a given MC—it lies to the left of several other least-cost studies referenced in the Stavins paper.[8] Newell and Stavins (2000) examine variations on Stavins' (1999) policy scenarios, including restrictions on the ability to harvest the afforested stands.

Plantinga et al. (1999) take a similar approach to the three studies described above by using econometric models to estimate C supply functions for Maine, South Carolina, and Wisconsin. They, like the authors of the other econometric studies referenced, found that the earlier, normative studies of C supply tended to underestimate the marginal cost of sequestration.

In summary, the body of research applying econometric methods to data on revealed land use behavior shows that landowners are not as responsive to the land rent differentials as the published studies using normative models assume they would be. Thus, future research should be aimed at better understanding why landowners respond weakly to seemingly profitable land use change opportunities and how this might be modified by policy design.

3.1.3 Models Linking the Land and Commodity Markets

Up to this point, the studies referenced in this section all ignore the interactions between the land market and the commodity markets for timber and agricultural products. But, as described earlier, changes in land allocation between the forest and agricultural sectors, especially at the large scale being discussed in these studies, will alter supply conditions in timber and agricultural markets, causing price changes that will feed back into the land allocation decision.

Perhaps the first study to address the link between the forest and agricultural sectors is by Richard Adams et al. (1993). This study links a normative spatial equilibrium model of the agricultural sector (ASM, Chang et al. 1992) with a positive econometric model of the U.S. timber markets (TAMM, Darius Adams and Haynes 1980) to capture the price feedback between the forest, agricultural, and land markets illustrated in figure 13.3. Adams et al. (1993) provide empirical evidence of the rise in agricultural prices, fall in timber prices, and changes in stakeholder welfare that could result from a large-scale afforestation program in the United States. Among the study's results is the finding that private landowners might need to be compensated to keep their land in forests. Otherwise, the commodity price effects induced by the policy provide an incentive for offsetting movements of land from forest to agriculture. Without using the exact term, this is the first study that explicitly addresses the C leakage effect, which is discussed in more detail below.

Development of the Forest and Agricultural Sector Optimization Model (FASOM, Adams et al. 1996a,b) pivots off of the work by Adams et al. (1993). FASOM explicitly and simultaneously connects forest and agricultural commodity markets with the market for land in alternative uses. FASOM is an intertemporal, price endogenous, spatial equilibrium model in which market solutions for the forest and agriculture sectors are obtained each decade in the entire projection period at one time. The model employs a joint objective function, maximizing the present value of producers' and consumers' surplus in the markets of the two sectors, with restrictions on the disposition of the land base that is suitable for use in either sector. The optimizing spatial equilibrium market model simulates prices, production, consumption, and management actions in the two sectors. Endogenous variables in FASOM include:

- Land transfers between the forest and agriculture sectors;
- Forest management investment activity (e.g., conversion of natural forests to plantations, increased intensification of plantation silviculture);
- Timber harvest quantities and log prices for nine U.S. regions, two species groups, and three classes of products; and
- Agricultural prices and production in 11 regions for 40 primary and 46 secondary commodities.

The FASOM model also includes a C accounting component, based in part on the FORCARB model of Birdsey and Heath (1995).

To gauge the potential impact of a targeted forest C policy, Alig et al. (1997) use the FASOM model to examine the total cost and C consequences of a one-time forcing of 4.9 million ha (12 million acres) of agricultural land into forests. The connection between this land movement and a particular policy is not defined. Therefore, the simulations conducted are more accurately characterized as the modeled response to an exogenous shift in land from agriculture to forest, not the response to a particular afforestation subsidy program. Because land market equilibrium in FASOM is determined by relative demands in the agricultural and forest sectors, forcing acres from agriculture to forest—and in this article's case, no restrictions on land reverting back to agriculture—is found to induce a large offsetting movement (i.e., leakage) of land from forest to agricultural after 50 years. The baseline scenario has forest area declining 5 million ha from 1990 to 2040. The simulated decline under the afforestation scenario is 3.3 million ha. Therefore, the 4.9 million ha afforestation program in the first decade produces only 1.7 million ha more forest area by 2040 than the base case.

This substantially mitigates the first-order C sequestration effects of the initial afforestation effort.

A subsequent paper by the FASOM authors (Adams et al. 1999) examines a more complex set of policies than the simple afforestation target modeled in Alig et al. (1997). The later study evaluates a wide range of quantitative annual C flux targets that vary with respect to magnitude and timing. As the C target is a binding constraint to the model, changes in land use and forest management intensity combine to generate the additional C.

Perhaps the most significant improvement of the linked market model studies relative to the land use studies is that they can be used to derive producer and consumer surplus-based social welfare measures. Thus, they provide the only theoretically consistent measure of the full cost to society of programs that cause intersectoral changes in resource allocation. In principle, the social cost measures associated with the change in producer and consumer surplus can be compared to social benefits measures of climate change mitigation to estimate the net social welfare benefits of the policy.

4. SPECIAL ISSUES RELATED TO THE ECONOMICS OF FOREST C SEQUESTRATION

Of the various complicating issues related to the economics of forest C sequestration, two deserve special attention: (1) leakage of C benefits, and (2) the impermanence of terrestrial C sequestration.

4.1 Leakage

As just demonstrated by the example of land and commodity market interactions, economic behavior can mitigate, perhaps substantially, the direct objectives of policy interventions to sequester C. In the discourse of climate change mitigation policy, this phenomenon has been placed under the mantle of *leakage*. Leakage refers to situations in which targeted C emission reductions or sequestration in one place lead to offsetting C emission increases elsewhere. This might be problematic, for instance, in the accounting of a C trading system if the targeted sequestration is credited without the offsetting emissions being debited. This would occur when a specific site-limited project to sequester C shifts extractive activity (and C emissions) to land not affected by the project. Then the credit assigned directly to the sequestration achieved by the project overstates the net effect on the atmosphere, unless some adjustment for leakage is rendered.

A similar problem surfaces if a country obligated to meet a GHG net emission target meets that target in part with sequestration actions that ultimately displace emission-generating activity to countries not obligated to meet emission reduction targets. While compliance has been achieved in this case, the benefits of compliance are diminished by the shifting of emissions to unregulated sources.[9] One way to eliminate leakage is to ensure that all sources and sinks are covered under a comprehensive accounting system. The only pure comprehensive accounting system would be global; unfortunately, political and institutional barriers to such a system are likely to persist. In lieu of that, some efforts have been made at the individual project-site level to better account for the leakage that occurs as a result of the project and to design policies to minimize leakage upon implementation. Chomitz (2002) compares leakage propensity in forest C projects versus energy sector GHG mitigation projects. Murray et al. (2002) estimate the size of leakage for different types of activities and regions in the United States.

4.2 Impermanence

A critical issue that has been raised with regard to terrestrial C sequestration as a mitigation strategy is the duration of the benefits. For instance, a forest acts as a C sink throughout its growth phase; however, when growth culminates in a mature forest, there may be relatively little additional net sequestration, because growth and mortality tend to counterbalance. Thus the forest sink can saturate. Meanwhile, the mature forest serves as a reservoir of C in trees, other vegetation, litter, and soils, but these reservoirs are a potentially large emissions source by harvest or natural disturbance such as wildfire or pest infestation. As a result, some investigators question the long-term benefit of C sequestration as a mitigation strategy, or at least propose methods that discount C credits on the grounds of their impermanence (Lashof and Hare 1999).

In essence, impermanence is simply *temporal* leakage, in contrast to the *spatial* leakage described earlier. As with spatial leakage, one solution for temporal leakage is an accounting system that is comprehensive over time. Sequestration is credited as it occurs, and emissions are debited when they occur. In that way, nothing leaks from the system over time. However, given the uncertainty of future commitments and the ability to collect C debits from landowners, some have proposed a system wherein the amount of credit given to C as it is sequestered is discounted using a method called *ton-year accounting* (e.g., McCarl and Murray 2001, Moura-Costa and Wilson 2000). Ton-year accounting grants partial credits for tons sequestered during a finite time period. The amount credited grows as the sequestered C is left intact over time. Ton-year accounting is built on the premise that different time paths of sequestration and release have different consequences for the residency of CO_2 in the atmosphere and the corresponding effect on climate. Thus it might be considered an imperfect, but pragmatic, approach to making comparable C sequestration at different points in time.

5. LITERATURE CITED

ADAMS, D.M., AND R.W. HAYNES. 1980. The Timber Assessment Market Model: Structure, projections, and policy simulations. For. Sci. Monog. No. 22, Society of American Foresters, Bethesda, MD. 64 p.

ADAMS, D.M., R.D. ALIG, AND B.A. MCCARL. 1996a. An analysis of the impacts of public timber harvest policies on private forest management in the U.S. For. Sci. 42:343-358.

ADAMS, D.M., R.D. ALIG, B.A. MCCARL, J.M. CALLAWAY, AND S.M. WINNETT. 1996b. The forest and agricultural sector optimization model: Model structure and applications. Res. Pap. PNW-RP-495. Pacific Northwest For. and Range Exper. Station, USDA Forest Service, Portland, OR.

ADAMS, D.M., R.J. ALIG, B.A. MCCARL, J.M. CALLAWAY, AND S.M. WINNETT. 1999. Minimum cost strategies for sequestering carbon in forests. Land Econ. 75(3):360-374.

ADAMS, R.M., D.M. ADAMS, J.M. CALLAWAY, C-C. CHANG, AND B. MCCARL. 1993. Sequestering carbon on agricultural land: Social cost and impacts on timber markets. Contemp. Policy Iss. 11(1):76-87.

ALIG, R.J., D.M. ADAMS, B.A. MCCARL, J.M. CALLAWAY, AND S.M. WINNETT. 1997. Assessing effects of mitigation strategies for global climate change with an intertemporal model of the U.S. forest and agricultural sectors. Env. and Res. Econ. 9:259-274.

BIRDSEY, R.A., AND L.S. HEATH. 1995. Carbon changes in U.S. forests. in Productivity of America's Forests and Climate Change, L.A. Joyce (ed.), pp. 56-70. Gen. Tech. Rep. RM-271. USDA Forest Service, Ft. Collins, CO.

BIRDSEY, R.A. 1996. Carbon storage for major forest types and regions in the conterminous United States. P.23 – 29 and Appendix 2 in Forests and Global Change, R.N. Sampson and D. Hair (eds.), Vol. 2. Forest Management Opportunities. American Forests, Washington DC.

CHANG, C.C., B.A. MCCARL, J. MJELDE, AND J. RICHARDSON. 1992. Sectoral implications of farm program modifications. Am. J. Agric. Econ. 74:38-49.

CHOMITZ, K.M. 2002. Baseline, leakage and measurement issues: How do forestry and energy projects compare? Climate Policy 2(1):35-49.

CLINTON, W.J., AND A. GORE. 1993. The Climate Change Action Plan. Washington, DC: White House Office of Environmental Policy.

DAVIS, L.S., AND K.N. JOHNSON. 1987. Forest Management. Third Edition. McGraw-Hill, Inc., New York, NY. 790 p.

DIXON, R.K., K.J. ANDRASKO, F.G. SUSSMAN, M.A. LAVINSON, M.C. TREXLER, AND T.S. VINSON. 1993. Forest sector carbon offset projects: Near-term opportunities to mitigate greenhouse gas emissions. Water, Air, and Soil Pollution 70:561-577.

HARDIE, I.W., AND P.J. PARKS. 1997. Land use with heterogeneous land quality: an application of an area base model. Am. J. Agric. Econ. 79:299-310.

HARTMAN, R. 1976. The harvesting decision when a standing forest has value. Econ. Inquiry 14(Mar):52-58.

HOEN, H.F., AND B. SOLBERG. 1997. Carbon dioxide taxing, timber rotations, and market implications. P. S151-S162 in Economics of Carbon Sequestration in Forestry, R.A. Sedjo, R.M. Sampson, and J. Wisniewski (eds.). CRC Press, Boca Raton, FL.

HOEN, H.F. 1994. The Faustmann rotation in the presence of a positive CO_2-price. Scand. For. Econ. 35:278-88.

LASHOF, D., AND B. HARE. 1999. The role of biotic carbon stocks in stabilizing greenhouse gas concentrations at safe levels. Environ. Sci. and Pol. 2(2):101-110.

MCCARL, B.A., AND B.C. MURRAY. 2001. Harvesting the greenhouse: Comparing biological sequestration with emissions offsets. Unpublished Paper, Department of Agricultural Economics, Texas A&M University, College Station, TX.

MOULTON, R.J., AND K.R. RICHARDS. 1990. Costs of sequestering carbon through tree planting and forest management in the United States. General Technical Report WO-58. USDA Forest Service, Washington DC.

MOURA-COSTA, P., AND C. WILSON. 2000. an equivalence factor between CO2 avoided emissions and sequestration—description and applications in forestry. Mitigation and Adaptation Strategies for Global Change 5(1):51-60.

MURRAY, B.C. 2000. carbon values, reforestation, and perverse incentives under the Kyoto Protocol: An empirical analysis. Mitigation and Adaptation Strategies for Global Change 5(3):271-295.

MURRAY, B.C., B.A. MCCARL, H.C. LEE. 2002. Estimating leakage from forest carbon sequestration programs. Working Paper 02-06, Environmental and Natural Resource Economics Program, RTI International. Contact bcm@rti.org for copy.

NEWELL, R.G., AND R.N. STAVINS. 2000. Climate change and forest sinks: Factors affecting the costs of carbon sequestration. J. of Environ. Econ. and Manag. 40 (3):211-235.

PARKS, P.J., AND I.W. HARDIE. 1995. Least cost forest carbon reserves: cost-effective subsidies to convert marginal agricultural lands to forests. Land Econ. 71(1):122-136.

PARKS, P.J., AND I.W. HARDIE. 1996. Forest carbon sinks: Costs and effects of expanding the conservation reserve program. Choices 2nd quarter:37-39.

PFAFF, A.S.P., S. KERR, R.F. HUGHES, S. LIU, G.A. SANCHEZ-AZOFEIFA, D. SCHIMEL, J. TOSI, AND V. WATSON. 2000. The Kyoto Protocol and payments for tropical forest: An inter-disciplinary method for estimating carbon-offset supply and increasing the feasibility of a carbon market under the CDM. Ecol. Econ. 35:203-221.

PLANTINGA, A.J., T. MAULDIN, AND D.J. MILLER. 1999. An econometric analysis of the costs of sequestering carbon in forests. Am. J. Agric. Econ. 81:812-824.

SAMPSON, R.N., AND R.A. SEDJO. 1997. Economics of carbon sequestration in forestry: An overview. Critical Reviews in Environmental Science and Technology 27(Special):S1-S8.

STAVINS, R.N., AND A.B. JAFFE. 1990. Unintended impacts of public investments on private decisions: The depletion of forested wetlands. Am. Econ. Rev. 80(3):337-352.

STAVINS, R. 1999. The costs of carbon sequestration: A revealed preference approach. Am. Econ. Rev. 89(4):994-1009.

VAN KOOTEN, C.G., C.S. BINKLEY, AND G. DELCOURT. 1995. Effect of carbon taxes and subsidies on optimal forest rotation age and supply of carbon services. Am. J. Agric. Econ. 77:365-374.

WATSON, R.T., I.R. NOBLE, B. BOLIN, N.H. RAVINDRANATH, D.J. VERARDO, AND D.J. DOKKEN (EDS.) 2000. Special Report on Land Use, Land Use Change, and Forestry. Intergovernmental Panel on Climate Change, Geneva, Switzerland: Cambridge University Press, Cambridge, UK.

---

[1] Other trace GHGs may be affected by forest clearing, especially biomass burning in the tropics. However, CO_2 is the primary GHG constituent relevant to forest C sequestration.

² C can also be enhanced on nonforested land by changes in management intensity, for example, by changes in agricultural practices. Analyses of land management intensity outside the forest sector are beyond the scope of this chapter.

³ The underlying growth and yield functions used in the analysis are from Birdsey (1996) and are expressed in C pool components by 5-year increments for stand ages 0 to 120 years. The 5-year increments are interpolated to 1-year increments. Optimal values are computed numerically by evaluating the discrete age with the highest bare land value.

⁴ Any corresponding restrictions on harvesting timber would diminish the expansion in timber supply from afforestation and weaken the price feedback.

⁵ Pfaff et al. (2000) look at aggregate C sequestration in a developing country setting. However, their work describes an integrated methodology which is similar to that used in the studies cited here, though customized to Costa Rica, and does not present empirical results.

⁶ Though not specifically stated in the Parks and Hardie article, this yield would appear to be the average annual carbon sequestered per acre for a typical forest rotation on the established lands. Thus, they are essentially steady state levels, and would not necessarily reflect near term yields, say, in the early commitment periods of the Kyoto Protocol.

⁷ Parks and Hardie (1996) do not provide details on the probit estimation. That detail is provided in a prepublication version of the paper obtained from the authors.

⁸ One should note that the Stavins study uses the data from the sample of 36 counties in the U.S. Delta region to develop national estimates. This could be a rather strong assumption in light of the wide regional variability of C yields and costs shown by Moulton and Richards (1990), Parks and Hardie (1995), Plantinga et al. (1999), and Murray (2000).

⁹ Of course, leakage is not limited to C sequestration. The same thing can occur when one entity installs equipment or technology to reduce CO_2 emissions, thereby raising costs of production and shifting emitting activities to uncontrolled entities.

Chapter 14

Timber and Amenities on Nonindustrial Private Forest Land

Subhrendu K. Pattanayak, Karen Lee Abt, and Thomas P. Holmes
Research Triangle Institute, USDA Forest Service, and USDA Forest Service

Economic analyses of the joint production timber and amenities from nonindustrial private forest lands (NIPF) have been conducted for several decades. Binkley (1981) summarized this strand of research and elegantly articulated a microeconomic household model in which NIPF owners maximize utility by choosing optimal combinations of timber income and amenities. Most follow-up attempts have been limited to either simulations based on stylized characterization of joint production (Max and Lehman 1988) or to empirical representations hampered by data limitations—particularly with regard to measuring amenity production (Hyberg and Holthausen 1989). In attempting to redress this gap, Holmes (1986) was limited to binary representations of timber and amenities and did not get conclusive results. In this chapter, we use data from North Carolina that includes timber output and amenity indices to illustrate a method for empirically characterizing Binkley's household model.

By accounting for heterogeneous motives for forest ownership through explicit attention to amenity demand, the approach described in this chapter begins the process of developing more comprehensive forest management models that look beyond timber supply. Such efforts can improve the predictions of national and regional timber market models that have assumed timber profit maximization. Moreover, we can provide a clearer understanding of household production of socially desirable forest amenities that lie outside the reach of markets.

Sills and Abt (eds.), Forests in a Market Economy, 243–258. ©Kluwer Academic Publishers. Printed in The Netherlands.

1. TIMBER AND AMENITY MODELS

There are two major schools of timber supply modeling, optimal harvest age simulations and timber supply microeconometrics (Pattanayak et al. 2002). The optimal harvest age approach focuses on the timber harvesting decision for forest landowners who consider amenity benefits from standing forests. The main objective is to determine the optimal rotation length, given a set of parameters such as prices, biological technology, and preferences, and to simulate changes in the optimal length in response to changes in these given parameters (see chapter 8). Therefore, these studies are successors to Hartman's (1976) description of why and how amenity considerations lengthen optimal rotation. These studies draw their theory from Binkley (1981) or Hartman (1976). Simulations are based on assumed functional forms and parameters drawn from empirical studies, expert opinion, and informed conjectures regarding supply and demand of forest products.

A second set of literature uses microeconometrics to estimate parameters of timber supply behavior by private landowners. These positive analyses describe how landowners do behave instead of how they should or would behave. Binkley's (1981) work has been the starting point and conceptual template for this literature. A representative private landowner is assumed to maximize utility by consuming goods and amenities, where utility is separable over time and commodity space and is subject to an income and a production constraint. The technical production constraint links the landowner's scarce inputs, e.g., land or capital, to multiple products, e.g., timber and amenity. Amenity is conceptualized as self-produced recreation proxied by some form of forest inventory; most timber supply studies do not estimate amenity services. Timber supply is derived using first-order conditions of a typical constrained maximization problem to be a function of prices, interest rates, sociodemographics (income, occupation, and education) and biophysical factors (tract size, species mix, and inventory characteristics). Survey data are typically used to estimate the timber supply model with some direct or indirect accommodations for amenity services. A detailed description of these studies is presented in Pattanayak et al. (2002).

The strength of the optimal harvest age models lies in explicitly addressing the choice of forest age or structure. The major shortcoming of the optimal harvest age studies is the lack of empirical underpinnings. Moreover, the optimal harvest age studies tend to rely on Fisherian separation of consumption and production even in specifications with amenity services; therefore, attention to owner characteristics is inadequate. In comparison, the primary contribution of the microeconometric utility maximization tradition is the recognition of the role of owner characteristics on timber supply (because of uncertainty in prices and interest rates, nontimber amenities, imperfect capital markets, and forest taxation). Their main problem is a lack of connection with the biological aspect of forests.

An approach that combines the strengths of the two traditions and fills the gap between theory and empiricism in timber supply modeling, is identified by Binkley (1987) and Wear and Parks (1994). This chapter develops an empirical nontimber index derived from the biological attributes of forest stands and integrates it with ownership characteristics as described in Pattanayak et al. (2002). Our interest is in the practical aspects of timber supply modeling, particularly measures of amenity flows. In the next section, we present a stylized household production model of timber supply and amenity demand that draws on the literature and previous data explorations to define the conceptual basis for our empirical analysis.

2. A CONCEPTUAL MODEL OF HOUSEHOLD JOINT PRODUCTION

The model presented in this section modifies Holmes's (1986) and Hyberg and Holthausen's (1989) interpretations of Binkley's (1981) household model in which owners produce amenities and timber income. A typical landowner maximizes utility comprising income (π) and amenities (a).[1] Acknowledging nonseparability, presumably due to incomplete amenity markets, utility will be conditioned by preference parameters θ that measure the shape of the utility curve and account for the risk characteristics and bequest motives of landowners. Studies by Kuuluvainen and Salo (1991), Kuuluvainen et al. (1996), and Dennis (1989) use sociodemographic data on age, occupation, and income to proxy for preferences.

As in Binkley (1981), utility is maximized subject to two constraints. The first constraint is a multi-input, multi-output production function that is twice-differentiable, continuous, and convex; y and a are vectors of timber and amenity outputs, and production possibilities will be conditioned by ecological factors, Z. This joint production function is the core of our analysis of simultaneous timber and amenity choices. Because amenities flow from the resulting forest condition, landowners are described as self-producing the amenities. The shape of the production function, including cross-effects, reflects increasing marginal productivity and decreasing second-order effects.[2] The second constraint is a budget constraint where income must be no greater than the sum of exogenous plus timber income. The Lagrangian for this problem, in which μ and λ are the Lagrangian multipliers, is presented in equation 14.1. μ is the marginal utility of jointly produced timber and amenities, and λ is the marginal utility of income.

$$\ell_{\pi,a,y} = U(\pi, a; \theta) + \mu[G(y, a; Z)] + \lambda[\pi_e + p \cdot y - \pi] \quad \quad 14.1$$

The first-order conditions of this utility maximization are presented in equations 14.2 to 14.4. Simultaneous solution of these first-order conditions determines optimal allocation of resources and consumption levels. Resources are allocated so that marginal opportunity costs are equal to marginal utility of consumption generated by that resource.

$$\ell_\pi = 0 = U_\pi - \lambda \quad \Rightarrow \quad \pi^*(p; \theta, \pi_e) \quad \quad 14.2$$

$$\ell_a = 0 = U_a + \mu \cdot G_a \quad \Rightarrow \quad a^*(p; \theta, \pi_e) \quad \quad 14.3$$

$$\ell_y = 0 = \lambda \cdot p + \mu \cdot G_y \quad \Rightarrow \quad y^*(p; a^*, \pi_e, Z) \quad \quad 14.4$$

By manipulating equations 14.2, 14.3, and 14.4, we get

$$U_a = U_\pi p \cdot \frac{G_a}{G_y} \quad \quad 14.5$$

This implies that the marginal utility of amenity is equal to the marginal utility of the timber income forgone to self-produce amenity. In other words, at the margin, benefits equal the costs of self-producing amenities. By isolating the U_π in this same condition, we see that the marginal benefits of income equal the marginal cost measured in terms of forgone amenity benefits. We can totally differentiate equations 14.2 to 14.4 to obtain comparative static results for timber and amenities with respect to price (equations 14.6 and 14.7).

$$\frac{\partial a}{\partial p} = \frac{y \cdot U_{\pi a} \cdot {G_y}/{G_a} + y \cdot p \cdot U_{\pi\pi} + U_\pi}{U_{aa} \cdot {G_y}/{G_a} + U_a \left[{G_{ya}}/{G_a} - {G_y}/{G_{aa}} \right] + p \cdot U_{\pi a}} \quad \quad 14.6$$

$$\frac{\partial y}{\partial p} = \frac{y \cdot U_{\pi a} \cdot {G_y}/{G_a} + y \cdot p \cdot U_{\pi\pi} + U_\pi}{p \cdot U_{\pi a} \cdot {G_y}/{G_a} + U_a \left[{G_{yy}}/{G_a} - {G_y}/{G_{ya}} \right] + p^2 \cdot U_{\pi a}} \quad \quad 14.7$$

The signs of these expressions are ambiguous because the relative strengths of offsetting income and substitution effects of price changes are unknown. For example, consider a landowners' amenity choice in response to higher timber prices. From equation 14.5 we can see that higher timber prices raise the opportunity cost of consuming amenity and therefore impart a negative effect on amenity consumption (substitution effect). However, higher timber prices also increase the landowners' overall income through higher valued timber sales. Assuming amenity is a normal good, the landowner will wish to consume more amenities (income effect). In addition, the landowner must consider the timber production impacts of self-producing more amenities (G_{ya}) and declining marginal utility of amenities (U_{aa}) (equation 14.6). Similar considerations will influence timber supply responses to higher prices (see equation 14.7). For additional comparative statics, see Binkley (1981) and Hyberg and Holthausen (1989). The ambiguities created by the offsetting consumption and production responses strengthen our case for conducting empirical studies to determine how landowners actually behave.

There are three considerations in establishing an empirical timber supply model from these first-order conditions. First, because all consumption and production commodities are linked to the budget constraint either directly or indirectly (i.e., linked to market commodities through a joint production technology), all prices influence all consumption and production allocations, including self-produced/consumed amenities. Second, the supply of timber will depend on the actual level of amenities produced/consumed because there is no amenity price. That is, the quantity of amenity demand conditions the supply of timber. Third, in contrast to traditional timber production variables, landowner choice of amenity will be a function of prices and sociodemographic factors, not quantities of timber. The inclusion of timber prices in the amenity model accounts for the economic opportunity costs of producing amenities and, therefore, makes the inclusion of timber quantity redundant. See Ebert (1998) for a discussion of market and nonmarket choices of this nature. To the extent that sociodemographic factors proxy for preferences and attitudes and therefore influence the level of amenities chosen, they will be an argument of the amenity function. Although these factors are not in the timber production functions, they indirectly influence the timber allocations through their effect on amenity choice. The model presented in this section is a utility-theoretic characterization of supply of timber and amenity demand. We estimate amenity demand (equation 14.2) and timber supply (equation 14.3) to explore the tradeoffs between timber and amenity.

3. NORTH CAROLINA DATA

The primary data used in this analysis were obtained from USDA Forest Service plot surveys (USDA Forest Service, various years) and from Timber Mart-South price surveys (Norris Foundation, various issues). Neither the timber nor amenity data are collected directly; these values were calculated from the primary data (for details see Lee 1997). In addition, we used county-level sociodemographic data from the decennial census. Descriptive statistics are reported in table 14.1. Overall we have more than 4400 observations in this data set.

Table 14.1. Descriptive statistics for timber and nontimber data used in empirical model

Symbol	Variables	Units	Mean
a	Amenity index	See equation A-2	-0.002
H	Dummy for harvest	1 = harvest, 0 = not harvest	0.164
P	Timber price index	See equation A-1	3.288
D	Dummy for NIPF owner	1 = NIPF, 0 = otherwise	0.358
D	% of county with bachelor's degree	Percentage	12.22
D	Ln (median household income)	Ln ($)	10.04
Z	Dummy for operability	1 = no problems; 0 = otherwise	0.630
Z	Site class index	1= most growth to 6 = least growth	3.680
Z	Dummy for mountain ecoregion	1 = mountain, 0 = otherwise	0.162
Z	Timber inventory	See equation A-3	5.263

3.1 Timber Price and Quantity (P, y)

Timber harvesting information is collected at the plot and tree level and enables the determination of the product class and year of harvest. There are five product classes based on diameter of the tree and species class: pulpwood, chip and saw, and sawtimber in softwoods, and pulpwood and sawtimber in hardwoods. Although the details on the product class can be useful in measuring product substitution, our interest is in the broader timber-amenity tradeoff. Consequently we focus on the composite timber harvest data. Recognizing that there is no harvest for approximately 85% of the observations, our empirical tests use a discrete choice model of timber supply—that is, whether or not the owner-manager has harvested timber in any of the five product classes. We use the data on the year of harvest to match the relevant price information.

Price is the most critical variable in an economic model of timber and amenities. Our price data are from Timber Mart South (Norris Foundation, various years), which had three regions in North Carolina. The annual stumpage price for each of five product classes is measured in U.S. dollars per cubic foot. We assign prices to plots based on the year of harvest; for plots that were not harvested at all, the 1990 price is used because that was the last price information considered by landowners in choosing not to harvest. We use principal component analysis (PCA) to construct a price index. The five prices are highly collinear, and we are interested in the overall opportunity cost of not harvesting (see equation A-1 in the appendix). PCA is a data reduction method that seeks patterns of variation among observed variables, and, therefore, simplifies subsequent analyses (see Hamilton 1992).[3] The first principal component of the five prices explains 72% of the price variance and has a reliability coefficient of 0.9.

3.2 Amenity Characteristics (a)

Amenity characteristics are classified as either indices or raw data. Because most of the ecology data underlying amenity characterizations are either in discrete measures or in complex estimates of cover, occupancy, and species, Lee (1997) developed several indices to measure the nontimber amenity attributes. Our interest is in modeling joint production of timber and nontimber broadly, and so we focus on the indices that comprise measures of tree diversity index, scenic beauty, and deer and bird habitat. The formulas for these indices are taken from the literature on ecology, wildlife, and scenic beauty. We use the Rudis et al. (1988) index to estimate scenic beauty, which has the highest values in the mountains and higher values on harvested (before harvest) than on nonharvested sites. An index of tree diversity was developed using the Shannon-Weaver index, which will weight rare species more heavily (Shannon 1948). Habitat suitability for deer is based on Crawford and Marchington (1989) and for wild birds on Sheffield (1981).[4]

To focus on the timber-amenity tradeoff, we develop a composite amenity index using PCA. We first create a faunal index that is based on the first principal component of the habitat indices for the deer and nine birds.[5] We then apply PCA to the faunal, tree diversity (Shannon-Weaver), and scenic beauty (Rudis) indices. We use only the first principal component that (1) is the only component with an eigen value > 1, (2) explains 61% of the combined variation, and (3) has a reliability coefficient of 0.67. For the amenity index or scores derived from this process (see appendix equation A-2), a positive value implies a site rich in faunal habitat, floral diversity, and scenic beauty. The computed scores range from −4.4 to 3.0.

3.3 Ecosystem Characteristics (Z)

Timber production is influenced by a number of site characteristics. Of the variables recorded in the Forest Inventory and Analysis (FIA) data set, we found four to be particularly relevant. The first is a measure of harvest operability. We code this as a dummy variable that is equal to 1 when the site is described as having "no problems." Sixty-three percent of our sites are in this category. The second is a measure of inherent capacity to grow crops of industrial wood using six site productivity classes that identify the average potential growth in cubic feet/acre/year and is based on the culmination of mean annual increment of fully stocked natural stands. Sites with the highest average potential growth are assigned a value of 1, and the lowest growth sites are assigned a value of 6. The percentage of the county covered by national forests is the third explanatory variable. To the extent that national forests are logged less frequently, there may be diseconomies of logging in counties that have a greater share in national forests. Finally, the volume of timber inventory can be a critical determinant of timber production. We have information on initial volume for five product classes and we use prices for the product classes to create a weighted average of the inventory.[6] We take the natural log of the weighted average to reduce scale differences (which can cause convergence problems in maximum likelihood estimation), improve linearity, and pull in outliers (see appendix equation A-3).

3.4 Sociodemographic Characteristics (D)

The final layer of data is socio-demographic information from census files. Although applying broad averages to plot-level ecological data is less than ideal, it is the closest we can get to measuring sociodemographic heterogeneity. See King (1997) for a discussion of inference from aggregated data of this nature. We use income and education at the county level as a first approximation. Income, proxied by the median household income in the counties, has a mean and standard deviation of $23,000 and $4000, respectively, in our data set. We use the log of the median household income. Education, proxied by the percentage of the population with a graduate degree, has a mean and standard deviation of 12 and 6. The FIA data provide us with the last bit of sociodemographic data: the owner type. We start with the simplest distinction, using a dummy variable to identify NIPF owners. Approximately 36% of our sample are other private owners.

4. EMPIRICAL MODEL

At the core of the joint production choice are equations 14.3 and 14.4 on amenity demand and timber supply.[7] We focus on this two-equation system as a first approximation of the tradeoffs of timber and amenity production.

4.1 Estimating Equations

We specify amenity demand as a function of timber prices and landowner sociodemographics (proxied by county aggregates). An innovation of this research is the characterization of amenity demand as an amenity index, which is constructed using the first principal component of three nontimber indices (see section 3). Because amenity is self-produced, we account for the associated endogeneity by regressing our amenity index on timber price and landowner sociodemographics, with the latter proxying the preference parameter θ from equation 14.3. The estimated form of amenity demand is

$$a = \beta_p P + \beta_d D + \varepsilon_a \qquad 14.8$$

where P is timber price; D is a vector of sociodemographics including education, income, and dummy variable for ownership; ε_a is the error term; and β is a vector of parameters to be estimated. The predicted value from this regression ($\hat{a}$) is used in the probability model of timber harvest, described in equation 14.9.

Timber supply is specified as a function of timber price, site characteristics, and amenity demand. Given that our data contain a large number of observations, approximately 85%, for which there is no timber harvest, we use limited-dependent variable methods to model timber supply. A probit model, which is based on a marginal utility discrete choice motivation, is employed to estimate the probability of a nonzero harvest.[8]

$$Prob(H = 1) = \Phi(\gamma_P P + \gamma_Z Z + \gamma_a \hat{a}) \qquad 14.9$$

where H is an indicator variable that is equal to 1 when $y^*>0$ and zero otherwise; $\Phi(\cdot)$ is the cumulative distribution function of the unobservable in the timber supply equation; P is timber price; Z is a vector of site characteristics such as site index, operability, timber inventory, ecoregion, and national forest percentage; and γ are parameters to be estimated.[9]

4.2 Estimation Results: Least Squares and Probit

The results from estimating equations 14.6 and 14.7 are reported in tables 14.2 and 14.3, respectively. Although the overall performance of our models is not very strong ($R^2 = 0.06$ and pseudo $R^2 = 0.18$), presumably because county aggregates serve as weak proxies for individual owner characteristics, we can detect some clear statistical signals in this otherwise noisy data set. Based on the signs and statistical significance of the estimated parameters, we find that both models generate plausible results.

Table 14.2. Amenity demand (index): robust least-square regression

Independent Variables	β (estimated coeff.)	p-value
Timber price index	-0.160	0.000
Dummy for NIPF owner	0.165	0.000
% of county with bachelor's degree	0.018	0.000
Log (median household income)	-0.324	0.020
Constant	3.491	0.011
R^2	0.06	
Root mean square error	1.31	
N = 4403		

Timber price has the expected negative influence on amenity demand, suggesting that the opportunity costs of preserving forests for nontimber uses (i.e., the substitution effect) dominate landowner choices. This finding also validates the utility-theoretic foundation of our approach. The coefficient of the ownership dummy indicates that NIPF landowners have a higher demand and self-production of amenities than other groups. That is, in comparison to other owners such as forest industry, NIPF landowners are less focused on timber management, with interest in a broad range of forest-based goods and services. We see higher amenity demand in counties with higher education levels, a result that is consistent with the logic that education is likely to be correlated with forest conservation or broader forest management goals. Finally, income is negatively correlated with the forest amenity index. One interpretation of this seemingly counterintuitive result is that the county level income variable is a proxy for urbanization such that more urban counties are positively correlated with poorer quality forests as measured by our habitat and scenic beauty index.

In the timber supply equation, we see the positive influence of timber prices. This finding that landowner-managers are more likely to harvest at higher prices is a critical piece of evidence in support of rational economic behavior and is consistent with our utility-theoretic model scenario in which the substitution effect dominates the income effect.

Table 14.3. Timber harvest: probit model

Independent Variables	γ (estimated coeff.)	p-value
Timber price index	0.272	0.000
Predicted nontimber index	-0.430	0.044
Timber inventory	0.314	0.000
Dummy for operability	0.562	0.000
Site class index (1 = most to 5 = least productive)	-0.054	0.115
Dummy for mountain region	1.236	0.000
% of county acres in national forests	-1.461	0.000
Constant	-4.025	0.000
Wald χ^2 (7)	567.59	0.000
Pseudo R^2	0.182	
N = 4403		

We find a positive coefficient on the timber inventory index, a positive coefficient on the operability dummy, and a negative coefficient on the site class index, which is significant only at the 11% confidence level. Collectively, these three results show a higher likelihood of harvests on sites with better silvicultural potential. We see lower harvests in counties with greater percentage of national forests, suggesting that there are some diseconomies of being in areas that are infrequently logged. The positive coefficient for the mountain region dummy indicates greater harvest in this ecoregion. While this may seem puzzling at first, we must recognize the multiple-regression nature of this result. That is, controlling for price, silvicultural potential, and externalities of national forests—which are typically cited as reasons for less logging in the mountains—the data suggest that there is a positive effect of being in the mountain region. Finally and most importantly, we see a negative coefficient on the predicted amenity index. This offers support for the hypotheses regarding the tradeoff between timber and amenities within a joint bioeconomic production process. The statistical significance of the amenity coefficient in the timber supply model is the core result of our study.

5. DISCUSSION

Using a data set from North Carolina that contains information on timber output and amenity indices, this chapter illustrates an approach to empirically characterize joint household production. At the heart of our model is a utility-maximizing NIPF landowner whose choices of timber and amenities depend on timber prices, site conditions, and individual preference characteristics. The unique features of this data set are the use of several nontimber indices to develop principal components to serve as measures of amenities. Census data is combined with the ecological and economic data to proxy for landowner characteristics. We estimate a two-step probit model of amenity demand and timber supply. Based on the signs and statistical significance of the estimated parameters, we find that both models generate plausible results.

5.1 Methodological Contribution

Although formal economic modeling of multiple forest production by private agents is now 20 years old, we extend Binkley's (1981) logic by explicitly identifying timber and amenity outputs within a household production framework. In particular, we focus on the amenity demand function and offer insights on how to empirically specify joint amenity demand and timber supply models. In this context, the mix of plot-level ecological data, county-level sociodemographics, and market-level price data offer unique opportunities to test hypotheses on joint forest production. While the resolution of the sociodemographic data somewhat limits our ability to rigorously test hypotheses, it presents a useful step towards combining these kinds of information. Future studies could consider surveying a sample of households from the FIA study region to enrich the socioeconomic story. From an empirical perspective, our use of principal component analysis to summarize information on amenities as an amenity index illustrates a parsimonious method to identify the core issue of joint production—the tradeoff between timber and amenities. Finally, the two-step probit estimator used to model timber supply and amenity demand illustrates the application of an equation systems method to account for endogeneity arising out of the joint production nature of this problem. While these conceptual, data, or estimation methods are not individually unique to this study, collectively they constitute a useful kit of analytical tools to address forest economics questions of this kind.

5.2 Policy Implications

The statistical significance of the amenity index in the timber supply equation suggests that models that exclude nontimber outputs will generate biased timber supply parameters. In our data set, we found that in three specifications of the probit model, the model using the predicted value of amenities generates the smallest price coefficient. The three specifications are (1) excluding the amenity index, (2) including the amenity index, and (3) including the predicted value of the amenity index to account for its endogeneity. The mean elasticity of harvest probability with respect to price, defined as the percent change in harvest probability for a percent price change, is 1.88, 1.87, and 1.50 in the three models. While it is premature to draw generalizations regarding the size and sign of this bias, this suggests the potential for errors and over-estimates in policy simulations that are based on timber supply characterization in which timber is the sole product.[10] In addition, we found that in two specifications of the probit model using amenity indices, the model using the predicted value of amenity generates the larger amenity coefficient.

Perhaps the more important policy contribution of the methods described in this chapter lies in providing a conceptual and empirical framework for better understanding how and why private agents manage forests. In particular, the model and the data allow us to investigate whether landowner-managers are jointly producing timber and amenities. Understanding how private agents can supplement public supply of forest amenities is of particular interest to policy makers because amenities (1) lie outside the purview of markets and (2) provide social benefits, in addition to private benefits to owner-managers. We can also identify the characteristics of owner-managers who are more likely to jointly produce amenities and timber. The empirical evidence supports the existence of joint production and its positive correlation with NIPF ownership and counties characterized by higher education and lower incomes. Clearly, future research that uses richer socioeconomic data can more fully characterize the socioeconomics of joint production. In the meanwhile, the methods and empirical results presented in this chapter provide a stepping-stone to the development of analytical tools for studying emerging forest policy questions such as the household joint production of timber and amenities.

6. LITERATURE CITED

BINKLEY, C. 1981. Timber supply from non-industrial forests: A microeconomic analysis of landowner behavior. Yale University Press, New Haven, CT.

BINKLEY, C. 1987. Economic models of timber supply. P. 109-136 *in*: The global forest sector: An analytical perspective, M. Kallio, D. Dykstra, and C. Binkley (eds.). John Wiley and Sons, New York, NY.

BOLLEN, K., D. GUILKEY, and T. MROZ. 1995. Binary outcomes and endogenous explanatory variables. tests and solutions with an application to demand for contraceptive use in Tunisia. Demography 32 (1):111-131.

CRAWFORD, H., and R. MARCHINGTON. 1989. A habitat suitability index for white-tailed deer in the Piedmont. So. J. App. For. 13(1):12-16.

DENNIS, D. 1989. An economic analysis of harvest behavior: integrating forest and ownership characteristics. For. Sci. 35(4):1088-1104.

EBERT, U. 1998. Evaluation of nonmarket goods: Recovering unconditional preferences, Am. J. Agri. Econ. 80(May): 241-254.

HAMILTON, L. 1992. Regression with graphics: A second course in applied statistics. Duxbury Press: California. 363 p.

HARTMAN, R. 1976. The harvest decision when a standing forest has value. Econ. Inq.14:52-58.
HOLMES, T. 1986. An economic analysis of timber supply from nonindustrial private forests in Connecticut. Unpublished Ph.D. Dissertation. University of Connecticut. Storrs, CT.
HYBERG, B., and D. HOLTHAUSEN. 1989. The behavior of nonindustrial private forest owners. Can. J. For. Res. 19:1014-1023.
KING, G. 1997. A solution to the ecological inference problem: Reconstructing individual behavior from aggregate data. Princeton University Press, Princeton, NJ. 346 p.
KUULUVAINEN, J., H. KARPPINEN, and V. OVASKAINEN. 1996. Landowner objectives and non-industrial private timber supply. For. Sci. 42 (3):300-309.
KUULUVAINEN, J and J. SALO. 1991. Timber supply and life cycle harvest of nonindustrial private forest owners: An empirical analysis of the Finnish case. For. Sci. 37:1011-1029.
LEE, K.J. 1997. Hedonic estimation of nonindustrial private forest landowner amenity value. Unpublished Ph.D. Dissertation. North Carolina State University, Raleigh, NC. 71 p.
MADDALA, G. 1983. Limited dependent and qualitative variables in econometrics. Cambridge University Press. New York, NY. 401 p.
MAX, W., and D. LEHMAN. 1988. A behavioral model of timber supply. J. Environ. Econ. Manage. 15: 71-86.
NORTON, E.C., R.C. LINDROOTH, and S.T. ENNETT. 1998. Controlling for the endogeneity of peer substance use on adolescent alcohol and tobacco use. Health Econ. 7:439-453.
PATTANAYAK, S. K., B.C. MURRAY, and R.C. ABT. 2002. How joint is joint forest production? An econometric analysis of timber supply conditional on amenity values. For. Sci. 48(3):479-491.
RUDIS, V., J. GRAMANN, E. RUDDELL, and J. WESTPHAL. 1988. Forest inventory and management based visual preference models of southern pine stands. For. Sci. 34:846-863.
SHANNON, C.E. 1948. A mathematical theory of communications. Bell System Technical J. 37: 379-423, 623-656.
SHEFFIELD, R. 1981. Multiresource inventories: techniques for evaluating non-game bird habitat. Res. Pap. SE-218. USDA Forest Service Southeastern For. Exper. Station, Asheville, NC. 28 p.
WEAR, D., and P. PARKS. 1994. The economics of timber supply: An analytical synthesis of modeling approaches. Nat. Res. Model. 8(3):199-223.

7. APPENDIX: FORMULAE USED

Price Index (P)

$$P = \left(\sum_{j}^{5} \kappa_j \cdot \left(\frac{p_j - \bar{p}_j}{\sigma_{pj}} \right) \right) + 3.297 \qquad \text{A-1}$$

where j product classes include three softwoods (pulp, chip-saw, and sawtimber) and two hardwoods (pulp and sawtimber); p_j is the per unit price; 2.96 is a scaling constant to ensure that the sum within the brackets is non-negative; and κ_j is the scoring coefficient generated by the first principal component. The coefficients are 0.37, 0.48, 0.52, –0.44, and –0.42, respectively, suggesting that a positive factor score measures a plot with relatively high softwood prices and low hardwood prices.

Nontimber Index (*A*)

$$A = \sum_{i}^{3} w_i \cdot \left(\frac{A_i - \overline{A}_i}{\sigma_{Ai}} \right) \qquad \text{A-2}$$

where *i* includes fauna, tree diversity (Shannon-Weaver), and scenic beauty (Rudis); A_i is the amenity index; and w_i is the scoring coefficient generated by the first principal component. The coefficients are 0.66, 0.54, and 0.56, respectively, suggesting that a positive factor score measures a site rich in faunal habitat, floral diversity, and scenic beauty.

Timber Inventory Index (*I*)

$$I = Ln \left[\sum_{j}^{5} \frac{p_j}{\sum p_j} I_j \right] \qquad \text{A-3}$$

where *j* product classes include three softwoods (pulp, chip-saw, and sawtimber) and two hardwoods (pulp and sawtimber); p_j is the per unit price; and I_j the initial inventory.

---

[1] The utility function is assumed to be concave, continuous, and twice-differentiable with the following properties: $U_a > 0$, $U_\pi > 0$, $U_{aa} < 0$, $U_{\pi\pi} < 0$, and $U_{\pi a} > 0$. The first four properties are usual assumptions (see Binkley 1981). The last condition cannot be extracted from standard assumptions. It says that a landowner will value amenities more at higher incomes than at lower incomes—the normal good argument for amenities. The validity of this can only be determined empirically.

[2] The production function is assumed to be convex, continuous, and twice-differentiable with the following properties: $G_a > 0$, $G_y > 0$, $G_{aa} < 0$, $G_{yy} < 0$, and $G_{ya} < 0$. The first four properties are usual assumptions (see Binkley 1981). The last condition says that it is more difficult to generate amenities at higher timber production than at lower timber production—a diseconomies of scale argument for amenities. The validity of this assumption can only be determined empirically.

[3] PCA can result in more parsimonious models, improved measurement of indirectly observed concepts, and may avoid multicollinearity. This method relies on the linear relationship among sets of measured variables using simple transformation of the data. That is, each variable is a linear transformation of the principal components using factor loadings. Alternatively, each principal component is the linear combination of the variables using score coefficients. Typically, the first principal component explains the most and represents the "best possible" (maximum variance) combination, and its contributions are evaluated in terms of a reliability coefficient or explained variance. The reliability

coefficient = $(k/k-1)*(1-1/\lambda_1)$, and the explained variance = (λ_1/k), where k is the number of variables combined and λ_1 is the eigen value. See Hamilton (1992) for details.

[4] White-tailed deer are ubiquitous in North Carolina, so deer habitat on an individual plot may have little impact on deer population in general or even on deer presence on that site. However, there are limited means to measure the quality of habitat other than suitability indices for a plot. Similar concerns and conclusions exist for the six species of wild birds.

[5] The species (and the associated scoring coefficient) included in this index are white-tailed deer (0.175), wood thrush (0.432), red-eyed vireo (0.451), pileated woodpecker (0.375), downy woodpecker (0.361), prothonotary warbler (0.133), brown-headed nuthatch (0.020), pine warbler (0.021), prairie warblers (–0.428), and eastern bluebird (–0.327). The first principal component has an eigen value of 2.8 and a reliability coefficient of 0.72, and is used as an index of faunal habitat.

[6] Price weights reflect the relative value of the types of inventory, which is appropriate in an economic model, and result in an inventory measure that is a value index.

[7] For the remainder of the chapter, we refer to the level of amenity self-produced or consumed as "amenity demand." As pointed out by Steve Swallow (personal communication), the level of amenities demanded or consumed is conditioned by what is available for the existing shadow price. To the extent that landowners manage their forest lands to obtain specific forest landscapes and structures, they self-produce amenity flows. This level of self-production is conditioned by the opportunity costs of creating these forest conditions or its shadow price, and therefore measures landowner amenity demand.

[8] Consider the choice facing a landowner-manager when deciding whether to harvest. The decision clearly depends on the net utility with and without harvesting (EU_i*). This net utility is given by $EU_i* = \alpha_1 P_i + \alpha_2 Z_i + \alpha_3 a_i + e_i$, where e_i is a random disturbance and P_i, Z_i, and a_i are as defined in the text. Note that while EU_i* is not directly observable, the owner-manager's decision outcome is. Let H_i be an indicator of whether the owner i harvests or not. Then $H_i = 0$ if $EU_i* \leq 0$ and $H_i = 1$ if $EU_i* > 0$. The structural relationship presented above can be estimated using a probit model (Maddala 1983).

[9] We use a standard two-step estimator instead of more complicated methods (FIML and GMM) because Monte Carlo experiments (Bollen et al. 1995) and other studies (Norton et al. 1998) have shown no gain in performance in using more complicated methods.

[10] As suggested by Steve Swallow (personal communication), using the elasticity of harvest probability, we calculate the expected supply for a plot in response to a 10% increase in price. The expected supply is equal to the product of the plot timber inventory and the change in harvest probability associated with the 10% price change. Comparing the model without amenity indices (model 1) with the model with the predicted amenity index (model 3), we see that model 1 will over-predict expected timber supply by 14 feet^3 for a plot with the mean level of inventory. If we extrapolate to North Carolina levels using the volume expansion factors for each of the 4403 in our analysis, model 1 will over-predict supply by 220 million feet^3 of timber.

Chapter 15

Nontimber Forest Products in the Rural Household Economy

Erin O. Sills, Sharachchandra Lele, Thomas P. Holmes, and
Subhrendu K. Pattanayak
North Carolina State University, Centre for Interdisciplinary Studies in Environment and Development, USDA Forest Service, and Research Triangle Institute

Among the multiple outputs of forests, the category labeled nontimber forest products, or NTFPs, has drawn increased policy and research attention during the past 20 years. NTFPs have become recognized for their importance in the livelihoods of the many relatively poor households who live in or near forests, especially in the tropics. Policy concern about NTFPs takes two forms. On the one hand, collection of relatively high-volume, low-value NTFPs, such as fuelwood, fodder, and mulch, has raised concerns about degradation of the forest resource, potentially resulting in hardships for households and negative environmental externalities. On the other hand, collection of relatively high-value, low-volume NTFPs, such as specialty food products, inputs to cosmetics and crafts, and medicinal plants, has drawn interest as an activity that could raise standards of living while being compatible with forest conservation. Addressing these policy concerns requires an improved "understanding of how households interact with natural resources and how one can affect household behavior in desired ways" (Ferraro and Kramer 1997: 207).

In this chapter, we show how both types of NTFPs and related concerns can be understood and evaluated in the household production framework. We illustrate this with two case studies, from the distinct cultural and historical contexts of the Western Ghats of India and the Brazilian Amazon. Our approach is first to clarify objectives, constraints, and conditioning factors using household production theory, and then to estimate econometric

Sills and Abt (eds.), Forests in a Market Economy, 259–281. ©Kluwer Academic Publishers. Printed in The Netherlands.

models consistent with that theory and feasible given available data. This raises modeling issues such as the implications of missing or incomplete markets, the relation of other household activities to NTFP collection, and the representation of heterogeneity across households. Appropriately specified models can provide insight into the role of NTFPs in the rural household economy (Pattanayak and Sills 2001), identify policy levers (Lele 1993), and serve as the building blocks for valuation of local forest access (Pattanayak et al. [forthcoming]) and policy simulations (Bluffstone 1995).

1. NTFP LITERATURE

NTFPs include a wide range of subsistence and commercial products (Neumann and Hirsch 2000, Pérez and Arnold 1995). Although much of the literature focuses on products collected from natural forests in developing countries, NTFPs are also produced in plantations and agroforestry systems (see chapter 16) and in developed countries (Jones et al. 2002). Fuelwood is probably the NTFP collected in greatest volume. In fact, fuelwood and charcoal are often placed in a category of their own, with other NTFPs relabeled as nonwood forest products (NWFPs). These include rattans and bamboos; edible fruits, nuts, and other foods; medicinal plants; resins and latex; wildlife and derivative products; and cultural, religious, and aesthetic commodities (Thandani 2001). The Food and Agriculture Organization of the United Nations estimates that approximately 150 of these NWFPs are "significant in terms of international trade" (FAO 2002), some as traditional commodities (e.g., rattan) and some as "green" products marketed as environmentally friendly (e.g., Brazil nuts). While products that enter formal international markets are easiest to quantify, NTFPs are also known to play a critical role in household subsistence and local and regional markets. For example, FAO (2002) asserts that "80% of the population of the developing world use NWFPs for health and nutritional needs." Byron and Arnold (1999) emphasize that the exact nature and degree of forest dependence varies widely across regions and households. Here, we review three prominent strands of the literature on forest dependence.

1.1 Local Value of NTFPs

Many researchers have sought to quantify the value of NTFPs. Tewari (2000) reviews the motivations and policy implications of these valuation efforts, and Wollenberg (2000) reviews the methodological challenges of obtaining accurate data on quantities and prices. NTFP value can be calculated per hectare of forest (returns to land) or per household (returns to labor). For the first, researchers typically combine botanical or ethnobotanical information with market price data to find the potential value of NTFP production (Godoy and Bawa 1993, Peters et al. 1989). For the second, researchers (a) track small samples of households with frequent visits to record quantities and prices, (b) rely on respondent recall of quantities and prices in household surveys, or (c) elicit values directly with stated preference methods (see Shyamsundar and Kramer [1996] and chapters 17 and 18). Recent studies that carefully tracked household income conclude that NTFPs contribute between 10% and 60% of full income (Cavendish 2000, Kvist et al. 2001, Reddy and Chakravarty 1999). This contribution varies substantially across households, which is the theme of the next strand of literature.

A common hypothesis is that poorer households are more dependent on the forest (Godoy et al. 1995, Reddy and Chakravarty 1999). The relationship between NTFP collection (quantities or gross value) and socioeconomic characteristics including income or wealth has been analyzed most often with cross-tabulations and graphical methods (Bahuguna 2000, Cavendish 2000, Godoy et al. 1995, Hegde and Enters 2000, Lele 1993, Takasaki et al. 2000). Econometric approaches are discussed in section 2.4. Many of these studies find that poor households depend *relatively* more on NTFPs, conditional on an array of other socioeconomic and geographical characteristics. Findings regarding absolute dependence on NTFPs vary, as does the pattern across middle- and high-income households. Takasaki et al. (2000) use participatory rural assessment methods to categorize Amazonian households by specific types of wealth. They contend that not only the level but also the type of wealth determines how people use the forest and hence their dependence on NTFPs.

The common finding that the poor depend relatively more on NTFPs raises questions about the role of NTFPs in economic development. In the early 1990s, there was great interest in NTFPs as a basis for sustainable development (e.g., Nepstad and Schwartzman 1992, Plotkin and Famolare 1992). More recently, the economic potential of NTFPs has been sharply debated in the literature (Pérez and Byron 1999), with some authors arguing that the role of NTFPs as engines of local development has been greatly exaggerated (Southgate 1998, Wunder 2001). Much of the empirical literature concludes that NTFPs are neither the main driver nor an impediment to development, but rather that they play an important supplemental or fallback role (Godoy et al. 2000, Pattanayak and Sills 2001). In this capacity, NTFPs are seen as supporting the economic development process by serving as a safety net for households entering new economic activities and markets (Byron and Arnold 1999). To better understand this role, we turn next to a conceptual framework of household behavior.

2. HOUSEHOLD PRODUCTION THEORY

Household production theory has been used to model the economic activities of rural households in a wide variety of cultural contexts, especially where households' time endowments are their primary factor input, and households consume most of their own production outputs. Singh et al. (1986) remains the basic reference for agricultural household production models. Hyde and Amacher (1996, 2000) argue for wider application to forestry issues and report such applications to fuelwood. The basic theory posits a household that combines the time endowments of its members with other variable and fixed inputs (including available forest resources) to produce a utility-maximizing bundle of goods, subject to technology, budget, and time constraints.

2.1 Agrarian Households on the Forest Edge

For purposes of this chapter, we present a generic model of a typical agrarian household living on the forest margin (equation 15.1). This household engages in agriculture (A) and collection of NTFPs (F). We assume complete markets for agricultural products and for market goods (M), but incomplete markets for NTFPs and labor. Thus, the amount of labor and leisure available are constrained by household time (T), and cash expenditures are constrained by the value of agricultural output plus any exogenous income (I) such as remittances. The household seeks to maximize a single utility function, which depends on consumption of agricultural goods (A_H), market goods (M_H), forest goods (F_H), and home time (T_H, including leisure, child care, etc.). Household utility is conditioned on preferences (Φ).

$$\max U(A_H, M_H, F_H, T_H; \Phi)$$

s.t.

$(1) T \geq T_H + T_A + T_F$

$(2) A = a(T_A, F_A, M_A; \Psi)$

$(3) F = f(T_F; B, H, \Psi)$

$(4) F \geq F_H + F_A$

$(5) P_A(A - A_H) + I \geq P_M(M_H + M_A)$

15.1

where the constraints apply to (1) household time, (2) agricultural production, (3) nontimber forest production, (4) forest output allocation, and (5) budget. The choice variables are T_H, T_F, M_A, M_H, F_H, F_A, and A_H.

Agricultural production is a function (a) of household time allocated to agriculture (T_A) and other inputs collected from the forest (F_A) or purchased in the market (M_A), conditioned on fixed household production endowments (e.g., land, livestock) and technology (Ψ). Forest production (f) is also conditioned on fixed production endowments. However, we assume that it does not compete with agriculture for land, but rather takes place in public forest, conditioned by its biophysical state (B) and household knowledge of the forest (H). The only variable input in forest product collection is household time (T_F). Forest products are either consumed (F_H) or used as inputs to agriculture (F_A).

To write the Lagrangian function, we combine constraints 3 and 4, resulting in four constraints with four Lagrangian multipliers ($\mu, \gamma, \delta, \lambda$) or shadow values (equation 15.2).

$$\ell = U(A_H, M_H, F_H, T_H; \Phi) + \mu(T - T_H + T_A + T_F) + \\ \gamma(a(T_A, F_A, M_A; \Psi) - A) + \delta(f(T_F; B, H, \Psi) - F_H - F_A) \\ + \lambda(P_A(A - A_H) + I - P_M(M_H + M_A))$$

15.2

The seven choice variables (equation 15.1) and four constraints result in eleven first-order conditions (FOC). To conserve space, we only present the FOC with respect to the three choice variables directly related to forest production and consumption decisions, T_F, F_H, and F_A:

$$\frac{\partial \ell}{\partial T_F} = -\mu + \delta \frac{\partial f}{\partial T_F} = 0$$

$$\frac{\partial \ell}{\partial F_H} = \frac{\partial U}{\partial F_H} - \delta = 0 \qquad 15.3$$

$$\frac{\partial \ell}{\partial F_A} = \gamma \frac{\partial a}{\partial F_A} - \delta = 0$$

Algebraic manipulation of the FOC yields the following results (equation 15.4). First, households allocate their time such that the shadow value of time (μ) is equal to the marginal utility of NTFPs obtained by allocating more time to collecting. This is the familiar proposition that marginal cost equals marginal benefit applied to forest collection. Second, the marginal utility of increased agricultural production arising from inputs of forest products must equal the marginal utility of household consumption of forest products. This condition indicates how households allocate forest production between household consumption and agricultural inputs. Finally, this second condition implies that the shadow value of time must also equal the marginal utility of increased agricultural production due to forest inputs obtained with more time spent collecting.

$$\mu = \delta \frac{\partial f}{\partial T_F} = \frac{\partial U}{\partial F_H} \frac{\partial f}{\partial T_F}$$

$$\frac{\partial U}{\partial F_H} = \gamma \frac{\partial a}{\partial F_A} \Rightarrow \mu = \gamma \frac{\partial a}{\partial F_A} \frac{\partial f}{\partial T_F}$$

15.4

Note that the shadow value of time depends on the parameters of both the utility function and the production functions. Further, the other FOCs would show that the marginal utility of increased agricultural production (γ) is related to the shadow value of income (λ) and consequently to prices and exogenous income. In fact, the shadow values, which are internal to each household, depend on the full set of exogenous variables. As a result, collection, consumption, and the derived demand for labor are also functions of all exogenous variables in the system. This dependence of production decisions on preferences and endowments is termed nonseparability in the household production literature and results whenever key markets are missing or incomplete (Sadoulet and de Janvry 1995).

2.2 Incomplete Markets

To further explore the relationship between nonseparability and incomplete markets, we turn to a graphical treatment. Consider panel A of figure 15.1, representing an individual household's demand (WTP) for a NTFP and two possible household supply curves, or marginal costs of production. If a market for the good exists, the price (P) is exogenous to the household. If the household has high production costs (MC″) relative to P, it will not produce and will purchase the amount Qd. If the household has low production costs (MC′), it will produce Qp′, of which it will consume Qd and sell Qp′ − Qd. In either case, demand is set where WTP equals P. The production level is independent of demand and is established where MC equals P (the profit- maximizing solution). Of course, demand depends on the income generated by production. However, household decisions can be modeled recursively, with production decisions treated as if they were made prior to and independent of consumption decisions (separability).

The usual case for households living on the forest margin is somewhat incomplete or imperfect markets (see Carter and Yao 2002 for argument that this is generally true of rural markets and Behrman 1999 for empirical tests). Following Sadoulet and de Janvry (1995), this can be conceptualized as price bands for the sale and purchase of goods. That is, households can usually purchase at some—perhaps very high—"buyer price" (P_b). Likewise, households can usually sell at some "sale price" (P_S), although it may be so low that it is irrelevant. There are both spatial (e.g., distance to market) and household-specific (e.g., connections to traders) reasons for the transactions costs (t) that create these price bands. As a result, each household faces different price bands in addition to having unique demand and supply functions.

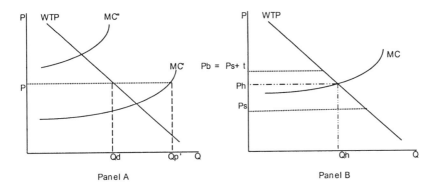

Figure 15.1. Household production and consumption of NTFPs

Consider a household producing an NTFP with the WTP and MC curves shown in panel B of figure 15.1. When the intersection of WTP and MC occurs above P_S and below P_b, the household decision about the quantity to collect and consume (Q_h) is determined jointly with a household-specific shadow price P_h (δ in the conceptual model). Both depend on the parameters of household utility (Φ) and household production technology (Ψ, B, H). The same holds true for inputs whose demand derives from this household market for NTFPs, such as labor or collection trips. Finally, note that variation in household supply and demand means that some households may still base their production and consumption decisions on the market price if the intersection of WTP and MC falls either below P_S or above P_b. Thus, even when some households are observed to participate in the market, it may be incomplete for others.

2.3 Dynamics of Forest Collection

Thus far, we have presented a static model, with only one time period. However, much of the interest in NTFPs stems from the link between current behavior and future resource conditions. Consider a two-period model, in which households maximize the sum of current and expected future utility, discounted by ϕ (equation 15.5). In the second period, forest production depends on household knowledge of the forest, which in turn depends on time spent in the forest (learning) during the first period (Pattanayak and Sills 2001). In addition, the quality of forest stocks in period two are affected by the aggregate amount collected by all households during period one (ΣF_1). If access to forest land is privately (or community) controlled, the household (or community) can set ΣF_1.

$$\max U_1(A_{H_1}, M_{H_1}, F_{H_1}, T_{H_1}; \Phi) +$$
$$\varphi E\{U_2(A_{H_2}, M_{H_2}, F_{H_2}, T_{H_2}; \Phi)\} \qquad 15.5$$
s.t.

$$(3.2)\ F_2 = f_2[T_{F2}; B_2(\sum F_1), H_2(T_{F1}), \Psi]$$

The household is also subject to constraints 1 through 5 for period 1 and constraints 1, 2, 4, and 5 for period 2 from equation 15.1.

If we redrew figure 15.1 for period two, the marginal cost curve (MC_2) would shift up if forest stocks had been degraded by collection in the first period.[1] Conversely, if increased forest knowledge more than offsets any forest degradation, MC_2 would shift down. In either case, P_{H2} and Q_{H2} would also adjust. Consideration of these impacts changes the marginal conditions for the first period. For example, time would be allocated such that its shadow value is equal to the *net* contribution of collection time to utility, through increased knowledge as well as increased production (equation 15.6). Households will also consider impacts of current collection on future biophysical conditions when they control access to the forest and hence can determine ΣF_1.

$$\mu_1 = \delta_1 \frac{\partial f_1}{\partial T_{F1}} + \delta_2 \frac{\partial f_2}{\partial H_2} \frac{\partial H_2}{\partial T_{F1}} \qquad 15.6$$

2.4 Model Specification

The household production framework described above gives the analyst multiple options for empirical estimation: the dependent variable may be NTFP production, consumption, marketed surplus, or labor allocation. Estimation results can provide insight into the behavior of households, including the determinants of forest use, the distribution of forest use across households, and responses to potential policy interventions. In the case of complete markets, specification would follow standard production or consumption theory, with prices and income playing key roles. However, as argued above, the more common case is imperfect markets. Here, we review the specifications and findings of previous empirical work.

2.4.1 Market Assumptions

In much of the literature on fuelwood, either the labor or the product market is assumed to be complete. When only the labor market is incomplete (i.e., a product market exists), a shadow wage can be estimated as the value of the marginal product of labor (Jacoby 1993). For example, Amacher et al. (1999), Köhlin and Parks (2001), and Mekonnen (1999) calculate shadow wages as the value of the marginal product of labor in fuelwood collection. When only the labor market is complete, the reverse approach is possible: the shadow price of fuelwood can be estimated as the value of the time required to collect a unit of fuelwood. For example, Cooke (1998) and Bardhan et al. (2001) use time to collect a kilogram of fuelwood multiplied by the household wage. In a third approach, Pattanayak et al. (forthcoming) and MacDonald et al. (2001) use the cost of a collection trip (wage multiplied by time) in travel cost models of fuelwood collection. Finally, the collection time itself is used in some studies as a proxy biophysical variable, representing scarcity of the forest resource (Edmunds 2002, Heltberg et al. 2000). While the direction of influence varies, almost all studies find that household behavior is significantly influenced by the productivity of labor in fuelwood collection, whether that is interpreted as a factor in the shadow wage, a factor in the shadow price, or a measure of scarcity.[2]

If multiple markets are incomplete, as they are for our two case studies, analysts usually resort to reduced form models. This approach has been applied both to specific NTFPs (e.g., fodder, game, rattan) and to the NTFP category in general, represented either as total gross income from NTFPs or total household time allocated to collection of NTFPs (Godoy 2001, Gunatilake 1998, Wickramasinghe et al. 1996). Household behavior is modeled as a function of socioeconomic and environmental characteristics reflecting preferences, technology, and input endowments, rather than prices. In this case, "none of the original parameters and hence the constraints that they are supposed to identify can be identified. There is no justification for any specific restrictive form for the system" (Sadoulet and de Janvry 1995: 160). In general, linear or log-linear functional forms are used, and models are assessed based on their explanatory power and ability to identify determinants of NTFP collection and consumption.

2.4.2 Determinants of Behavior

In studies of both fuelwood and other NTFPs, household size is one of the most common explanatory variables. The number of people in a household affects both production possibilities (as a measure of available labor) and demand (as a measure of cooking and other consumption needs). Size has been found positively and significantly related to collection time, gross income from NTFPs, and production and consumption quantities. Household size and other demographic variables, such as age, may be included in quadratic form to represent non-linear family life cycles. Size may also be combined with or disaggregated into the gender distribution within a household to better represent the labor endowment. Other sociocultural factors (e.g., education and caste) are also hypothesized to reflect production abilities and preferences regarding NTFPs.

Household assets affect production capabilities and preferences, and many studies include some measure of household wealth, such as landholdings (Edmunds 2002, Amacher et al. 1999) and livestock ownership (Gunatilake 1998, Joshee et al. 2000). The effect of wealth varies across studies, even for the same region and NTFP (compare Amacher et al. 1999 and Edmunds 2002). Some studies use wealth as a proxy for permanent income, while others use expenditures or exogenous income from nonforest sources (often negatively related to NTFP collection). These income variables may also reflect the opportunity cost of household labor. Some studies represent income sources as dummy variables, especially when the accuracy of reported income levels is in doubt. Another household variable often included in fuelwood models is ownership of a substitute fuel technology, such as a kerosene stove. Substitutes generally have the expected negative effect on collection and consumption.

Regional characteristics commonly found in these models include measures of the forest stock, which is generally positively related to NTFP collection time and quantities. Other studies emphasize distance to forest (generally negatively related to collection) and/or distance to market (generally positively related to fuelwood collection, but negatively related to gross NTFP income). Despite these commonalities across models, the specific combination of variables and specific measurement of those variables differs substantially among studies (*cf.* chapter 16).

To summarize, recall that collection (quantities and time allocation) is a function of all exogenous factors, $h(\Phi, \Psi, P, I, H, B)$, in nonseparable household models. In practice, these factors are typically represented by variables drawn from the following categories:
1. Household demographics
2. Wealth or assets (physical and human capital)

3. Income opportunities or sources
4. Substitutes
5. Regional resource and market characteristics.

The choice of specific variables depends on the particular NTFP, the socioeconomic and ecological context, the available data, and the objectives of the analysis. For reduced form estimations of nonseparable models, representation of the variables is governed by hypothesized relationships, data quality, and goodness-of-fit, rather than any theoretical restrictions.

3.   CASE STUDIES: INDIA AND BRAZIL

The richness of the household production model as a tool for examining economic behavior in subsistence and low-income economies is illustrated by two case studies. Each study provides insight regarding the interplay between agricultural and forest-based activities in areas where access to markets for labor and production outputs is limited. Policy concerns in both areas arise from the role of NTFP collection activities in the sustainable use of forest resources.

In the first case, we model fuelwood collection (quantity of fuelwood supplied) in the Malnaad region of the Western Ghats of India. These forests have been used heavily for grazing, mulch, fodder, and fuelwood. Forest land tenure falls under two regimes: private access forests (*soppinbettas*) for which usufruct rights are held by particular households, and *de facto* open access forests in which all community members can collect NTFPs. Fuelwood serves as an input to agriculture (processing areca nut and sugarcane) and household services (cooking and bathing).

In the second case, we model the collection of multiple NTFPs (quantity of labor demanded for collection) in the Tapajós region of the Brazilian Amazon. This forest is federally owned and officially designated for timber production, although no large-scale harvesting had taken place at the time of this study. Local households had informal access to the forest under a *de facto* open access tenure regime, effectively regulated only by community norms. The NTFPs collected in this case are both consumed by the households (as food and medicine) and occasionally sold in the market.

Data were collected through surveys of 260 Malnaad households in 1992 and 324 Tapajós households in 1997 (Lele 1993, Pattanayak and Sills 2001). Although the specific questions varied, both surveys provide information on household demographics (family size, age, proportion male), human capital (education, years in local area), and physical assets (land, cattle). Malnaad and Tapajós households are similar in that they both rely on agriculture as their primary economic activity, harvest only small quantities of timber on a sporadic basis, and have limited opportunities for wage labor.[3] While they collect different NTFPs, both rely primarily on household labor for collection and face significant transactions costs that serve as barriers to market participation.

The case studies demonstrate alternative approaches to two methodological issues. First, there were missing values in both data sets due to household nonresponse to particular questions. For Malnaad, we use class-based imputation to interpolate missing fuelwood quantities, based on median values for households from the same socioeconomic class, in the same village, collecting from the same source. In the Tapajós, we use community averages for several explanatory variables, which then represent community, as opposed to individual household, conditions.

Second, we illustrate two approaches to quantifying household wealth or socioeconomic class: cluster and principal components analyses. These are two of the most common methods for grouping observations by attributes when there is no *a priori* classification scheme (Hand 1981). They incorporate more of the available information than previous studies that have proxied wealth with individual variables such as land or livestock ownership. Malnaad society is distinctly stratified, with wealthier households generally having higher income and more assets of all kinds (Lele 1993). We therefore cluster households using measures of both physical and human capital. In the Tapajós, previous cluster analysis of households (Sills et al. 2000) was found to mask different relationships between specific types of assets and forest use, consistent with the asset specialization among Amazonian households found by Takasaki et al. (2000). We therefore summarize different types of household wealth using principal components.

3.1 Fuelwood Collection in India

In Malnaad, all but a few households collect and consume fuelwood. The largest proportion (47%) collects only from open access forest, while a somewhat smaller proportion collects only from private access forest (42%), and 11% collect from both ownership types. To place this in the context of the household production model developed above, consider first the 58% of households that control private access forests and hence have two supply options. Private access forests are generally more accessible and have better, or at least better known, stocks of fuelwood. Therefore, the marginal cost of collection from private access is likely to be lower than the marginal cost of collection from open access for initial units of fuelwood. However, collection from private forests entails an additional cost, in terms of reduced future fuelwood stocks and thus higher future collection costs ($\partial f_2 / \partial F_1 < 0$). Users of open access forests do not consider this opportunity cost of reduced fuelwood stocks. Hence, the marginal cost of collection from open access is likely to rise more slowly than the marginal cost of collection from private forests. If household demand is sufficient, collection will eventually switch to open access forest. These relationships are consistent with the greater quantities of fuelwood collected from private access forest (mean = 3993 kg per household per year, st.dev. = 2508) than from open access (mean = 1317, st.dev. = 1913). Of course, for the 42% of households who do not control private access forest, the marginal cost of collecting from open access forest is the only relevant supply curve.

3.1.1 Empirical Specification

To investigate whether there are differences in collection behavior, we specify separate models for open and private access forests (*cf.* Joshee et al. 2000 and Mekonnen 1999). We include all households in the model of collection from open access forests and use a Tobit estimator to account for households that do not exercise this option. Only households with private access forest are included in the second model, which is estimated with OLS. Note that in the Malnaad context, control of private access forest is not a choice but rather an inherited endowment.

Explanatory variables are drawn from the five categories listed in section 2.4.2 (table 15.1). To represent wealth and income, we divide households into three classes through nonhierarchical cluster analysis based on all available measures of physical and human capital.[4] The method initially divides the households into three clusters and then reassigns them to the closest cluster, as defined by Euclidean distance to the cluster median (Johnson and Wichern 1998). This k-medians clustering produces reasonably balanced cluster sizes and classifies households in a manner consistent with field observations on social stratification. As expected, the clusters are quite distinct in terms of both land ownership (e.g., 100% in the wealthiest cluster and 13% in the poorest cluster control access to private forest) and other socioeconomic characteristics (e.g., 34% in the wealthiest cluster and 82% in the poorest cluster have less than a fifth-grade education).

While previous analysis suggested that class variables capture most differences among households (Lele 1993), it is possible that the individual variables used in the cluster analysis could affect household production abilities or preferences independent of their relationship to class. To test the null hypothesis that the class variables capture all of these influences, we include both the cluster variables and their component variables in the estimations.

Table 15.1. Characteristics of Malnaad, India (n=255)

Variable	Definition	Mean	St. dev.
HH size	Number of household members	7.00	3.71
% Male	Percent of household who are males of working age	0.37	0.18
% Low educ.	Percent of household with less than 5th-grade education	0.57	0.34
Off-farm job	Dummy = 1 if household has job outside of village	0.08	0.27
Livestock	Number of cattle and buffalo owned	5.45	5.78
Sugarcane	Dummy = 1 if cultivate sugarcane	0.03	0.10
Areca	Dummy = 1 if cultivate areca	0.69	0.46
Private forest	Hectares of private access forest	9.15	8.26
Substitute	Dummy = 1 if own substitute fuel technology	0.35	0.48
Forest access	Accessibility of open access forest, on scale of 1 to 10	4.73	3.12
Road access	Dummy = 1 if better road access	0.67	0.47

3.1.2 Results[5]

Table 15.2 presents estimation results for fuelwood collection from open access and private access forests. The only common result across the two models is that larger households collect more fuelwood. In the case of private access, the square of size is negative and significant, suggesting diminishing marginal productivity of labor due to the fixed area of private forest. In general, there are more statistically significant variables in the open access model, possibly due to the larger number and greater variation of households in that sample. The class variables are significant only in the case of open access: all else equal, poorer households collect the most and middle-class households the second most from open access forest.[6]

The null hypothesis that the component variables have no effect is clearly rejected. Households with little education who are not employed outside the village collect significantly more fuelwood from open access forest. This may reflect their lower opportunity cost of time. The number of cattle and buffalo owned is negatively related to private and positively related to open access fuelwood collection. Households may use cattle to transport fuelwood from the more distant open access forests, and/or they may herd cattle and collect fuelwood at the same time (joint production). If livestock grazing diminishes forest productivity, households may prefer to limit joint grazing and fuelwood collection on private access forest.[7]

Sugarcane and areca cultivation are associated with fuelwood collection from different forest types. The coefficients on sugarcane indicate that jaggery production is a major factor in fuelwood consumption from open access forests. The negative coefficient on this variable in the private forest specification suggests that these families either contract out sugarcane processing or obtain the necessary fuelwood from open access forests, encouraging greater overall reliance on these alternative fuel strategies. The coefficients on areca have the opposite signs, perhaps reflecting the historical relationship between areca orchards and allocation of private access forest. As expected, acres of private access forest is positively correlated with collection from those forests and negatively correlated with collection from open access forest. Finally, ownership of a substitute has the expected negative effect only on collection from private access forests. Regional variables only affect collection from open access forest: households that have better access to the forest and to the road collect more fuelwood.

Table 15.2. Estimates of fuelwood collection in Malnaad, India

Variable	Open Access^a		Private^b	
	Coefficient	p-value	Coefficient	p-value
Intercept	-6536.9	0.000	-32.88	0.984
HH size	363.32	0.053	361.50	0.003
HH size^2	-15.08	0.198	-7.70	0.054
% Male	408.61	0.628	1291.77	0.282
Poor class	2486.85	0.019		
Middle class^c	1239.34	0.143	-389.05	0.504
% Low educ.	1060.44	0.076	-843.62	0.344
Off-farm job	-1283.7	0.108	246.12	0.726
Livestock	103.54	0.015	-82.05	0.048
Sugarcane	4743.79	0.003	-2431.8	0.042
Areca	-2081.6	0.000	1943.18	0.002
Private forest	-92.98	0.103	87.74	0.025
Substitute	6.95	0.986	-1156.9	0.003
Forest access	545.27	0.000	110.49	0.435
Road access	1980.04	0.003	-37.10	0.969
σ	1803.58	0.000		

^{a} Tobit model; sample size = 255; log-likelihood = -1104.5.
^{b} OLS, corrected for heteroskedasticity; sample size = 151; adjusted R^2 = 0.224.
^{c} Definition of this variable differs across the two regressions: for open access forest, it designates the 38% of households (out of 255) in the middle class, while for private access forest, it designates the 60% of households (out of 151) in either the middle or poor class.

3.2 Collection of Multiple NTFPs in Brazil

In the Tapajós region, 84% of households collect NTFPs, including vines, honey, nuts, fruits, and medicinal products. On average, these households reported collecting five products (counting all fruits as one product). In contrast to Malnaad, households in the Tapajós rarely collect fuelwood from the forest but rather rely on dead wood from agricultural fields and fallows. Most households spend less than 10% of their time collecting NTFPs, and only a very few households indicated that forest products are a primary source of income. Nevertheless, NTFP collection is important to these households, as evidenced by community action and negotiation to defend *de facto* access rights to the Tapajós National Forest. To better understand the role of NTFPs in the livelihoods of these households, we focus on their time allocation, and, specifically, on their derived demand for forest collection trips.

We use the annual number of typical collection trips as an index of labor allocation to NTFP collection largely because it was relatively easy for households to recall. As is often the case with high-value, low-volume NTFP collection from tropical forests, the Tapajós households had difficulty remembering precise quantities and time allocated to collection of specific products. To place our measure of collection effort in the context of the household production model described earlier, consider figure 15.1 relabeled with number of trips on the horizontal axis and cost or return to trips on the vertical axis. The demand (WTP) for trips derives from the household demand for forest products, as both subsistence and commercial goods, while the supply (MC) depends on the effort required per trip and the opportunity cost of household time. Households do not hire others to take forest collection trips, probably due to the difficulties of monitoring effort and the particular human capital (knowledge of forest) required. Therefore, this is another case of household production with missing markets.

3.2.1 Count Data Model

The number of NTFP collection trips in the survey year is a non-negative integer variable, best modeled using the count data approach described in chapter 19. In this case, 31% percent of households report zero trips. Among those who report positive trips, the mean is 8.6, and the standard deviation is 12.4. One explanation for this distribution is that (a) some households are not collectors and make zero trips, and (b) among the collecting households, a few also make zero trips in the survey year, many make a small number of trips, and a few make a large number of trips. Pattanayak and Sills (2001) conclude that the best fit to these data is a zero-inflated-tau negative binomial model. The negative binomial accounts for the overdispersed count of trips (variance greater than mean). The probability of being a collector is modeled as a multiplicative function of the variables explaining the count, with tau as the single additional parameter (Cameron and Trivedi 1998).

With the count of trips as the dependent variable, the independent variables are drawn from the five categories listed in section 2.4.2. As in the Malnaad, household wealth or assets are key explanatory variables, affecting consumption preferences, ability to sell products, and the opportunity cost of collection. To represent wealth in this case, we create linear combinations of different types of household assets, using principal component analysis (see chapter 14).[8] The model includes the first principal component of each set of variables, which are all correlated in the same direction with the first component because they are selected to represent a like set of assets.

Finally, we return to the observation that even though NTFPs comprise only a small part of income and labor effort, the households claim they are important and have actively sought to protect access to them. Pattanayak and Sills (2001) suggest that this is because NTFP collection serves as a form of natural insurance, providing a backstop source of income to households who have access to the forest and know how, where, and when to find NTFPs. This suggests that forest collection trips provide valuable on-the-job education about the spatial and temporal distribution of NTFPs ($\partial f_2/\partial T_{F1} > 0$) and that households facing greater risks should take more trips. We test this hypothesis by relating trips to two variables representing risk and shortfall at the community level: variability in production of the main agricultural crop (manioc) and percent of households who reported a worse than usual harvest. For details on how these variables are measured and incorporated into the household production framework, see Pattanayak and Sills (2001).

3.2.2 Results

As in the Malnaad case, the estimation results reported in table 15.3 show that household demographics affect NTFP collection patterns: middle-age households with more men make more forest collection trips. Informal education about the local forest, represented by percent of life spent in the local area, is positively related to number of collection trips. The results support our hypothesis that NTFPs serve as natural insurance, since agricultural risk and shortfalls are both positively correlated with collection trips. Contrary to expectations, neither the number of children living outside the forest (potential source of remittance income) nor distance to forest (resource access) is statistically related to collection trips. The former may indicate that the backstop possibilities provided by the immediately available forest and by children who live far from home are not good substitutes. The latter may indicate simply that we have not captured the relevant travel cost variable, which depends on the specific collection site preferred by households and is therefore difficult to measure.

Turning to the principal components of household assets, we see that NTFP collection is positively associated with several measures of wealth, contrary to fuelwood collection from open access forest in the Malnaad and to much of the literature reviewed in section 1.1. For example, trips are positively related to garden production assets (poultry, orchards, gardens) and to domestic assets (clocks, radios, sewing machines). The coefficient on agricultural wealth also has a positive sign, although it is not significant at the 10% level.

Table15.3. Household demand for NTFP collection trips in Tapajós, Brazil

Variable	Definition	Coeff.	p-value	Mean (st. dev.)
Intercept		-3.95	0.000	
Age	Age of head of household in decades	0.74	0.028	4.55 (1.61)
Age²	Square of age	-0.09	0.006	na
Men	Number of males	0.12	0.101	1.90 (1.34)
Local	Percent of life spent in current village	0.75	0.001	0.72 (0.37)
Children out	Number of children living outside national forest	-0.04	0.282	1.62 (2.33)
Distance	Walking distance to forest	0.02	0.818	1.03 (0.91)
Risk	Coefficient of variation of manioc	14.01	0.000	0.17 (0.02)
Shortfall	Dummy = 1 if poor crop	1.71	0.023	0.17 (0.10)
Principal Components				
Agric. PC	PC of agricultural assets	0.11	0.138	0.22 (1.00)
Livestock PC	PC of ranching assets	-0.29	0.017	0.04 (1.30)
Fishing PC	PC of fishing assets	0.02	0.893	0.97 (0.60)
Garden PC	PC of garden assets	0.29	0.005	1.31 (0.78)
Home PC	PC of domestic assets	0.46	0.006	1.04 (0.53)
Alpha		1.54	0.000	
Tau		-1.55	0.135	
Sample size = 308	Log-likelihood = -803.3		Vuong Statistic = 2.94	

This positive correlation with wealth may reflect the fact that these NTFPs are not necessities for day-to-day survival, like fuelwood in the Malnaad, but rather add some variety to consumption possibilities. On the other hand, collection trips are negatively associated with ranching assets, where head of cattle owned has the greatest weight in that principal component. Two possible explanations for the different signs on ranching and other types of wealth are that cattle may represent an alternative way to mitigate risk (i.e., an alternative form of insurance) and that investment in ranching may reflect a more modern orientation and choice of a wealth accumulation pathway that reduces reliance on NTFPs. These different relationships between asset categories and forest collection would be obscured by either a single principal component or cluster variables based on all assets. Thus, the use of principal components to represent the diversity of household wealth provides a better understanding of how forest dependence varies across household.

4. SUMMARY

Dependence on NTFPs varies across households, even within relatively small geographic areas that are often perceived as homogeneous by policy makers. Understanding this heterogeneity is key for projects that seek to reconcile conservation and development on the forest margin. For example, we find significant effects of household wealth in both Malnaad and the Tapajós, whether represented as socioeconomic classes or asset categories. The direction of these effects, however, differs across forest type (in Malnaad) and asset categories (in the Tapajós). Livestock, which is often used as a proxy for wealth, may play a much more complex role, as a complement to open access fuelwood collection in the Malnaad and as a culturally distinct alternative risk-mitigation strategy in the Tapajós. Rather than seeking general principles of NTFP use, such as "poor households are more forest-dependent," researchers should build models that account for the particular socioeconomic and environmental context, as well as the type of NTFP. Fuelwood in Malnaad is a good example of a relatively high-volume, low-value NTFP, whose collection depends on labor availability and demands for domestic uses and agricultural processing. NTFPs in the Tapajós are collected much less frequently, in smaller volumes, and we find that determinants are related more to the abilities of the household (age and local knowledge) and to risk and shortfalls in the primary agricultural activity.

Since markets are incomplete in both Malnaad and Tapajós, a wide range of household attributes, rather than an exogenous market price, determine household supply and demand behavior. The household production framework provides a structure for specifying and interpreting models in this context. It also helps identify clues to households' dynamic behavior from typical cross-sectional data. For example, we find that the determinants of collection from private and public access forests in Malnaad differ. In particular, ownership of a fuel substitute substantially reduces collection from private access forest, consistent with the premise that private resources are treated with greater care. In the Tapajós, we find that households facing greater agricultural risks take more forest collection trips, possibly because of a desire to maintain NTFP collection as a fallback option. By granting local households access to public forests, the government could facilitate this natural insurance. Thus, insight into the highly heterogeneous role of NTFPs in rural household economies around the world can be obtained with microeconometric modeling in the household production framework.

5. LITERATURE CITED

AMACHER, G.S., W.F. HYDE, AND K.R. KANEL. 1999. Nepali fuelwood production and consumption: Regional and household distinctions, substitution and successful intervention. J. Dev. Stud. 35(4):138-163.

AMACHER, G.S., W.F. HYDE, AND K.R. KANEL. 1996. Forest policy when some households collect and other purchase fulewood. J. For. Econ. 2(3): 273-288.

BAHUGUNA, V.K. 2000. Forests in the economy of the rural poor: an estimation of the dependency level. Ambio 29(3):126 – 129.

BARDHAN, P., J. BALAND, S. DAS, D. MOOKHERJEE, AND R. SARKAR. 2001. Household firewood collection in rural Nepal. Working Paper, Network on the Effects of Inequality on Development, University of California, Berkeley, CA.

BEHRMAN, J.R. 1999. Labor markets in developing countries, P. 2859-2939 *in* Handbook of Labor Economics, Ashenfelter, O. and D. Card (eds.). Elsevier Science, New York.

BLUFFSTONE, R.A. 1995. The effect of labor market performance on deforestation in developing countries under open access: An example from rural Nepal. J. Env. Econ. Manage. 29:42-63.

BYRON, N. AND M. ARNOLD. 1999. What futures for the people of the tropical forests? World Dev. 27(5):789-805.

CAMERON, C. AND P. TRIVEDI. 1998. Regression Analysis of Count Data. Cambridge University Press, New York. 411 p.

CARTER, M.R., AND Y. YAO. 2002. Local versus global separability in agricultural household models: The factor price equalization effect of land transfer rights. Am. J. Agr. Econ. 84(3):702-715.

CAVENDISH, W. 2000. empirical regularities in the poverty environment relationship of rural households: Evidence from Zimbabwe. World Dev. 28(11):1979-2003.

COOKE, P. A. 1998. Intrahousehold labor allocation responses to environmental good scarcity. Econ. Dev. Cult. Change 46(4):807-830.

EDMUNDS, E. 2002. Government initiated community resource management and resource extraction from Nepal's forests. J. Dev. Econ. 68(1):89-115.

FAO. 2002. Non-Wood Forest Products. Food and Agriculture Organization of the United Nations. Available at: http://www.fao.org/forestry/FOP/FOPW/NWFP/new/nwfp.htm.

FERRARO, P.J., AND R.A. KRAMER. 1997. Compensation and economic incentives, P.187-211 *in* Last Stand: Protected Areas and the Defense of Tropical Biodiversity, Kramer, R., C. van Schaik, and J. Johnson (eds.). Oxford University Press, New York.

GODOY, R.A. 2001. Indians, Markets, and Rainforests: Theory, methods, analysis. Columbia University Press, New York. 256 p.

GODOY, R.A., AND K.S. BAWA. 1993. The economic value and sustainable harvest of plants and animals from the tropical forest: Assumptions, hypotheses, and methods. Econ. Bot. 47(3):215-219.

GODOY, R., D. WILKIE, H. OVERMAN, A. CUBAS, G. CUBAS, J. DEMMER, K. MCSWEENY, AND N. BROKAW. 2000. Valuation of consumption and sale of forest goods from a Central American rain forest. Nature 406:62-63.

GODOY, R., N. BROKAW, AND D. WILKIE. 1995. The effect of income on the extraction of non-timber tropical forest products. Human Ecology 23:29-52.

GUNATILAKE, H.M. 1998. The role of rural development in protecting tropical rainforests: evidence from Sri Lanka. J. Env. Manage. 53:273-292.

HAND, D.J. 1981. Discrimination and Classification. John Wiley, New York. 218 p.

HEGDE, R., AND T. ENTERS. 2000. Forest products and household economy: a case study from Mudumalai Sanctuary, Southern India. Env. Conserv. 27(3):-250-259.

HELTBERG, R., T.C. ARNDT, AND N.U. SEKHAR. 2000. Fuelwood consumption and forest degradation. Land Econ. 76(2):213-232.

HYDE, W.F., AND G.S. AMACHER. 2000. Economics of Forestry and Rural Development: an Empirical Introduction from Asia. University of Michigan Press, Ann Arbor. 287 p.

HYDE, W.F., AND G.S. AMACHER. 1996. Applications of environmental accounting and the new household economics: new technical economic issues with a common theme in forestry. For. Ecol. Manage. 83:137-148.

JACOBY, H. G. 1993. shadow wages and peasant family labor supply: an econometric application to the Peruvian Sierra. Rev. Econ. Stud. 60:903-921.

JOHNSON, R.A., AND D.W. WICHERN. 1998. Applied Multivariate Statistical Analysis. Prentice Hall, New Jersey. 816 p.

JONES, E.T, R.J. MCLAIN, AND J. WEIGAND. 2002. Nontimber Forest Products in the United States. University Press of Kansas, Lawrence, KS. 445 p.

JOSHEE, B.R., G.S. AMACHER, AND W.F. HYDE. 2000. household fuel production and consumption, substitution, and innovation in two districts of Nepal. P. 57-86 in Economics of Forestry and Rural Development: an Empirical Introduction from Asia, Hyde, W.F., and G.S. Amacher (eds.). University of Michigan Press, Ann Arbor.

KÖHLIN, G., AND P.J. PARKS. 2001. spatial variability and disincentives to harvest: deforestation and fuelwood collection in South Asia. Land Econ. 77(2):206-218.

KVIST, L.P., S. GRAM, A. CÁCARES, AND I. ORE. 2001. socio-economy of flood plain households in the Peruvian Amazon. For. Ecol. Manage. 150:175-186.

LELE, S. 1993. degradation, sustainability, or transformation? a case study of villagers' use of forest lands in the Malnaad Region of Uttara Kannada District, India. Ph.D. Dissertation. Department of Energy and Resource Economics, University of California, Berkeley.

MACDONALD, D.H., W.L. ADAMOWICZ, AND M.K. LUCKERT. 2001. Fuelwood collection in Northeastern Zimbabwe. J. For. Econ. 7(1):29-51.

MEKONNEN, A. 1999. Rural Household Biomass Fuel Production and Consumption in Ethiopia: A Case Study. J. For. Econ. 5(1):69-97.

NEUMANN, R.P., AND E. HIRSCH. 2000. Commercialisation of non-timber forest products: review and analysis of research. CIFOR, Indonesia, and FAO, Italy. 176 p.

NEPSTAD, D., AND S. SCHWARTZMAN (EDS.). 1992. Non-timber products from tropical forests: evaluation of a conservation and development strategy. Advances in Economic Botany 9. New York Botanical Garden, New York. 164 p.

PATTANAYAK, S.K., AND E.O. SILLS. 2001. Do tropical forests provide natural insurance? the microeconomics of non-timber forest product collection in the Brazilian Amazon. Land Econ. 77(4):595-612.

PATTANAYAK, S.K., E. SILLS, AND R. KRAMER. Forthcoming. seeing the forest for the fuel. Env. Dev. Econ.

PÉREZ, M.R., AND J.E.M. ARNOLD (EDS.). 1995. Current issues in non-timber forest products research. CIFOR, Indonesia. 264 p.

PÉREZ M.R., AND N. BYRON. 1999. A methodology to analyze divergent case studies of non-timber forest products and their development potential. For. Sci. 45(1):1-14.

PETERS, C., A. GENTRY, AND R. MENDELSOHN. 1989. Valuation of an Amazonian rainforest. Nature 339:655-656.

PLOTKIN, M., AND L. FAMOLARE (EDS.). 1992. Sustainable harvest and marketing of rain forest products. Island Press, Washington DC. 325 p.

REDDY, S., AND S. CHAKRAVARTY. 1999. Forest dependence and income distribution in a subsistence economy: evidence from India. World Dev. 27 (7):1141-1149.

SADOULET, E., AND A. DE JANVRY. 1995. Quantitative Development Policy Analysis. Johns Hopkins University Press, Baltimore. 397 p.

SILLS, E., S. PATTANAYAK, AND T. HOLMES. 2000. Living on the edge: collecting tropical forest products to mitigate risk in the Brazilian Amazon. Presented at the Western Economics Association Meetings in Vancouver, June.

SINGH, I., L. SQUIRE, AND J. STRAUSS (EDS). 1986. Agricultural Household Models. Johns Hopkins University Press, Baltimore. 335 p.

SHYAMSUNDAR, P., AND R.A. KRAMER. 1996. Tropical Forest Protection: An Empirical Analysis of the Costs Borne by Local People. J. Env. Econ. Manage. 31:129-144.

SOUTHGATE, D.D. 1998. Tropical forest conservation: an economic assessment of the alternatives in Latin America. Oxford University Press, New York. 175 p.

TAKASAKI, Y., B. BARHAM, AND O. COOMES. 2000. Rapid rural appraisal in humid tropical forests. World Dev. 28(11):1961-1977.

TEWARI, D.D. 2000. Valuation of non-timber forest products (NTFPs): models, problems, and issues. J. Sust. For. 11(4):47-68.

THANDANI, R. 2001. International non-timber forest product issues, P. 5-24 in Non-timber Forest Products: Medicinal Herbs, Fungi, Edible Fruits and Nuts, and Other Natural Products from the Forest, Emery, M., and R. McLain (eds). Haworth Press, New York.

WICKRAMASINGHE, A., M. PÉREZ, AND J.M. BLOCKHUS. 1996. nontimber forest product gathering in Ritigala Forest (Sri Lanka): household strategies and community differentiation. Human Ecology 24(4):493-519.

WOLLENBERG, E. 2000. Methods for Estimating Forest Income and their Challenges. Society Nat. Res. 13:777-795.

WUNDER, S. 2001. Poverty alleviation and tropical forests – what scope for synergies? World Dev. 29(11):1817-1833.

---

[1] Households could also augment the forest stock by planting trees on their land. For simplicity, we do not consider household decisions about tree planting in this model. See chapter 16 for further discussion.

[2] The relationship between productivity and labor allocation differs across studies. The Nepalese households modeled by Amacher et al. (1996) respond to higher productivity (interpreted as a higher shadow wage) by supplying more labor, and likewise households in Indonesia and Zimbabwe modeled by Pattanayak et al. (forthcoming) and MacDonald et al. (2001) respond to higher productivity (interpreted as lower cost of collection trips) by taking more trips. On the other hand, the households in Nepal and India modeled by Cooke (1998), Bardhan et al. (2001), Köhlin and Parks (2001), and Heltberg et al. (2000) respond to higher productivity (interpreted as a lower price for fuelwood or higher shadow wage) by supplying less labor. This could be due to differences in data and estimation procedures, different substitution possibilities or market conditions across regions, or a backward-bending aggregate fuelwood labor supply curve.

[3] Only 15% of Malnaad households include someone who worked for wages outside of the home (half of those outside the village), and only 33% of Tapajós households include someone who participated in wage labor for at least one day in the survey year.

[4] The variables are acres of areca, rice paddy, private access grassland, private access forest, coconut, and sugarcane; caste; number of household members with less than fifth-grade education and with high school degree; number of household members who hold off-farm jobs (in and outside their home village); and number of cattle and buffalo owned.

[5] We report probability values for coefficients, allowing readers to apply their own preferred level of significance. We consider the 10% level to indicate statistical significance, while the 15% level suggests the possibility of a statistical relationship.

[6] While it is insignificant in the specification reported in table 15.2, the class variable combining the poor and middle clusters is negative and significant when the component variables are not included. There may not be enough variation in the private forest sample to separate the effect of class from the effects of the component variables in the cluster analysis. A third possible specification would include interaction terms between cluster variables and other independent variables, but specification testing indicated that most of these interaction terms are insignificant.

[7] Another possibility is that better-off households have fewer but higher quality cattle.

[8] These include 25 categories of assets, some measured simply as dummy variables (dummy = 1 if household owns asset), and others as quantities (e.g., head of cattle). To ensure that the quantity variables do not dominate, we first standardize these variables by subtracting the mean and dividing by the standard deviation.

Chapter 16

Agroforestry Adoption By Smallholders

D. Evan Mercer and Subhrendu K. Pattanayak
USDA Forest Service and Research Triangle Institute

Agroforestry is a joint forest production system whereby land, labor, and capital inputs are combined to produce trees and agricultural crops (and/or livestock) on the same unit of land. Although existing for centuries (maybe millennia) as an array of traditional land use practices in the tropics, agroforestry emerged in the late 1970s as a modern, improved tropical land use system suitable for scientific study, replete with its own international research center, the International Center for Research in Agroforestry (ICRAF) and journal, Agroforestry Systems. During the 1990s, interest in agroforestry in temperate regions increased rapidly when the scientific community discovered the complex land management systems developed by rural landowners in North America and Europe, including forest farming, alley cropping, shelterbelts, riparian buffers, and silvopastural systems (Lassoie and Buck 1999).

Despite some impressive scientific and technological advances, agroforestry rural development efforts in the 1980s and 1990s were frequently unsuccessful (Nair 1996). Although agroforestry projects failed for a number of different reasons, one common factor was the inadequate attention given to socioeconomics in the development of systems and projects (Current et al. 1995). Beginning in the mid 1990s, agroforestry leaders argued for increased emphasis on research to understand the agroforestry adoption decision process (Mercer and Miller 1998, Sanchez 1995). For example, former Director General of ICRAF, Pedro Sanchez (1995: 24), stated that "the need to develop predictive understanding of how farm households make decisions regarding land use is as essential as developing a predictive understanding of the competition between tree and crop roots."

As a result, agroforestry adoption studies have proliferated recently (Pattanayak et al. [forthcoming]). Most of these studies use dichotomous choice (logit or probit) regression models to explain how various characteristics of farmers, farms, and development projects influence the adoption decision. Unfortunately, many of these recent studies fail to link their empirical analyses to underlying theory. Rather, they often just report a number of factors that are correlated with adoption of specific technologies in specific locations, which does little to promote a general predictive understanding of the farm household decision-making process.

In this chapter, we develop a model of the adoption decision process using microeconomic theory and illustrate its econometric application with two case studies in the Philippines and Mexico. These case studies examine different types of agroforestry systems in sites that contrast ecologically, socially, and culturally.

1. REVIEW OF EMPIRICAL STUDIES ON AGROFORESTRY ADOPTION[1]

A large and growing literature addresses the adoption of agricultural technologies and technological change as engines for economic development. Examples include the seminal survey by Feder et al. (1985) and a recent study of sustainable agricultural intensification by Clay et al. (1998). More recently, the study of agroforestry adoption has intensified as governments, donor agencies, and scientists search for technologies that will be adopted by farmers to generate economic growth while protecting ecological capital. Consequently, we draw on the general technology adoption literature to identify clusters of factors that empirically explain adoption behavior and compare these to recent empirical analyses of agroforestry adoption.

The broader literature reveals five categories of determinants of technology adoption: economic incentives, biophysical conditions, risk and uncertainty, household preferences, and resource endowments. These are not mutually exclusive because of correlation and complementarity between factors within categories and because different empirical applications often use the same variables to proxy different factors. In addition, researchers have developed conflicting conceptual and empirical arguments regarding the way in which several variables (e.g., plot size and tenure) influence adoption. For each of the five general categories, we discuss the expected direction of influence, based on our review and summary of 26 empirical analyses of agroforestry (for details see Pattanayak et al. [forthcoming]). We then use this information to develop a stylized economic model of farmer adoption in section 2.

1. Market incentives (I) include factors that explicitly lower costs and/or produce higher benefits from technology adoption and as such are the standard economic determinants of adoption. The empirical literature suggests that adoption is positively influenced by variables such as expected yield increases and share of income from farming. Unfortunately, explicit market data such as prices are lacking in most analyses. Often market data are absent because the studies focus on subsistence economies where markets are thin and proxies for price are usually not available, or the studies are so limited geographically and/or temporally that there is little variability of the available market data among respondents.

2. Biophysical conditions (Z) such as soil quality, steepness of farmland, and plot size influence the physical production process. Since these conditions directly impact production costs and returns, they are implicitly economic determinants of adoption. Our review shows that adoption is more likely on farms with steep plots. As in much of the broader literature, the correlation with plot size is ambiguous, perhaps because of the confounding influence of scale economies and resource constraints. Based on the few studies that have included soil quality variables, we see a similar ambiguity. Again, this might reflect the desire to protect good-quality land being confounded by complacency because of sufficient soil resources.

3. Risk and uncertainty (R) relates to the market and institutional environment under which decisions are made. Short-term risk (e.g., fluctuations in commodity prices and rainfall) and long-term risk (e.g., tenure insecurity) influence the adoption decision and process. Our review shows an unambiguous and consistent result for the tenure variable: landowners are more likely than tenants to adopt agroforestry and other conservation technologies. We also found that previous experiences and familiarity with agroforestry/conservation investment projects, possibly because of information disseminated through extension services or community group membership, were positively correlated with adoption.

4. Household preferences (H) are a placeholder for the broad category of household-specific influences such as risk tolerance, intrahousehold homogeneity, and conservation attitude. Since preferences are difficult to measure explicitly, they are usually proxied with sociodemographic variables such as age, gender, and education. Our assessment of the literature shows adoption is more likely in a household with a higher education level and greater proportion of males. The male effect could

reflect the endowment effect discussed next. By and large, age is an insignificant explainer of adoption.

5. Resource endowment (L) measures the decision makers' abilities to employ resources necessary for implementing the technology. Asset holdings and wealth measures such as land, labor, livestock, and savings are examples of resource endowments. Our review shows the consistent and unambiguous positive influence of wealth on agroforestry adoption and conservation investments.

2. THEORY OF FARMER ADOPTION

Using household production theory as a conceptual framework (Amacher et al. 1993, Pender and Kerr 1998) and the five broad determinants of adoption discussed above, we develop a model of agroforestry adoption as an investment choice. Consider a representative farm household that maximizes its utility, U, which is assumed to be a concave, continuous, twice-differentiable function of agricultural commodities, Q_C, (e.g., rice/corn) and household time inputs, Y_C (e.g., leisure). The function is conditioned by household preferences that are proxied by sociodemographics, H. Utility maximization is subject to three constraints (time input endowment, technology, and cash income). The household time input constraint implies that the sum of own input supply of time, Y_P (labor), and own input consumption of time, Y_C (leisure), cannot exceed the household time endowment, Y_E, which is conditioned by household characteristics, H.

Agricultural outputs, Q_P, are assumed to be a convex, continuous production function, F, of Y_P. Productivity depends on household resource endowments, L, such as land, tools, money, human capital, and economic incentives provided by the government, such as subsidies. The biophysical characteristics of the farm, Z, also mediate the production technology. A typical cash constraint requires household expenditures on agricultural commodities and inputs to be less than or equal to the sum of agricultural profits, π, which depend on market prices, P_Y, and exogenous income, E. The household's budget constraint combines a typical cash income constraint with the endowment constraint such that expenditures are equal to the sum of the monetary equivalent of the household input endowment, agricultural profits, and exogenous income; this sum is the "Beckerian" full income (Strauss 1986).

Following Amacher et al. (1993), adoption of agroforestry requires joint investments of money, labor, and land to acquire agroforestry capital. That is, labor and money are collectively embodied in the amount of land dedicated to agroforestry. As described above, this joint investment is conditioned by the resource endowments and biophysical conditions faced by the household. Agroforestry (L_{AF}) can therefore be conceived as one among many sets of coordinated investments that produce an annual rate of return, r, to enhance overall well-being. Since the returns to agroforestry occur in the future, households consider the expected stream of income net of consumption (I) or the market-based incentives, in choosing between alternate investments. These expectations are based on the household's assessment of the relative importance of agroforestry income to total farm income, which depends on risks and uncertainty, R, in the short and long terms. Mathematically, the household's utility-maximization problem is expressed with the Lagrangian in equation 16.1.

$$Max\ E\ \{[U(Q_C, Y_C; H) + \lambda(\pi + E - P_Y Y_P - rL_{AF}) + \mu(Q_P, Y_P; L_{AF}, Z) + \eta(Y_E - Y_C - Y_P)], R\} \quad 16.1$$

The objective is to maximize expected utility by choosing levels of inputs (including land) and outputs. The first-order conditions with respect to Q_C, Y_C, Q_P, Y_P, and L_{AF} have the standard Marshallian equimarginal interpretations when households choose the level of agroforestry technology that maximizes total utility.

Consider the choice facing household i when deciding whether to adopt agroforestry. The utility-maximizing household compares its expected net utility with and without adoption (EU_i^*). A reduced form version of this net utility is given by equation 16.2:

$$EU_i^* = \alpha_I I_i + \alpha_L L_i + \alpha_R R_i + \alpha_Z Z_i + \alpha_H H_i + \varepsilon_i \quad 16.2$$

where I_i, L_i, R_i, Z_i, and H_i are as defined above. Note that I_i captures market incentives because net income is a function of explicit and implicit prices of outputs and inputs of the agroforestry process. Since the true net utility function is unknown, we treat the estimated function as random by including the error term ε_i.[2] Although EU_i^* is not directly observable, the researcher can observe the owner-manager's adoption decision. Let L_{AFi} be an indicator of whether the household i adopts agroforestry ($L_{AFi} = 1$) or not ($L_{AFi} = 0$), so that

$$L_{AFi}^* = 0\ if\ EU_i^* \leq 0 \quad and \quad L_{AFi}^* = 1\ if\ EU_i^* > 0 \quad 16.3$$

Depending on the assumptions regarding the distribution of the error term in equation 16.2, this structural relationship can be estimated using a variety of methods. In most analyses of binary choice data, probit or logit models are estimated assuming either a normal or logistic distribution, respectively, for the error term (Maddala 1983). That is,

$$\text{Prob}(L^*_{AFi} = 1) = \Phi(\alpha_I I + \alpha_L L + \alpha_R R + \alpha_Z Z + \alpha_H H) \qquad 16.4$$

where $\Phi(.)$ is the cumulative distribution function and I, L, R, Z, and H are the explanatory variables in equation 16.2 and α is a vector of parameters to be estimated. Although one might expect different predictions from the logit and probit models for samples with very few positive responses for the dependent variable ($y = 1$), or very few nonresponses ($y = 0$) and very wide variation in important independent variables, usually the two models produce similar results. In fact, little theoretical justification exists for choosing between the probit and logit models (Greene 1997). To investigate the determinants of agroforestry adoption, our two case studies empirically estimate equation 16.4 with binary adoption data from the Philippines and Mexico with probit (Philippines) and logit (Mexico) regression models.

3. EMPIRICAL ANALYSES OF AGROFORESTRY ADOPTION

We present two case studies of adoption of tree planting by small farmers in the Philippines and Mexico as examples of using empirical analysis to test the predictions of the theory of adoption. These case studies provide ecological, social, and cultural contrasts as well as contrast in the types of agroforestry systems being promoted. For example, land is a major constraint for Filipino farmers, with an average farm size of 2.63 ha on steeply sloped land, resulting in high erosion rates under traditional agricultural systems. In Mexico, however, the absolute amount of land is not a constraint because the average farm size is 48 ha on relatively flat slopes. Inadequate and highly variable rainfall and very thin, poor soils are major ecological constraints to corn-based agricultural production in the Mexico site. Labor, seeds, seedlings, and fertilizers constrain production in both Mexico and the Philippines. Thus, the main objective for the agroforestry projects in the Philippines was erosion control to facilitate long-term agricultural production from small, steeply sloped plots, and in Mexico, the primary objectives were to develop systems to reduce farm production risk, improve total farm productivity, and reduce pressures on natural forests.

The Philippines case study examines the factors influencing farmers' decisions to adopt contour hedgerow systems, a form of alley cropping, to reduce the negative impacts of soil erosion. Contour hedgerows are a set of agroforestry practices in which food crops are planted between hedges of woody perennials established along the contours of sloping upland farm plots. Prunings from the hedgerow trees or shrubs are placed at the up-slope base of the hedges to trap the eroding soil so that over time natural terraces are formed. The hypothesized benefits of contour hedgerows are erosion control, enhanced soil nutrient availability, weed suppression, and enhanced fuel and fodder production. However, the hedgerows may also produce increased demand for scarce labor and skills, loss of annual cropping area, difficulty in mechanizing agricultural operations, and excessive competition with the crops for soil nutrients, light, and water (Nair 1993).

In the Mexico case study, low and sporadic corn production with traditional *milpa* (slash and burn) systems on thin soils and inadequate rainfall led to the search for tree-based systems to improve long-term productivity and reduce the risk of catastrophic agricultural failure. Hence, this study examines factors influencing farmers' decisions to plant trees in a variety of different systems. Projects offered timber trees and/or fruit trees to farmers who agreed to plant the trees in association with agricultural crops in 1 hectare agroforestry plots (Snook and Zapata 1998). The projects' objectives were to provide short-, medium-, and long-term production, starting with annual crops, followed by fruits and finally timber.

3.1 The Philippines: Leyte in the Eastern Visayas

Two villages, Visares and Cagnocot, on the island of Leyte, Eastern Visayas, were the sites for the Philippines case study. Both sites are hilly and subject to significant erosion. Visares has a pronounced maximum rainy period in December but no dry season, while Cagnocot receives even rainfall throughout the year except for the dry months of February to April. The soils are acidic, varying from sandy loam to extremely clayey. Ipil-ipil (*Leucaena leucocephala*) and kakawate (*Gliricidia sepium*) are the two primary tree species used as hedgerows. Both communities engage in fishing, carpentry, and other nonfarm activities, and Visares has a rudimentary rattan furniture industry introduced by a U.S. Agency for International Development project. The Philippines data were collected in 1993 and 1994 through a socioeconomic survey of 277 agricultural households. Two questionnaires, one on socioeconomic and agronomic characteristics and the other on farm budgets, were administered to each household. Descriptive statistics for the Philippines case study are in table 16.1. The average farm covered 2.63 ha sitting on a 28% slope and produced an annual income of 10,500 pesos (US$402), accounting for 58% of total household income. The average education across household members was 2 years. The head of the average farm household had lived in the village for 33 years. While 66% of respondents had planted trees on their farms, only 31% had previously constructed contour hedgerows. Sixty two percent of the respondents reported owning the plots they farmed.

Table 16.1. Descriptive statistics for Philippines case study (n = 277)

Variable Description	Mean	Standard Deviation
Average education (years)	1.90	1.17
Number of labor days in farming	83.30	88.2
Annual agricultural profits (US$)	402.00	617
Percent of income from farm agriculture	57.70	34.0
Farm size (ha)	2.63	3.47
Made contour hedgerows on your farm? (yes = 1, no = 0)	0.31	0.47
Ever planted trees on farm? (yes = 1, no = 0)	0.66	0.48
Heard of contour hedgerow farming? (yes = 1, no = 0)	0.77	0.42
Received training in contour hedgerows (yes = 1, no = 0)	0.19	0.39
Extent of assistance from project official [a]	0.10	0.20
Steepness of farmland (degree)	28.6	15.7
Length of residency in the village (years)	33.0	15.5
Distance from home to fields (minutes)	18.5	23.5
Tenant? (yes = 1, no = 0)	0.38	0.43
Member of farmer or community development group (yes = 1, no = 0)	0.50	0.66

[a] The extent of assistance is measured as the normalized sum of dummy variables where each dummy measures the receipt of one of four types of assistance (cash, technical information, labor, and seeds) from project staff.

3.2 Mexico: Calakmul Biosphere Reserve, Campeche

This case study site is located in the buffer zone of the 723,000-ha (1.7 million-acre) Calakmul Biosphere Reserve in southeastern Campeche, Mexico. Contiguous with the Maya Biosphere Reserve in Guatemala, the Calakmul Biosphere Reserve was created in 1989 to protect the last great frontier in Mexico to which Mexicans continue to migrate in search of land for farming. The Calakmul region comprises a municipality (Calakmul), the core bioreserve area where settlement is prohibited, a buffer zone of 72 communities (15,000 inhabitants), and a few privately owned properties (Bosque Modelo 1997). In the communities, called *ejidos,* each member family has equal rights to the use of communal forest and agricultural lands. *Ejidos* vary in size from 100 to 50,000 ha and from 10 to 150 members. The allotment for each family's agricultural use varies from 25 to 50 ha. Communal forest areas vary from 250 to 25,000 ha per *ejido.*

Data were collected in winter of 1998 via in-person interviews of a stratified random sample (by *ejido*) of farmers in the buffer zone of the Calakmul Biosphere Reserve. The final sample consisted of 176 farmers in 15 separate *ejidos*. Casey et al. (1999) provide details on field logistics and data gathering. Descriptive statistics for the variables used for the analysis of the Mexican case study are in table 16.2. The average farmer is 38 years old with 4 children and an annual income of US$1,510. The education level of the farmers is very low; 60% had not finished primary school, only 29% had finished primary school, and only 11% had finished secondary school. Ninety-four percent of respondents immigrated to Calakmul from outside the state of Campeche, with the average farmer having immigrated to Calakmul 11 years ago. The average farmer received 48.2 ha of land on joining the *ejido*, 39.7 ha of which was originally under primary forest cover and 8 ha under secondary fallow. Farmers had harvested an average 9.9 ha of forests with an average 28 ha currently under forest cover, 19 ha under fallow, and 4.8 ha in *milpa*. While 67% of respondents had established an average of 1.27 ha of nonagroforestry tree plantings since joining the *ejido,* only 31% (55) reported establishing agroforestry systems on an average plot size of 1.15 ha. Forty-seven percent of respondents had previous experience with an agricultural or forestry development project, and 79% reported an interest in participating in future agroforestry development projects.

Table 16.2. Descriptive statistics for Mexican case study (n = 176)

Variable description	Mean	Standard Deviation
Age of farmer (years)	38.31	13.76
Secondary education (1 = yes; 0 = no)	0.11	0.31
Total farm income (US$/year)	$1510	$1638
Timber income (US$/year)	$118	$486
Immigrant from outside Campeche (1 = yes; 0 = no)	0.94	0.23
Length of residency in Calakmul (years)	10.97	6.33
Distance to fields from house (km)	2.81	2.22
Farm size (ha)	48.16	25.25
Non-agroforestry tree plantings (ha)	1.27	2.54
Forestry experience (1 yes; 0 = no)	0.29	0.46
Previous project experience (1 = yes, 0 = no)	0.47	0.50
Interest in planting more trees (1 = yes; 0 = no)	0.79	0.40

3.3 Empirical findings

3.3.1 Leyte, Philippines: Building Contour Hedgerows

The results of the probit analysis for contour hedgerow adoption are presented in table 16.3. The dependent variable is the probability of being a contour hedgerows adopter (0 = nonadopter, and 1 = adopter); 86 respondents (31%) had constructed hedgerows (y = 1). The overall model fit the data well as indicated by the high χ^2, Veall-Zimmerman pseudo R^2 statistics, and the percentage of correct predictions (84%). The signs of statistically significant regressors have theoretical and intuitive appeal. Statistical significance of variables can be determined by studying the probability values (p-values) reported in column 3. The marginal effects (or probabilities) on adoption from a unit increase in independent variables (calculated at the means) are reported in column 4.

Table 16.3. A probit model of agroforestry adoption in Leyte, the Philippines (n = 277)

Variable	Coefficient	P-Value	Marginal Effect
Constant	-2.83	0.000	-0.721
Ever planted trees on farm? (yes = 1, no = 0)	1.87	0.000	0.365
Heard of contour hedgerow farming? (yes = 1, no = 0)	0.46	0.137	0.104
Percent of income from farm agriculture	0.53	0.143	0.134
Distance from home to fields (minutes)	-0.01	0.034	-0.003
Steepness of farmland (degree)	0.01	0.092	0.003
Length of residency in the village (years)	-0.02	0.031	-0.004
Tenant? (yes = 1, no = 0)	-0.67	0.029	-0.169
Labor in farming (days)	0.002	0.087	0.001
Average education of household head	-0.01	0.905	-0.003
Received training (yes = 1, no = 0)	1.72	0.000	0.569
Member of community group (yes = 1, no = 0)	0.28	0.106	0.072
χ^2 (11) statistic	159	0.000	
Veall-Zimmerman pseudo R^2	0.66		
% Correctly predicted	84		

Those households that have historically planted trees on their own farms and are familiar with agroforestry are more likely to adopt. Households which earn a greater percentage of their income from agriculture and which live closer to their agricultural fields are more likely to adopt. Agroecological needs also influence the adoption choice, and we find that households farming steeper lands are more likely to adopt. The length of residency indicates that households that have lived in the area for a long time are less likely to adopt. The coefficient on the tenant variable shows that tenants are less inclined to make long-term soil conservation investments. Labor endowments, proxied by the number of days in farming, are positively correlated with the adoption choice.

We do not find a significant relationship for education, possibly due to limited statistical variation in our sample. If a household received training in making hedgerows, it is very likely to adopt agroforestry. The variable indicating membership in community organizations is positively related with adoption. To the extent that community organizations provide information on new technologies and infrastructural support, membership in such groups and direct training should encourage adoption. Finally, as in many rural development projects, greater project assistance appears to have a substantial impact on the adoption of agroforestry technology. Unfortunately, project assistance is correlated with other variables, such as tenant and training, and causes interpretation problems due to multicollinearity.

We do not report the model with the assistance variable (see Pattanayak and Mercer 2002), but do find that it is positively correlated with adoption. In other results not reported here, prices of outputs (banana and corn) and inputs (labor and seed) were not statistically correlated with the adoption decision. This may be due to inadequate variability in prices. Looking across the column of marginal probabilities, we find that familiarity with the technology, either through previous personal experience with planting or through training by extension agents, has the highest impact on the probability of adoption. This suggests that efforts to minimize the uncertainty regarding new technologies may have significant payoffs.

3.3.2   Campeche, Mexico: Planting Timber and Fruit Trees

Table 16.4 presents the results of the maximum likelihood estimation of the logit regression model. The dependent variable, whether or not the farmer had established an agroforestry system, is regressed against the list of explanatory variables in table 16.2; 55 of the 176 respondents (31%) had established an agroforestry system (y = 1).

The χ^2 (12) statistic and pseudo R^2 suggest that the estimated model fits the data reasonably well; 74.43% of all responses were predicted correctly. Statistical significance of variables is identified by the p-value (probability value) reported in column 3. The effects of the independent variables on the logit or log odds of adopting agroforestry are reported as odds ratios in column 4.[3]

Six variables are significant at or below the 5% level (farm income, distance to fields, immigrant, previous project experience, nonagroforestry tree plantings, and interest in planting more trees). Age, education, and farm size are significant at the 6% to 11% level. Timber income is significant at the 18% level. The length of residency in Calakmul is not significantly different from zero. Signs for all variables are intuitively credible, with higher probabilities of adoption being positively correlated with age, education, income, length of residency, forestry experience, previous agricultural or forestry project experience, and interest in more tree planting. The greater the distance that farmers have to walk to their fields, the larger the income from timber harvesting, the larger the farm size, and the more hectares in tree plantations, the less likely that farmers will adopt agroforestry. Immigrants from outside Campeche also have a lower probability of adopting agroforestry than those immigrating from inside Campeche.

Table 16.4. Logit regression model of agroforestry adoption in Campeche, Mexico

Variable	Coefficient	P-value	Odds Ratio
Constant	-0.119	0.927	—
Age of farmer (years)	0.026	0.078	1.026
Education (1 = secondary; 0 = no secondary)	1.039	0.106	2.827
Immigrant (1 = yes; 0 = no)	-2.380	0.004	0.092
Length of residency (years)	0.0113	0.727	1.011
Total farm income	0.0004	0.029	1.000
Timber income	-0.0002	0.179	0.999
Size of farm (ha)	-0.018	0.073	0.982
Distance to fields (km)	-0.244	0.018	0.784
Non-agroforestry tree plantings (ha)	-0.161	0.046	0.851
Interest in planting more trees (1 = yes; 0 = no)	1.073	0.051	2.924
Forestry experience (1 = yes; 0 = no)	0.638	0.123	1.893
Previous project experience (1 = yes; 0 = no)	0.878	0.024	2.407
χ^2 (12) statistic	41.20	0.0000	
Pseudo R^2	0.189		
% Correctly predicted	74%		
N	176		

4. CASE STUDY FINDINGS IN CONTEXT

Next, we examine the results from the two case studies in the context of the categories of factors influencing adoption identified in the literature in section 1 and theory in section 2.

1. Market Incentives. In both the Philippines and Mexico studies, higher farm incomes are positively correlated with adoption. Likewise, in both case studies, increasing distance from home to fields results in lower probability of adoption, since increasing distance to fields increases the cost of adopting the new technology. Although only significant at a p-value of 18%, increasing income from timber harvests in Mexico is negatively correlated with agroforestry adoption. Unfortunately, in both case studies as in most agroforestry adoption studies, market data was either unavailable or insufficiently variable to be included in the analyses.

2. Biophysical Conditions. In the Philippines study, the steepness of farmland is significantly and positively correlated with adoption. In the Mexico case study, since there is very little variation in the biophysical conditions facing respondents, they are not included in the empirical modeling.

3. Risk and Uncertainty. Tenure is usually a very strong predictor of adoption, because lack of tenure suggests a risk of not being able to reap the long-term benefits from installing agroforestry systems. In the Philippines, respondents who did not own the land they were farming were significantly less likely to adopt agroforestry than owners. In Mexico, since all farmers operate under the same community (*ejido*) tenure system, tenure was not a variable. Several variables are associated with reducing uncertainty of new technologies. In the Philippines these variables include previous experience, knowledge, membership in community groups, training, and assistance, all of which significantly raised the probability of adoption. In Mexico as well, previous experience with rural development projects and with forestry are significant and positive as predicted.

4. Household Preferences. Different cultural, educational, and life experiences are expected to lead to different preferences for investing in new production methods, and we observe stark differences between the Philippines and Mexico in this category. In the Philippines, education is insignificant, but the length of residency in the village is significant and positive. In Mexico, however, education is significant at the 10% level and positively associated with adoption, while length of residency was insignificant. In addition, the age of Mexican farmers is significant and positively related to adoption. Immigrants from outside of Campeche are significantly less likely to adopt a new technology like agroforestry, perhaps reflecting increasing risk aversion for more distant immigrants. The impact of immigrant status on adoption suggests that the people of the Campeche share a knowledge base of the local soils, plants, and climate and generally adopt a modified version of the autochthonous natural resource management system common throughout the area.

5. Resource Endowments. In the Philippines, the proxy for labor endowment is positively and significantly correlated with adoption. Moreover, assistance, which included cash allowances, also increases the probability of adoption. In the Mexico study, two measures of land endowments—the total size of the farm and the size of nonagroforestry tree plantings—are both significant and

negatively related to likelihood of adoption. Farmers who control more land and who have planted more hectares in trees appear to perceive less need to adopt more intensive land use systems like agroforestry.

5. CONCLUSIONS

In this chapter, we develop a theoretical model of agroforestry adoption based on neoclassical economics, household production theory, and a review of the literature of empirical studies of adoption of agricultural and agroforestry innovations. The literature review reveals five general categories of factors shown to influence agroforestry adoption: market incentives, biophysical conditions, risk and uncertainty, household preferences, and resource endowment. We test these factors with two empirical case studies. In the Philippines study, we use a probit model to examine adoption of contour hedgerows on steep slopes to reduce the high erosion rates associated with traditional agricultural practices. In the Mexican case, we use a logit regression model to analyze adoption probabilities on flat land with very thin, poor soils and inadequate and sporadic rainfall. The results of the case studies confirm many of the predictions from the theory developed in section 2 and the literature review in section 1.

The literature review and our case studies also reveal several concerns with the recent studies on agroforestry adoption behavior. First, the majority of empirical studies examine agroforestry adoption as a snapshot in time, whereas agroforestry adoption is a dynamic process that occurs over a long time period as farmers experiment with agroforestry, incorporate it into their farming enterprise, or abandon it all together. Seldom will all households adopt any technology, even over the longest time period. Therefore, the adoption process might be modelled as a logistic or sigmoid function of time, with the rate of adoption and the final level of adoption as critical variables. The adoption studies to date (including the case studies in this chapter) only provide information on which farmers adopt agroforestry early on and can only show us which households to begin with when introducing new projects or programs. However, understanding the rate of adoption over time and the final expected level of adoption is necessary for assessing the potential benefits of new projects or programs. Unfortunately, we are unaware of any empirical research that rigorously analyzes the time dimension of adoption for agroforestry. This is likely due to the lack of data sets with adequate time series.

Second, the majority of adoption studies are limited in scope geographically as well as temporally. Like our case studies, most adoption studies are limited to a small number of sites within a limited geographic area. This is one of the factors limiting the inclusion of market data in many analyses, such as the case studies presented here. To really understand the adoption process, we need studies that compare adoption across a wide variety of communities that vary culturally, ecologically, and economically. Unfortunately, conducting surveys over extended geographic areas is costly and time consuming. An alternative approach is to use meta-analytic methods to analyze a large group of previous studies that as a group provide the needed variation (Cook et al. 1992).

Third, returning to a point made by Feder et al. (1985), if the technology is nondivisible, it takes on a dichotomous form at the individual level and a continuous form at the aggregate. For a divisible technology, the measure of adoption will likely be continuous and include examples such as the amount or share of farm area utilizing the technology. However, except for a study by Caviglia and Kahn (2001) that employs a Heckman selection model of agroforestry adoption, most studies typically estimate dichotomous data models and do not tackle the extent of adoption.

Fourth, the majority of recent studies (including our two case studies) apply standard logit or probit methods to the binary adoption data. Yet, a variety of alternative econometric techniques are available that apply different estimation approaches (such as linear probability, generalized method of moments) and/or make fewer assumptions about the distribution of the error term or about the estimation process (such as semi- and nonparametric approaches). It remains to be seen whether the results of agroforestry adoption studies are robust to the econometric methods applied.

Finally, recent studies have explored the usefulness of stated preference methods such as conjoint analysis to examine how and why farmers choose different types of agroforestry systems (Casey et al. 1999, Zinkhan et al. 1997). While most studies use past behavior (or binary adoption data) as a predictor of future adoption, the stated preference methods are well positioned to evaluate ex ante plans for tree planting. Collectively, these points indicate that additional study is needed to analyze the robustness of agroforestry adoption models to the analytic and econometric methods employed.

6. LITERATURE CITED

AMACHER, G., W. HYDE, AND M. RAFIQ. 1993. Local adoption of new forestry technologies: An example from Pakistan's NW Frontier Province. World Dev. 21(3): 445-453.

Bosque Modelo. 1997. Visión Estratégica 1997-2000. Bosque Modelo para Calakmul, Ecologica Productiva. Bosque Modelo de Calakmul, Zoh-Laguna, Campeche, Mexico. 35 p.

Casey, J., E. Mercer, and A. Snook. 1999. Evaluating farmer preferences for agroforestry systems: Survey instrument design. P. 153-158 *in* Proceedings of the 1998 Southern Forest Economics Workers Conference, Abt, K.L. and R.C. Abt (eds.). NC State University, Raleigh, NC.

Caviglia, J., and J. Kahn. 2001. Diffusion of sustainable agriculture in the Brazilian tropical rain forest: A discrete choice analysis. Econ. Dev. Cult. Change 49 (2):311-334.

Clay, D., T. Reardon, and J. Kangasniemi. 1998. Sustainable intensification in the highland tropics: Rwandan farmers' investments in land conservation and soil fertility. Econ. Dev. Cult. Change. 46 (2): 351-378.

Cook, T., H. Cooper, D. Corday, H. Hartmann, L. Hedges, R. Light, T. Louis, and F. Mosteller. 1992. Meta-analysis for explanation: A casebook. Russell Sage Foundation, New York, NY. 378 p.

Current, D., E. Lutz, and S. Scherr (eds.). 1995. Costs, benefits, and farmer adoption of agroforestry: project experience in Central America and the Caribbean. World Bank Environment Paper 14. World Bank, Washington DC. 212 p.

Feather, P., and G. Amacher. 1994. Role of information in the adoption of best management practices of water quality improvement. Agr. Econ. 11: 159-170.

Feder, G., R. Just, and D. Zilberman. 1985. Adoption of agricultural innovations in developing countries: A survey. Econ. Dev. Cult. Change. 33 (2): 255-98.

Greene, W.H. 1997. Econometric analysis. Third Edition. Prentice Hall, Upper Saddle River, New Jersey. 1075 p.

Lassoie, J.P. and L.E. Buck 1999. Exploring the opportunities for agroforestry in changing rural landscapes in North America. Agroforest. Syst. 44:105-107.

Maddala, G.S. 1983. Limited-dependent and qualitative variables in econometrics. Economic Society Monographs No. 3. Cambridge University Press, Cambridge, United Kingdom. 401 p.

Mercer, E., and R. Miller. 1998. Socio-economic research in agroforestry: Progress, prospects, priorities. Agroforest. Syst. 38:177-193.

Nair, P.K.R. 1993. Introduction to Agroforestry. Kluwer, Boston, MA. 499 p.

Nair, P.K.R. 1996. Agroforestry directions and trends. P. 74-95 *in* The Literature of Forestry and Agroforestry. McDonald, P. and Lassoie, J. (eds.). Cornell University Press, Ithaca, NY.

Pattanayak, S.K., D.E. Mercer, E. Sills, and J.C. Yang. Forthcoming. Taking stock of the agroforestry adoption studies. Agroforestry Systems.

Pattanayak, S. K., and D.E. Mercer. 2002. Indexing soil conservation: Farmer perceptions of agroforestry benefits. J. Sustain. For. 15(2): 63-85.

Pender, J., and J. Kerr. 1998. Determinants of farmers' indigenous soil and water conservation investments in semi-arid India. Agr. Econ. 19: 113-125.

Sanchez, P.A. 1995. Science in agroforestry. Agroforestry Systems 30:5-55.

Snook, A., 1996. Annual Report. International Centre for Research in Agroforestry-Mexico. Chetmual, Quintana Roo, Mexico. 49 p.

Snook, A., and G. Zapata. 1998. Tree cultivation in Calakmul, Mexico: Alternatives for reforestation. Agroforestry Today 10(1):15-18.

Strauss, J. 1986. The theory and comparative statics of agricultural household models: A general approach. P. 71-94 *in* Agricultural Household Models: Extensions, Applications, and Policy, Singh, I., L. Squire, and J. Strauss (eds.). Johns Hopkins University Press, Baltimore, MD.

Zinkhan, F.C., D.E. Mercer, and B. Stansell. 1997. Assessing the market and non-market benefits of Southern agroforestry. P. 123-132 *in* Proceedings of the 26th Annual Southern

Forest Economics Workers Conference, Greene, J.L. (ed.). University of Tennessee, Knoxville, TN.

---

[1] This review draws from ongoing research conducted by the authors and collaborators. We started with a set of 56 articles on adoption of agricultural and forestry technology by smallholders. Ultimately, based on the criteria of (1) empirical analysis and (2) focus on agroforestry and soil-water conservation investments, we narrowed our list down to 26 studies. The details of our meta-analytic review are presented in Pattanayak et al. (forthcoming).

[2] Randomness exists potentially because of unobserved attributes, instrumental variables, measurement errors, and taste variations (see Feather and Amacher 1994).

[3] Each additional increment of the independent variable increases the odds of adoption by $e^{(\beta)}$. These values are calculated as odds ratios: the amounts by which the odds favoring adoption (y = 1) are multiplied with each 1-unit increase in that independent variable assuming all other independent variables remain constant.

Section Three

NON-MARKET VALUATION

Thomas P. Holmes
USDA Forest Service

In addition to commodities such as timber, forest ecosystems provide an array of goods and services that are not priced in markets but maintain, to a large degree, the characteristics of public goods (non-rivalry and non-excludability). Markets do not recognize scarcity of non-market resources and cannot be relied upon to allocate these resources to their highest and best use. In addition, the production of commodities can diminish the non-market values of forests by more than the gain in commodity benefits, leading to losses in social welfare. Non-market valuation methods, as illustrated in this section, have been developed and applied to address these problems of sub-optimal resource allocation.

The first two chapters in this section (17 and 18) discuss stated preference methods, which are able to quantify both use and non-use values related to both current forest conditions and conditions that could exist in the future. The next two chapters present revealed preference methods, illustrating non-market valuation of forest resources first from the perspective of the recreation consumer (chapter 19) and second from the perspective of the downstream agricultural producer (chapter 20). The four non-market valuation methods discussed in the chapters are illustrated with econometric analyses of survey data to estimate values of non-market goods and services from public forests in North Carolina, Maine, Brazil, and Indonesia.

Sills and Abt (eds.), Forests in a Market Economy, 301. ©Kluwer Academic Publishers. Printed in The Netherlands.

Chapter 17

Contingent Valuation of Forest Ecosystem Protection

Randall A. Kramer, Thomas P. Holmes, and Michelle Haefele
Duke University, USDA Forest Service, and Colorado State University

In recent decades, concerns have arisen about the proper valuation of the world's forests. While some of these concerns have to do with market distortions for timber products or inadequate data on non-timber forest products, an additional challenge is to uncover the economic worth of non-market services provided by forest ecosystems (Kramer et al. 1997). This has led to a growing number of publications addressing the valuation of forest ecosystem services, on topics such as carbon sequestration and endangered species habitat. In this chapter, we focus on the contingent valuation method (CVM) to assess the structure, health, and extent of forest ecosystems.[1]

Forest ecosystems generate a wide variety of use values, the most important of which are timber, non-timber products, recreation, wildlife habitat, and watershed services. While use values are important, and their provision was the primary objective of public land management in the past, increasingly public land managers are confronted with demands arising from passive use values such as the knowledge that specific ecosystems exist or will be available for future generations to enjoy. Although use and passive use values are both non-market, passive use values can only be measured using stated preference methods.

Researchers use one of two stated preference methods, CVM and Attribute Based Method (ABM), to uncover non-market values of forest quality. ABMs represent a merging of the hedonic method in economics that views the demand for goods as derived from the demand for attributes, with marketing research methods for determining perceived values of particular product features. Values are revealed through a series of questions that ask people to rate, rank, or choose among a set of alternatives with varying levels of each attribute. ABMs have been used in recent years for valuing particular attributes of forests, e.g., age, species variety, and watershed protection. The use of ABMs is described in more detail elsewhere in this volume (see chapter 18). CVM remains an important tool for forest resource economists because forest ecosystems present bundles of goods and services that cannot be easily separated. In fact the components of a forest ecosystem often move together. For example, a forest with greater levels of species diversity may also have higher levels of watershed services and aesthetic value than less diverse forests. Thus, one can think of contingent valuation as a tool that is appropriate for valuing complex environmental goods such as forest ecosystems precisely because it leads to a holistic approach rather than focusing on individual components. Also, contingent valuation is appropriate for valuing unique resources. Estimating economic values for forest ecosystems can improve the formation and implementation of policies to manage those ecosystems.

In this chapter, we first review studies that used contingent valuation to evaluate forest quality, health, and extent. We find strong evidence that forest ecosystem condition can be considered an economic good and is therefore a candidate for cost/benefit analysis of forest protection actions. We then present a case study on forest quality in the Southern Appalachian Mountains. The spruce-fir ecosystem in this area is undergoing rapid change due to environmental stress. Although spruce-fir forests in the Southern Appalachians currently provide little in the way of commercial or market amenities, they provide significant non-market values, including recreation, scenic beauty, and biodiversity protection. We present results from an earlier study of this ecosystem and a new analysis of the consistency of measured willingness to pay (*WTP*) values along with a discussion of the theoretical constructs that allow an economic interpretation of measured forest values.

1. LITERATURE REVIEW

1.1 Existence Values/Passive Use Values for Forests

Existence value—what people are willing to pay to protect resources they have no plans to use—has emerged as the most important non-use or passive use value associated with environmental resources. Krutilla's widely cited paper "Conservation Reconsidered," provided the first formal argument for including existence value in benefit estimates: "When the existence of a grand scenic wonder or a unique and fragile ecosystem is involved, its preservation and continued availability are a significant part of the real income of many individuals" (Krutilla, 1967:779). In a footnote, Krutilla stated that "These would be the spiritual descendents of John Muir, the present members of the Sierra Club, the Wilderness Society, National Wildlife Federation, Audubon Society and others to whom the loss of a species or the disfigurement of a scenic area causes acute distress and a sense of genuine relative impoverishment."

Some non-economists have raised ethical objections to monetizing existence values of resources (Adams 1990), but the resource economics profession has pushed ahead with its valuation agenda, arguing that failure to do so will result in significant under-valuation of environmental resources in policy and management decision making. A variety of motivations for existence value have been proposed in the literature (Boyle and Bishop 1985, Brookshire et al. 1986, Krutilla 1967, McConnell 1997, Randall and Stoll 1983). The motivations include several types of altruism as well as bequest motives. In addition, there is vicarious consumption derived from reading books or watching documentaries about nature. Although this vicarious consumption could be seen as indirect use, in practice it cannot be separated from pure existence value (Smith and Desvousges 1986). While Boyle and Bishop (1985) consider sympathy for other species and concerns about environmental linkages as part of altruistic behavior, McConnell (1992:3) argues that preference for the natural order is distinct from altruism, which he defines as *WTP* to preserve a resource because the resource "enhances the well-being of others." We view it as plausible that some forest areas would meet with Krutilla's criteria regarding existence values for "unique and fragile ecosystems." While there is a lack of consensus in the literature about exactly what types of preferences are represented by existence values, if existence values enter into the total value of an ecosystem (either singly or in combination with use values), it is generally agreed that their influence should not render total value estimates inconsistent with economic theory. Recently, the argument has been made that well-behaved preferences for existence goods can be evaluated by examining the consistency of measured values with signs of the first and second derivatives of the *WTP* function (Diamond 1996, Loomis and Larson 1994, Rollins and Lyke 1998). Complete specification of a consistency test is a major focus of this chapter, and is discussed in section 2 below.

1.1.1   Previous Non-market Forest Valuation Studies

The first studies using CVM to estimate values for forest protection appeared in the early 1990s and were generally concerned with forest degradation due to insect infestations and air pollution (table 17.1). Walsh et al. (1990) used the iterative bidding technique to estimate the value of protecting ponderosa pine on national forests in the front range of the Colorado Rocky Mountains. This study confirmed that the general public is willing to pay for forest protection programs. In addition, by asking respondents to decompose total value into four categories of value (recreational use, option, existence, and bequest), the authors concluded that use values accounted for 27.4% of total value, and non-use values (including option, bequest, and existence values) accounted for 72.6% of total value.

Table 17.1. Contingent valuation studies of forest protection

Author	Year	Type of experiment [a]	Type of activity	WTP value
Haefele, Kramer, and Holmes	1991	PC, DC	Protect high-elevation spruce-fir forest in southern Appalachian mountains from exotic insect and air pollution	PC = $21/yr. DC = $100/ yr.
Jakus and Smith	1991	DC	Increase in aesthetic quality of homeowner property due to gypsy moth; private use value	$238–$394 for private control; $295–$494 for a public control
Kramer and Mercer	1997	PC, DBDC	Creating national parks and protected areas to preserve 10% of tropical rain forests	PC = $31/ yr. DBDC = $21/ yr.
Li and Mattson	1995	DC	Continued access to the forest environment under the Swedish Right of Common Access	12,817 Swedish Kroner ($1,600), adjusted for preference uncertainty
Loomis, Gonzalez-Caban, and Gregory	1996	OE, DC	Reduce fire hazard to old-growth forests	OE = $33/yr. DC = $98/yr.
Miller and Lindsey	1993	DC	Protect homeowner property from gypsy moth by state-run control program; private-use value	$69/ yr.
Reaves, Kramer, and Holmes	1999	OE, PC, DBDC	Restore 75,000 acres of old-growth longleaf pine for red-cockaded woodpecker habitat	OE = $11/yr. PC = $8/yr. DBDC = $13/yr.
Walsh, Bjonback, Aiken, and Rosenthal	1990	Iterative bidding	Protect mixed-age ponderosa pine from mountain pine beetle	$47/ yr.

[a] PC = payment card, DC = dichotomous choice, DBDC = double-bounded dichotomous choice, OE = open ended

Haefele et al. (1991) reported a positive *WTP* for protecting high-elevation spruce-fir forests in the southern Appalachian Mountains from exotic insect infestation and air pollution damage. Using a decomposition approach, they found that non-use values (bequest and existence values) dominated use values as reasons for protecting these forests. Subsequent analysis of responses given by people who never had visited the study area and did not intend to visit the study area in the future confirmed that ecosystem existence values were substantial and empirically distinct from total ecosystem values (Holmes and Kramer 1996).

Jakus and Smith (1991) and Miller and Lindsay (1993) used CVM to estimate *WTP* for gypsy moth protection programs. These studies differed from the earlier studies in that private, not public, property was the focus of valuation.

Loomis et al. (1996) elicited preferences of Oregon households for reducing fire hazards to old-growth forests in the Pacific Northwest. They also used two CVM response formats and found that *WTP* estimated from dichotomous choice responses was greater than *WTP* estimated from open-ended responses, which is consistent with the Holmes and Kramer (1995) study. It is interesting to note that dichotomous choice estimates for protecting old-growth forests from fire ($98/year) are very close to dichotomous choice estimates for protecting spruce-fir forests from insect epidemics/air pollution damage ($100/year). Likewise, payment card estimates of protecting spruce-fir forests ($21/year) are similar to open-ended estimates of *WTP* to protect old-growth forests ($33/year). Kramer and Mercer (1997) evaluated the preferences of a random sample of U.S. citizens regarding the creation of protected areas to preserve 10% of tropical rain forests. In contrast to Holmes and Kramer (1995), they found that *WTP* computed from dichotomous choice responses were lower than *WTP* estimated from payment card responses. Payment card estimates of *WTP* for creation of rain forest preserves ($31/year) were similar to payment card estimates of protecting spruce-fir forests ($21/year) and open-ended estimates of *WTP* to protect old-growth forests ($33/year). The dichotomous choice estimates of *WTP* for creation of rain forest preserves were, in general, lower than *WTP* values estimated using the dichotomous choice method in other studies.

Finally, one study estimated the value of restoring old-growth longleaf pine forests in South Carolina (Reaves et al. 1999). These forests were severely damaged by a natural event (hurricane) and provided habitat for an endangered species (the red-cockaded woodpecker). *WTP* estimates were quite similar for three valuation methods used. In addition, restoration *WTP* values were somewhat lower for this resource than for protection activities in old-growth forests in the Pacific Northwest, spruce-fir forests in the Southern Appalachian Mountains, and tropical rain forests.

2. CONCEPTUAL MODEL OF CVM CONSISTENCY

A major focus of this chapter is to evaluate the consistency of values measured using the CVM with constructs of neoclassical economic theory. We begin our discussion with the proposition that consumer preferences for the condition of a forest ecosystem can be represented by a utility function. Neoclassical economic theory states that utility is quasi-concave with respect to quantity or, equivalently, that preferences (indifference curves) are convex with respect to the origin (e.g., see Johansson 1987).[2] Thus, the first increment in quantity of an economic good should have a positive value. A second increment in quantity should also have a positive value (non-satiation), but the increase in value should be less than the first increment (diminishing marginal value).

This proposition implies that people are willing to make substitutions among bundles with varying levels of market goods and forest conditions. Some authors have argued that goods that embody existence values, such as endangered wildlife species, may invoke lexicographic preferences based on ethical concerns (Edwards 1986, Edwards 1992, Stevens et al. 1991). The lexicographic rule always ranks one characteristic of a decision problem above another. In the present context, a lexicographic decision rule would always rank improvements to ecosystem condition above other considerations, such as changes in household expenditures. That is, a household with lexicographic preferences would never be indifferent between various combinations of forest ecosystem conditions and expenditures on other goods and services. Lexicographic preferences are not well-behaved from an economic perspective.

If forest condition can be represented by a well-behaved utility function, then increments in the forest area protected will increase utility at a diminishing rate. A measure of the economic value of an increment in forest condition is the amount of money an individual is willing to pay to attain the increment and which leaves the individual just as well off as if there were no increment in forest protection and no payment. This measure is known as the compensating surplus and can be written using the expenditure function, which minimizes household expenditure subject to the constraint that utility equals or exceeds some reference level (Freeman 1993). In particular, the compensating surplus is written as the difference between two expenditure functions:

Contingent Valuation of Forest Ecosystem Protection

$$\text{Compensating surplus} = e(p, q^0(a^0), u^0) - e(p, q^1(a^1), u^0) \qquad 17.1$$

where p is a vector of market prices; q is a measure of forest condition, which, in turn, is a function of the area protected a; and u is utility. The superscript 0 refers to the *status quo*, and the superscript 1 refers to the changed condition. The compensating surplus is a welfare theoretic measure of *WTP* for a specific increment to forest ecosystem condition.

A positive *WTP* for an initial increment in forest condition beyond the status quo suggests that forest ecosystem condition can be considered an economic good and is therefore a candidate for cost/benefit analysis of forest protection actions. The first hypothesis to be tested is whether *WTP* for an initial increment is statistically different than zero. The null hypothesis is:

$$H_0^1 : WTP_a = 0 \qquad 17.2$$

where a is a measure of the area protected. The null is tested against the alternative hypothesis that $WTP_a > 0$.

Second, if preferences for forest ecosystem condition are consistent with consumer theory, then people will be willing to pay more for greater levels of protection. This suggests a second null hypothesis: people gain utility from protecting a core area of a forest ecosystem, but do not gain marginal utility from protecting more than the core area. To test this hypothesis, we establish the null hypothesis that incremental *WTP* for incremental gains in forest condition is equal to zero:

$$H_0^2 : \int_{q^0}^{q^1} \Theta(p, q(a), u^0) \, dq = 0 \qquad 17.3$$

where marginal *WTP*, $\Theta(p, q(a), u^0)$, is a partial derivative of the expenditure function:

$$\Theta(p, q(a), u^0) = \frac{-\partial e(p, q^1(a^1), u^0)}{\partial q} = \frac{\partial WTP}{\partial q} \qquad 17.4$$

The partial derivative represents the slope of the individual's indifference curve at the point of evaluation, and marginal *WTP* is integrated over the incremental change $q^0(a^0) \rightarrow q^1(a^1)$. Failure to reject the second hypothesis would imply that indifference curves are flat with respect to changes in forest protection—a violation of the neoclassical assumption regarding non-satiation. Letting the increment in the area of forest protection be represented by b, the second hypothesis can be rewritten:

$$H_0^2 : WTP_{a+b} = WTP_a \qquad 17.5$$

If the second null hypothesis is rejected, and the alternative hypothesis that marginal *WTP* is positive is accepted, then consistency of measured *WTP* with economic theory requires that the second derivative of *WTP* with respect to area protected be negative. The *WTP* curvature condition can be evaluated by comparing the average slope of two segments of the *WTP* function with respect to the forest area protected. In particular, the third null hypothesis is:

$$H_0^3 : \frac{WTP_{a+b} - WTP_a}{b} = \frac{WTP_a}{a} \qquad 17.6$$

In equation 17.6, the numerator in each expression represents incremental *WTP*, where it is implicitly assumed that *WTP* for no protection is zero, and the denominator represents the change in forest area protected. If the third null hypothesis is rejected, and the alternative hypothesis is accepted (the second derivative of *WTP* with respect to area protected is negative) then consumer preferences regarding forest ecosystem protection are consistent with the constructs of economic theory.

3. EXPERIMENTAL SETTING

Our experiment focuses on protection of the high-elevation spruce-fir forest ecosystem in the Southern Appalachian Mountains. This ecosystem covers 26,610 ha of mountaintops and high ridges in Virginia, North Carolina, and Tennessee. About three-fourths of this ecosystem is located in the Great Smoky Mountains National Park. This park receives about 9 million visitors per year and is the most heavily visited national park in the country.

Since the 1950s, there has been a dramatic increase in spruce-fir mortality in this ecosystem. Using aerial photography, a recent inventory determined that in one-fourth of this area, greater than 70% of the standing trees were dead (Dull et al. 1988). Research also indicates a decline in the growth rate of red spruce (*Picea rubens* Sarg.) on Mt. Mitchell, the highest mountain east of the Mississippi River (Bruck 1988). Decline of the spruce-fir forest is highly visible from roads and trails. The cause of decline of Fraser fir (*Abies fraseri* Poir.) is generally attributed to the balsam woolly adelgid, an exotic forest pest accidentally introduced from Europe. Also, some scientists have attributed the decline of these forests to air pollution impacts, through direct impacts on soils and foliage and indirect impacts on susceptibility to insect attacks (Hain 1987).

For our experiment, we considered the reference level of utility to be associated with the status quo forest condition. Because the entire ecosystem was at risk of degradation, reference utility was associated with protecting none of the existing forest area. Then, the first increment of forest protection was specified to occur along road and trail corridors, spanning one-third of the entire ecosystem at risk. This level of protection may be particularly appealing to people who value the ecosystem principally for recreational use. The second level of protection was for the entire ecosystem. It was thought that this level of protection may be appealing to people who value the ecosystem as a whole and may focus attention on the continued existence of the entire threatened ecosystem.

4. SURVEY INSTRUMENT

A contingent valuation mail-out mail-back survey was used to gather information about *WTP* for protection of the remaining healthy spruce-fir forests, along with information about socio-economic and other characteristics of the respondents. The format of the survey and its implementation closely followed the Dillman (1978) method. The sampling frame was people living within a 500-mile radius (approximate one day's drive) of Asheville, North Carolina. This sampling frame was used because we wanted a large share of our respondents to have some familiarity with the study area prior to receiving the questionnaire.

A sheet of color photographs representing three stages of forest decline and a map identifying the study area were included with the survey along with information about forest damage and forest protection programs. Two *WTP* response formats were used: payment card and dichotomous choice. A comparison of *WTP* models and estimates from the two response formats is reported elsewhere (Holmes and Kramer 1995). In this chapter, we only use responses to the dichotomous choice questions.

Response rate to the single-bounded dichotomous choice version of the questionnaire was 51% and resulted in 221 usable observations. Of those people responding to the questionnaire, 4% did not respond to the dichotomous choice questions.

Two sequential dichotomous choice questions were posed. The first question provided information to test the first hypothesis—that people have a positive *WTP* (WTP_a) for forest ecosystem protection—and asked whether or not people would be willing to pay a specified annual amount in higher taxes to protect spruce-fir forests along roads and trails (about one-third of the remaining forest area). The second question provided information to test the second and third hypotheses—incremental WTP (WTP_{a+b}) increases at a decreasing rate—and asked whether or not people would be willing to pay a specified annual amount in higher taxes to protect the entire ecosystem.

Specific dollar amounts were randomly assigned across questionnaires. Identical amounts were used in both questions within a questionnaire. This assignment method was used because if dollar amount in the second question exceeded the dollar amount in the initial question, then respondents may have construed that an increasing incremental value was being sought. Further, a decreasing amount in the follow-up question may have been construed as illogical.

Four response patterns to the dichotomous choice WTP questions were observable: No-No (NN), No-Yes (NY), Yes-No (YN), and Yes-Yes (YY). The hypothesis test that WTP increases at a decreasing rate critically depends on the pattern of NY responses. A NY response would indicate that WTP for protecting forests only along road and trail corridors was less than the bid amount X, but that WTP equalled or exceeded X for protecting the entire forest ecosystem. Other responses would indicate either a constant or decreasing WTP as the area protected increased.

Respondents who answered yes to the second WTP question were asked a follow-up question to provide information about their WTP rationale. In particular, people were asked to decompose their total WTP into four categories, by percentage: (1) use of forests for myself, (2) use of forests for others (including future generations), (3) protection of the forests even if no one uses them, and (4) other. This question was designed to identify the importance of non-use values associated with forest ecosystem protection.

5. EMPIRICAL METHODS

Sequential presentation of WTP questions in our experiment suggests that responses to these questions are not independent if unobserved factors influence both responses. Single-equation models of WTP should not be used in cases where equation errors are correlated, because preference parameter estimates are inefficient, and standard errors of the preference parameters are upwardly biased (Greene 1997). In turn, this bias affects hypothesis testing, because standard errors of WTP values, and differences in WTP values, are computed based on parameter estimates. To obviate these problems, we used a bivariate probit model to estimate preference parameters and identify correlation in the unobserved factors influencing responses across the two WTP equations. In general, a bivariate probit model is specified as:

Contingent Valuation of Forest Ecosystem Protection

$$y_1^* = \beta_1'x_1 + \varepsilon_1, y_1^* = 1 \, if \, y_1^* > 0, \, 0 \, otherwise$$
$$y_2^* = \beta_2'x_2 + \varepsilon_2, y_2^* = 1 \, if \, y_2^* > 0, \, 0 \, otherwise$$
$$E[\varepsilon_1] = E[\varepsilon_2] = 0 \qquad \qquad 17.7$$
$$Var[\varepsilon_1] = Var[\varepsilon_2] = 1$$
$$Cov[\varepsilon_1, \varepsilon_2] = \rho$$

where y_j is the response to WTP question j, the β_j's are vectors of preference parameters, the x_j's are vectors of explanatory variables, and the ε_j's are the equation errors. To simplify the interpretation of the model results, we use the same set of explanatory variables in both equations ($x_1 = x_2$).

WTP values should be nonnegative for economic goods. A non-negativity constraint can be imposed by assuming that the relationship between WTP_i and $\beta_j'X_i$ is log-linear. Median WTP is computed from the log-linear estimates as $exp(\beta_j'x_j/\mu)$, where μ is the parameter estimate on the bid amount Hanemann and Kanninen 1999:327. Mean WTP computed from a log-linear specification includes a term for the estimated variance of the model's error $(1/\mu^2)$: $WTP_{mean} = WTP_{median} \bullet [exp(1/\mu^2)]$. Thus, model specification errors can directly lead to inflated WTP_{mean} values with the log-normal model (Huang and Smith 1998). Further, because statistical tests using median WTP have greater statistical power than tests based on means (Mitchell and Carson 1989, Kealy and Turner 1993), we chose to use WTP_{median} in the tests below. For example, our second and third hypothesis tests were conducted using incremental WTP computed as $exp(\beta_1'x_1) - exp(\beta_2'x_2)$.

Hypothesis tests were conducted using the Krinsky-Robb (1986) bootstrap technique. This technique is used more often than the traditional bootstrap technique in estimating WTP confidence intervals because of its relative efficiency. This is because the traditional bootstrap resamples the raw data, and the model must be re-estimated for each draw (Efron and Tibshirani 1993). In contrast, the Krinsky-Robb procedure uses random draws from estimation results. [3]

The first hypothesis, $WTP_1 = 0$, was tested using the estimation results from the first equation in the bivariate probit model and the achieved significance level (ASL), which is defined as "the probability of observing at least that large of a value when the null hypothesis is true" (Efron and Tibshirani 1993:203). The ASL using the bootstrap percentile method ($ASL_\%$) for the first hypothesis test is written as:

$$ASL_\% = Pr(WTP_a = 0) = \frac{\#(WTP_a^{median} < 0)}{B} \qquad \qquad 17.8$$

where # is the number of times the condition is true and B is the number of bootstrap replications. The ratio on the right-hand side of equation 17.8 is a percentage indicating the significance level of the test.

The second hypothesis, $WTP_{a+b} = WTP_a$, was tested using the ASL:

$$ASL_\% = \Pr(WTP_{a+b} = WTP_a) = \frac{\#(WTP_{a+b}^{median} < WTP_a^{median})}{B} \qquad 17.9$$

This test was based on the results of both equations in the bivariate probit model. In this case, the distributions of median WTP values were not sorted before conducting the test.[4]

The expression for the third hypothesis, marginal WTP does not diminish as protected forest acres increase, can be written in a simplified form using the relationship specified in our experiment regarding protected forest area: $b = 2a$ (where a = area along road and trail corridors and b = the remaining area). Substituting this relationship into equation 17.6 and simplifying yields the third hypothesis test using the $ASL_\%$:

$$ASL_\% = \Pr((WTP_{3a(=a+b)} - WTP_a) < WTP_a)$$
$$= \frac{\#((WTP_{3a}^{median} - WTP_a^{median}) > 2(WTP_a))}{B}. \qquad 17.10$$

6. RESULTS

Descriptive statistics for the variables used in the empirical analysis are shown in table 17.2. Offer amounts were based on results from an open-ended WTP question in a pre-test survey. Based on pre-test results, it was decided to use an approximate log-normal offer distribution that ranged from $2 to $500. Data on household income were obtained using a categorical variable representing a range of incomes. People were asked if they belonged to any environmental organization or gave money to any environmental organizations or causes (no = 0, yes = 1). People were also asked to indicate how important various reasons were to them to protect the Southern Appalachian spruce-fir forests. Response categories were (1) not important, (2) somewhat important, and (3) very important. These variables were coded as 1 for the very important category and 0 otherwise.

Contingent Valuation of Forest Ecosystem Protection

Table 17.2. Descriptive statistics of model variables

Variable name	Description	Mean	Standard deviation
Ln_bid	Natural logarithm of offer amount	3.83 (antilog = 46.06)	1.38
Ln_inc	Natural logarithm of household income	10.40 (antilog = 32,860)	0.77
Enviro	Member of an environmental organization	0.29 (dummy variable)	0.45
Rec_val	Recreational opportunities very important	0.38 (dummy variable)	0.49
Scenic_val	Scenic beauty very important	0.69 (dummy variable)	0.46

The first step in the analysis was to estimate a bivariate probit model explaining *WTP* for protecting part and all remaining spruce-fir forests in the Southern Appalachian Mountains. Results are shown in table 17.3. As can be seen, logarithm of the offer amount is negative and statistically significant at the 0.01 level in both equations. Because offer amounts are varied across individuals, the variation in binary responses conveys information about the variance of the equation error (σ_i^2). Cameron and James (1987) show that the coefficient on the bid amount is a point estimate of $1/\sigma_i$. Taking the antilog

Table 17.3. Bivariate probit *WTP* results (N = 205)

Equation[a]	Constant	Ln_bid	Ln_inc	Enviro	Rec_val	Scenic_val
WTP part	-2.67*	-0.57***	0.35**	0.50**	0.53**	0.41
	(1.62)	(0.10)	(0.16)	(0.23)	(0.25)	(0.29)
WTP all	-1.97	-0.45***	0.28*	0.61***	0.08	0.56**
	(1.73)	(0.08)	(0.16)	(0.23)	(0.24)	(0.25)

* = significant at the 10% level, ** = significant at the 5% level, *** = significant at the 1% level. Standard errors in parentheses.
[a] ρ=0.04(0.04)***

of the inverse of the parameter estimates on ln_bid in each equation, we compute σ_1 = 5.80 and σ_2 = 9.35. This result indicates that the standard error of the second equation (for protecting all remaining spruce-fir forests in the Southern Appalachian Mountains) is larger than the standard error of the first equation (for protecting spruce-fir forests along roads and trails only). Apparently, responses to the second *WTP* question contain more statistical noise than responses to the first equation.[5] Differences in error distributions for the two equations support the rationale for using median *WTP* values for hypothesis tests rather than mean *WTP* values.[6] We also note that the parameter estimate for the correlation coefficient is highly significant and close to one, justifying the use of the bivariate probit model.

Table 17.4 shows the median *WTP* estimates and the 95% confidence intervals computed using random draws from the multivariate distribution of the bivariate probit parameter estimates. In B = 1000 draws, median *WTP* for protecting part of the spruce-fir ecosystem along roads and trails always exceeded zero. Consequently, we conclude that people are willing to pay a positive amount to protect at least part of the forest ecosystem at risk.

Table 17.4. Empirical median *WTP* distributions using B = 1000 random draws

Protection level	Lower bound (0.05)	Median	Upper bound (0.05)
Roads and trails (WTP_a)	$11.81	$18.17	$24.84
All remaining (WTP_{a+b})	$18.02	$28.49	$40.96

Table 17.5 presents the results of the three bootstrap hypothesis tests described in section 5. Results indicate that incremental *WTP* for forest ecosystem protection is positive ($28.49 ≠ $18.17) and that incremental *WTP* increases at a decreasing rate ($28.49 − 18.17 = $10.32 < $18.17). Consequently, we conclude that preferences for forest ecosystem protection, as obtained in this study, are well-behaved and are consistent with economic theory.

We note that the bootstrap hypothesis testing procedure described here is preferable to the non-overlapping confidence interval criterion used in an earlier treatment of the problem (Park et al. 1991). Under that criterion, the null hypothesis of no significant difference is rejected if the $(1 - \alpha)$ confidence intervals for *WTP* do not overlap. As pointed out by Poe et al. (1994), the actual significance level is higher than the significance level indicated by the test. This is consistent with our results.[7]

Table 17.5. Bootstrap hypothesis test results using the percentile method (B = 1000)

Null hypothesis	ASL%	Result
H_0^1 : $WTP_a = 0$	0.000	Reject H_0; Accept H_a
H_0^2 : $WTP_a = WTP_{a+b}$	0.009	Reject H_0; Accept H_a
H_0^3 : $(WTP_{a+b} - WTP_a)/b = (WTP_a/a)$	0.001	Reject H_0; Accept H_a

Finally, we report results for the value components of *WTP* (table 17.6). We recognize that there is debate in the literature about the cognitive ability of individuals to decompose total value in this way. However, we found that people allocated the greatest proportion of *WTP* to existence value. These results suggest that non-use values are an important component of total value for protection of this forest ecosystem.

Table 17.6. Value components of *WTP*

Type of value	Proportion of *WTP*	Component value
Use	0.13	$3.70
Bequest	0.30	$8.55
Existence	0.57	$16.24
Total	1	$28.49

7. CONCLUSION

Full and accurate assessment of forest values is essential for appraising projects and policies affecting the use of forests. Under-valuation of forest ecosystems can bias land use policies in directions that are not consistent with maximizing economic welfare. By improving the understanding of the economic importance of the structure, health, and extent of forest ecosystems, more informed forest policy and management decisions can be made.

The multiple outputs of forest ecosystems make their economic valuation challenging. This is particularly true when there are significant passive use values associated with protecting or restoring forest ecosystems. Contingent valuation is part of the tool kit available to forest resource economists. It allows a holistic approach to valuing the complex environmental good that a forest ecosystem represents.

A variety of studies using contingent valuation to value forest ecosystems have been conducted. The applications have included changes in forest quality due to insect infestations and air pollution, protection of existing ecosystems, and forest restoration. The studies show consistent support for the hypothesis that protection and restoration of forest ecosystems is an economic good that people are willing to pay for. Our own application to spruce-fir ecosystems confirmed this result and showed that consumer preferences regarding forest ecosystems were well-behaved and consistent with the constructs of economic theory. Thus, estimated *WTP* values can be used in cost/benefit assessments of forest ecosystem protection programs. These results were robust despite the fact that when *WTP* was decomposed, we found that existence value accounted for the greatest proportion of reported forest value.

8. LITERATURE CITED

ADAMS, J. G. 1990. Unsustainable economics. International Environ. Policy. 14-21.

BOYLE, K. J., and R. C. BISHOP. 1985. The total value of wildlife resources: Conceptual and empirical issues. Invited paper, Assoc. of Environ. Res. Econ. Workshop on Recr. Demand Modeling, Boulder, CO, May 17-18.

BROOKSHIRE, D. S., L. S. EUBANKS, and C. F. SORG. 1986. Existence values and normative economics: Implications for valuing water resources. Water Resources Research 22(11): 1509-1518.

BRUCK, R. J. 1988. Decline of red spruce and Fraser fir. P.108-111.*in* Forest Decline: Cause-Effect Research in the United States of North America and Federal Republic of Germany. Krahl-Urban, R., H. E. Papke, K. Peters, and C. Schimansky (eds.). U.S. Environmental Protection Agency, Corvallis, OR.

CAMERON, T.A., and M.D. JAMES. 1987. Efficient estimation methods for closed-ended contingent valuation surveys. Rev. Econ. Stat. 69:269-276.

CAMERON, T.A., and J. QUIGGIN. 1994. Estimation using contingent valuation data from a dichotomous choice with follow-up questionnaire. J. Environ. Econ. Manage. 27:218-234.

DIAMOND, P. 1996. Testing the internal consistency of contingent valuation surveys. J. of Environ. Econ. and Manage. 30:337-347.

DILLMAN, D.A. 1978. Mail and telephone surveys: The total design method. John Wiley and Sons: New York, NY. 325 p.

DULL, C.W., J.D. WARD, H.D. BROWN, G.W. RYAN, W.H. CLERKE, and R.J. WALKER. 1988. Evaluation of spruce and fir mortality in the Southern Appalachian Mountains. Protection Report R8-PR 13. USDA Forest Service, Atlanta, GA. 92 p.

EDWARDS, S.F. 1986. Ethical preferences and the assessment of existence values: Does the neoclassical model fit? Northeastern J. Agric. Res. Econ. 15:145-150.

EDWARDS, S.F. 1992. Rethinking existence values. Land Econ. 68:120-122.

EFRON, B., and R.J. TIBSHIRANI. 1993. An introduction to the bootstrap. Chapman & Hall, New York, NY. 436 p.

FREEMAN, A.M. III. 1993. The measurement of environmental and resource values. Resources for the Future, Washington, DC. 516 p.

GREENE, W.H. 1997. Econometric Analysis (third ed.). Macmillan Publishing Co., New York, NY. 1075 p.

HAEFELE, M., R.A. KRAMER, and T.P. HOLMES. 1991. Estimating the total value of forest quality in high-elevation spruce-fir forests. P. 91-96. *in* The Economic Value of Wilderness: Proceedings of the Conference. Gen. Tech. Rep. SE-78, Southeastern For. Exper. Station. USDA Forest Service, Asheville, NC.

HAIN, F.P. 1987. Interactions of insects, trees and air pollutants. Tree Phys. 3:93-102.

HANEMANN, W.M., AND B. KANNINEN. 1999. The statistical analysis of discrete-response CV data. P.302-441 *in* Valuing Environmental Preferences – Theory and Practice of the Contingent Valuation Method in the U.S., E.U., and Developing Countries. Bateman, I.J. and K.G. Willis (eds.). Oxford University Press, Oxford.

HOLMES, T.P., and R.A. KRAMER. 1996. Contingent valuation of ecosystem health. Ecos. Health 2:56-60.

HOLMES, T.P. and R.A. KRAMER. 1995. An independent sample test of yea-saying and starting point bias in dichotomous-choice contingent valuation. J. Environ. Econ. Manage. 28:121-132.

HUANG, J.-C., and V.K. SMITH. 1998. Monte Carlo benchmarks for discrete response valuation methods. Land Econ. 74:186-202.

JAKUS, P., and V.K. SMITH. 1991. Measuring use and nonuse values for landscape amenities: A contingent behavior analysis of gypsy moth control. Dis. Pap. QE92-07. Resources for the Future, Washington, DC. 48 p.
JOHANSSON, P-O. 1987. The Economic Theory and Measurement of Environmental Benefits. Cambridge University Press, New York. 223 p.
KEALY, M. J., and R. W. TURNER. 1993. A test of the equality of close-ended and open-ended contingent valuation. Am. J. Agric. Econ. 75:321-331.
KRAMER, R. and E. MERCER. 1997. Valuing a global environmental good: U.S. residents' willingness to pay to protect tropical rain forests. Land Econ. 73:196-210.
KRAMER, R., R. HEALY, and R. MENDELSOHN. 1997. Forest Valuation. Chapter 10. in Managing the World's Forests, Narendra Sharma (ed.). World Bank Natural Resources Development Series, Arlington, VA.
KRINSKY, I., and A.L. ROBB. 1986. On approximating the statistical properties of elasticities. Rev. Econ. Stat. 68:715-719.
KRUTILLA, J.V. 1967. Conservation reconsidered. Am. Econ. Rev. 57(4):777-786.
LI, C-Z. and L. MATTSSON. 1995. Discrete choice under preference uncertainty: An improved structural model for contingent valuation. J. Environ. Econ. Manage. 28:256-269.
LOOMIS, J.B., and D.M. LARSON. 1994. Total economic values of increasing gray whale populations: Results from a contingent valuation survey of visitors and households. Marine Res. Econ. 9:275-286.
LOOMIS, J.B., A. GONZALEZ-CABAN, and R. GREGORY. 1996. A contingent valuation study of the value of reducing fire hazards to old-growth forests in the Pacific Northwest. Res. Pap. PSW-RP-229. Pacific Southwest Research Station, USDA Forest Service, Portland, OR.
MCCONNELL, K. E. 1997. Does altruism undermine existence value? J. Environ. Econ. Manage. 32:22-37.
MILLER, J.D., and B.E. LINDSAY. 1993. Willingness to pay for a state gypsy moth control program in New Hampshire: A contingent valuation case study. J. Econ. Entom. 86:828-837.
MITCHELL, R.C., and R.T. CARSON. 1989. Using surveys to value public goods: the contingent valuation method. Resources for the Future, Washington DC. 463 p.
PARK, T., J.B. LOOMIS, and M. CREEL. 1991. Confidence intervals for evaluating benefits estimates from dichotomous choice contingent valuation studies. Land Econ. 67:64-73.
POE, G. L., E. K. SEVERANCE-LOSSIN, and M. P. WELSH. 1994. Measuring the difference (x-y) of simulated distributions: A convolutions approach. Am. J. Agric.. Econ. 76:904-15.
POE, G.L., M.P. WELSH, and P.A. CHAMP. 1997. Measuring the differences in mean willingness to pay when dichotomous choice contingent valuation responses are not independent. Land Econ. 73:255-267.
RANDALL, A. and J. R. STOLL. 1983. Existence value in a total valuation framework. in Managing Air Quality and Scenic Resources at National Parks and Wilderness Areas, R.D. Rowe and L.G. Chestnut, (eds.). Westview Press, Boulder, CO.
REAVES, D.W., R.A. KRAMER, and T.P. HOLMES. 1999. Does question format matter? Valuing an endangered species. Environ. Res. Econ. 14:365-383.
ROLLINS, K., and A. LYKE. 1998. The case for diminishing marginal existence values. J. Environ. Econ. Manage. 36:324-344.
SMITH, V.K. and W.H. DESVOUSGES. 1986. Measuring water quality benefits. Kluwer Nijhoff Publishing, Boston, MA. 327p.
STEVENS, T.H., J. ECHEVERRIA, R.J. GLASS, T. HAGER, AND T.A. MORE. 1991. Measuring the existence value of wildlife: What do CVM estimates really show? Land Econ. 67:390-400.
WALSH, R.G., R.D. BJONBACK, R.A. AIKEN, AND D.H. ROSENTHAL. 1990. Estimating the public benefits of protecting forest quality. J. Environ. Manage.30:175-189.

[1] While some analysts have used CVM to value individual elements of forest ecosystems, e.g. carbon sequestration or endangered species habitat, the focus of this chapter is on entire ecosystems.

[2] Johansson (1987:11) states that "A utility function $U(x)$ is *well-behaved* if (i) it is continuous where finite on X, (ii) it is increasing (and $\partial U(x)/\partial x_i > 0$ for all i), (iii) it is strictly quasi-concave on X, and (iv) it generates at least twice continuously differentiable demand functions."

[3] For each parameter in the estimated CVM model, random draws are made from a multivariate normal distribution with mean values set equal to the vector of parameter estimates and distribution set equal to the estimated variance-covariance matrix. Given the bootstrap parameter vector, median *WTP* is computed and stored. Computing and storing B bootstrap replications of median *WTP* yields a bootstrap distribution of the median for each equation. Sorting the median *WTP* bootstrap distribution allows confidence intervals to be established, and hypothesis tests can be constructed.

[4] A similar method to test for difference in mean *WTP* for nonindependent dichotomous choice responses was used by Poe et al. (1997).

[5] This result has also been observed using the double-bounded dichotomous choice format (Cameron and Quiggin 1994). In addition, this result is consistent with rank-order studies that indicate cognitive burden, and therefore respondent fatigue, increases with increasing rank (see chapter 18).

[6] It may be recalled that, in the log-linear specification, estimates of mean *WTP* are influenced by equation error. If mean values were used in the current application, estimates of mean *WTP* would be inflated for the second equation relative to the first. This would, in turn, affect the efficacy of the hypothesis test concerning whether *WTP* is the same in the two equations.

[7] A review of table 17.4 shows that the 95% confidence intervals overlap for the two distributions. However, the bootstrap hypothesis procedure shows that median *WTP* values are statistically different at the 1% significance level.

Chapter 18

Stated Preference Methods for Valuation of Forest Attributes

Thomas P. Holmes and Kevin J. Boyle
USDA Forest Service and University of Maine

The valuation methods described in this chapter are based on the idea that forest ecosystems produce a wide variety of goods and services that are valued by people. Rather than focusing attention on the holistic value of forest ecosystems as is done in contingent valuation studies, attribute-based valuation methods (ABMs) focus attention on a set of attributes that have management or policy relevance (Adamowicz et al. 1998a, Bennett and Blamey 2001). The attribute set might include, for example, measures of biological diversity, areas designated for timber production or set aside for conservation, size of timber harvesting gaps, or watershed protection measures. If human-induced changes in forest ecosystems can be meaningfully represented by a set of attributes, choices made by survey respondents among sets of alternatives can provide resource managers and policy makers with detailed information about public preferences for many potential states of the environment. If price is included as an attribute of the problem, a multidimensional valuation surface can be estimated for use in cost/benefit analysis.

In this chapter, we show how forest management systems can be modeled as sets of management attributes and how value tradeoffs among forest management attributes can be measured using survey methods. Because increasing public concern with the sustainable use of forest resources has placed forest management in the spotlight, we suggest using ABMs to provide public agencies with information relevant to the design of forest

Sills and Abt (eds.), Forests in a Market Economy, 321–340. ©Kluwer Academic Publishers. Printed in The Netherlands.

practice codes and alternative management systems. By using scientifically based survey designs, sampling methods, and analytical techniques, ABMs can provide policy makers with information on broad-based citizen preferences that complements information gathered in public meetings and can provide a balanced assessment of how the general public values changes in forest management and conservation.

The state of Maine serves as a case study for our approach. Maine is more heavily forested than any other state in the United States, and the forest products industry provides significant income and employment to Maine residents. In addition, recreation, hunting, fishing, and wildlife viewing values associated with the woods provide significant contributions to the economy and quality of life in Maine. In 1989, the state legislature passed a Forest Practices Act that sets standards for timber harvests. However, public concern with some provisions of the Act, particularly regarding clearcutting, led to a number of initiatives to modify the Act.[1] Although none of these initiatives has succeeded to date, it is clear that many among the voting public are dissatisfied with status quo forest practices and are seeking alternatives that reduce timber-harvesting impacts on the goods and services provided by Maine forests.

To develop a better understanding of the tradeoffs that residents of Maine were willing to make regarding timber harvesting practices, we conducted a survey based on a random sample of the population. Our intentions in conducting the survey were twofold. First, we wanted to gain a clearer understanding of how much people were willing to pay (WTP) for alternative timber harvesting practices. These results could then be used in a cost/benefit analysis of policy alternatives. Second, we recognized that the use of WTP studies for evaluating policy alternatives is controversial, and we wanted to assess the validity of our survey responses. Although issues surrounding the validity of WTP surveys have many dimensions, a major issue is whether or not WTP values are affected by the design of the WTP response format. Consequently, we designed our survey instrument so that we could compare alternative response formats.

Choice models are becoming increasingly popular for measuring the value of environmental goods. The basic idea is that underlying preferences are revealed by the choices that people make. Choice methods are consistent with random utility theory; therefore, economic benefits associated with changes in environmental services can be estimated. As shown in section 5, ranking models are a special form of choice and can be derived from a model of random utility maximization. In this chapter, we compare responses to choice and ranking questions.

First, we review the literature on using ABMs to value forest ecosystems and summarize conclusions that can be drawn from it. Second, we briefly review random utility theory and show how it is connected to the choice model. Third, we describe how to test hypotheses regarding the parameters of choice models estimated on independent sub samples, and we highlight the importance of understanding the scale parameter when comparing choice model parameter estimates. Fourth, we present our forest management experiment and interpret the results. Finally, we draw conclusions about using ABMs to inform forest management and policy decisions.

1. LITERATURE REVIEW

Applications of ABMs to forest valuation are relatively new, first appearing in the literature in the late 1990s. Various response formats are available for conducting attribute-based experiments, and the most popular formats (rating, ranking, and choice) have all been used to conduct forest valuation studies.[2]

Garrod and Willis (1997) used a ranking study to estimate the benefits of enhancing forest biodiversity. Generic standards of increases in forest biodiversity were used (no increase, low-medium increase, medium-high increase, high increase). Alternatives were constructed using the area of forest managed according to each biodiversity standard, and the price variable was tax cost. The results showed that respondents preferred a balance between conservation and commercial timber production and, in general, were not willing to pay higher taxes for the greatest level of forest biodiversity restoration.

Hanley et al. (1998) used a choice experiment to estimate the value of alternative forest landscapes. Their experiment included four attributes: forest shape (straight edges versus organic edges); felling gaps (large versus small-scale clear cuts); species mix (evergreen only versus a combination of evergreen, larch, and broadleaf species); and tax (the price variable). They found that the respondents preferred forests with organically shaped edges, small-scale felling gaps, and a diverse mix of species.

Adamowicz et al. (1998b) used a choice experiment to estimate passive use values for woodland caribou habitat in Alberta (woodland caribou rely on old-growth forests). Attributes used in their experiment included four levels for each of the following attributes: woodland caribou populations, wilderness area, recreation restrictions, forest industry employment, and changes in provincial income tax. They found that utility and WTP increased with increased caribou populations and wilderness area and decreased as more severe restrictions were placed on recreation options. Changes in forest industry employment did not have a significant impact on WTP.

Similar in design to an Alberta moose hunting study (Adamowicz et al. 1997), Boxall and MacNab (2000) studied the preferences of wildlife recreationists in Saskatchewan for different aspects of boreal forest management. The sample was split to identify preferences of wildlife viewers and moose hunters. Attributes included three levels for each of the following: opportunity to see wildlife species, evidence of moose populations, encounters with other recreationists, access within the recreation area, evidence of forestry activity, and driving distance. For both hunters and wildlife viewers, large straight-edged clearcut areas with no residual trees generated large decreases in trip values. However, small (maximum width 440 m), irregular-shaped cutover areas with scattered patches of residual trees generated positive trip values for both wildlife viewers and moose hunters.

Holmes et al. (1998) used a paired comparison method to evaluate ecotourism options and estimate the value of a remaining remnant of Atlantic Coastal Forest in the Brazilian state of Bahia. The study attributes included amount of forest cover, lodging options, level of traffic congestion, nature park attractions, daily expenditures, and user fees. The study reported that Brazilian tourists had a positive willingness to pay for the protection of 7,000 km^2 of the Atlantic Coastal Forest ecosystem, and WTP increased with recreation options (such as canopy walks) in the nature park.

Schaberg et al. (1999) used the rating method to evaluate preferences for attributes of national forest management plans. The experimental design was based on priority levels (low, medium, and high) for the following attributes: forest recreation, hunting and fishing, timber harvesting, water quality, and native ecosystems. A cost variable was not included in the design. They found that the ideal management plan would place high emphasis on ecosystem restoration and water quality protection, low emphasis on timber harvesting, and moderate emphasis on recreational opportunities.

Haefele and Loomis (2001a,b) used the rating method to estimate the value of changes in forest health. They included attributes for the number of acres infested by various forest pests, the percentage change in commercial timber harvests, the possible risk of water contamination from pesticide spraying, and the expected percentage change in recreation use. Similar to the Schaberg et al. (1999) study, they found a high level of concern with the water quality impacts of forest operations. This study also found that people preferred pest control programs that had minimal impact on commercial timber harvests.

Overall, then, we find evidence in previous studies that the general public is willing to pay for changes in forest management and timber harvesting operations that reduce the biological and amenity impacts on forest ecosystems. However, because only a limited number of forest preference studies have been conducted, many dimensions of citizen preferences for forest management remain unexplored. Future research needs to consider, among other things, how the social, economic, and natural resource context may influence preferences for forest attributes.

2. A RANDOM UTILITY FUNCTION

Our presentation of the choice modeling approach to forest valuation begins with a random utility function that can be used to link utility (a theoretical construct) with actual choices. A random utility function considers individual preferences (subscript n) to be the sum of systematic (V_{in}) and random (ε_{in}) components:

$$U_{in} = V_{in}(x_{in}, p_{in}; \beta) + \varepsilon_{in} \qquad 18.1$$

where U_{in} is the true but unobservable utility associated with alternative i, x_{in} is a vector of attributes associated with alternative i, p_{in} is the cost of alternative i, β is a vector of preference parameters for the population, and ε_{in} is a random error term with zero mean.[3] In its simplest form, utility is represented as linear-in-parameters:

$$U_{in} = \sum_{k=1}^{l} \beta_k x_{ink} + \beta_p p_{in} + \varepsilon_{in} \qquad 18.2$$

Differentiation of equation 18.2 shows that preference parameter estimates (a vector of βs) can be interpreted as marginal utilities: $\beta_k = \partial U_i / \partial x_{ik}$. The negative of the parameter estimate on cost, β_p, is interpreted as the marginal utility of money. The marginal rate of substitution between any two attributes k and l is easily computed ($MRS_{kl} = \beta_k / \beta_l$), and the implicit price (or marginal WTP) of attribute k is β_k / β_p.

3. CHOICE MODELS

The stochastic term in the random utility function shown in equation 18.1 allows probabilistic statements to be made about actual choices. Consider a choice set C containing $J > 2$ alternatives (such as J different recreation sites). The probability that a consumer will choose alternative i from choice set C can be expressed as a function of random utilities (McFadden 1973):

$$P_n(i) = P(U_{in} > U_{jn}) = P[V_{in} + \varepsilon_{in} > V_{jn} + \varepsilon_{jn}], \forall j \in C \qquad 18.3$$

Various probabilistic choice models can be derived from equation 18.3, depending on the assumption made about the distribution of the random error ε. The assumption that ε follows an Extreme Value Type 1 (EV1) distribution is often used, and the resulting model is referred to as multinomial logit (MNL).[4] This assumption is made purely for analytical convenience, as the difference between two EV1 variables is logistically distributed, and the logit distribution has convenient closed-form properties (Ben-Akiva and Lerman 1985).

Given this assumption, the probability of individual n choosing alternative i from the set C is written (using matrix notation and including cost in the x_n vectors):

$$P_n(i) = \frac{\exp(\mu\beta' x_{in})}{\sum_{j \in C} \exp(\mu\beta' x_{jn})} \qquad 18.4$$

where μ is a scale parameter (see section 5). If we let N represent the sample size, then the likelihood function for the MNL model is:

$$L = \prod_{n=1}^{N} \prod_{i \in C} P_n(i)^{y_{in}} \qquad 18.5$$

where

$y_{in} = 1$ *if individual n chose alternative i*
$\phantom{y_{in}} = 0$ *otherwise.*

Substituting equation 18.4 into equation 18.5 and taking the natural logarithm, the MNL model is estimated by finding the values of the βs that maximize the log-likelihood:

$$\ln L = \sum_{n=1}^{N}\sum_{i \in C} y_{in}(\mu\beta' x_{in} - \ln \sum_{j \in C}\exp(\mu\beta' x_{jn})) \qquad 18.6$$

The major limitation of the MNL model is that data are subject to the Independence of Irrelevant Alternatives (IIA) property. This property requires "that for a specific individual the ratio of the choice probabilities of any two alternatives is entirely unaffected by the systematic utilities of any other alternatives" (Ben-Akiva and Lerman 1985:108). Simply stated, this property requires that equation errors are independent. That is, none of the unobserved factors influencing the choice of any alternative i can influence the choice of any other alternative j. This condition limits the substitution possibilities among alternatives.[5] However, if this property holds, it allows the analyst to estimate the probability of choosing new alternatives not included in the choice experiment simply by adjusting terms in the denominator of equation 18.4.

The goal of many ABM nonmarket valuation studies is to estimate welfare impacts so they can be used in management and policy analysis. ABMs provide quantitative measures of tradeoffs between attributes, including price. Thus, they can be used to estimate how much money would be required to make a person as well off after a change in attributes as they were before the change. The fact that ABMs provide estimates of the indirect utility function allows one to calculate welfare measures for improvements or decrements in utility.

Ben-Akiva and Lerman (1985) show that for a set of independent variables that are EV1-distributed with common scale (μ), the maximum is also EV1-distributed. As defined in equation 18.1, utility is characterized as the sum of a systematic and stochastic component. By assuming that the stochastic component is EV1-distributed, it can be shown that the expected value of maximum utility can be specified as

$$E(U) = \ln(\sum_{j=1}^{J}\exp(V_j)) + D \qquad 18.7$$

where D is Euler's constant, and the other term is known as the log sum or inclusive value (Hanemann 1999, Morey 1999). This expression forms the basis for welfare measurement when multiple alternatives are available.

In the most general situation, compensating variation is computed as the difference between two expected values of maximum utility divided by the marginal utility of money ($\lambda = -\beta_p$):

$$CV = \frac{1}{\lambda}[\ln \sum_{j=1}^{J} \exp(V_j^1) - \ln \sum_{j=1}^{J} \exp(V_j^0)] \quad\quad 18.8$$

where the 0 superscript refers to the base situation (policy off), the 1 superscript to the altered situation (policy on), and J to the number of sites or locations included in the utility function. An example of this situation is the computation of compensating variation for a change in attribute levels for a set of recreation sites from the base situation to some altered levels. A simpler situation, as computed in this study, is the compensating variation for a change in attributes for a single site or location. In this situation, equation 18.8 reduces to:

$$CV = \frac{1}{\lambda}[V^1 - V^0] \quad\quad 18.9$$

where V^0 and V^1 are the utility expressions for the base and altered cases. In the simplest situation, where interest focuses on the value of a change in a single attribute, and utility is linear-in-parameters as in equation 18.2, equation 18.9 reduces to the ratio of the attribute coefficient and the marginal utility of money.

4. RANKING MODELS

Ranking question formats offer a more complex form of choice responses, in which respondents are asked not simply to choose their most preferred alternative but to order alternatives from most to least preferred. This question format results in a series of responses from 1 to K for a set of K alternatives. In the standard model, the respondent is assumed to first choose the alternative that provides the greatest utility from the choice set. Then the second ranked alternative is chosen from the remaining choice set, and so forth until all alternatives are ranked. Marschak (1960) showed that this sequence of choices can be considered as the product of independent probabilities:

$$P[alt.1\,ranked\,1st, alt.2\,ranked\,2nd, ..., alt.K\,ranked\,last] = \\ P(1|1,2,3,...,K) \bullet P(2|2,3,...,K) \bullet ... \bullet P(K-1|K-1,K) \quad\quad 18.10$$

Then, if the IIA property holds, an MNL model can be substituted for each of the $K-1$ probabilities in equation 18.10, resulting in the standard rank-ordered logit model (Beggs et al. 1981):

$$P(U_k > U_l > ... > U_K) = \prod_{k=1}^{K-1} \frac{\exp(\mu\beta' x_k)}{\sum_{i=k}^{K} \exp(\mu\beta' x_i)} \qquad 18.11$$

This formula implies that an observation of K ranked alternatives can be "exploded" into $K - 1$ statistically independent choices, and that the probability shown in equation 18.11 is the product of the exploded choices.

The log-likelihood function for the rank-order model is the sum of ordinary MNL log-likelihoods over the exploded choices:

$$L = \sum_{k=1}^{K-1} \sum_{n=1}^{N} [\mu\beta' x_{kn} - \ln \sum_{i=k}^{K} \exp(\mu\beta' x_{in})] \qquad 18.12$$

where μ is typically set equal to one.

Ranking models ostensibly offer the advantage of providing more information than a standard choice model because of the additional information contained in the sequence of choices. From a statistical perspective, the additional information provided by rankings should lead to smaller standard errors for parameter estimates. However, experience has shown that error variance increases (scale decreases) as respondents proceed down through the sequence of choices.

5. UNDERSTANDING THE SCALE PARAMETER

Equations 18.4 and 18.11 show that in the MNL model the scale factor and the preference parameters are always represented in multiplicative form $\mu\beta$, so it is not possible to identify scale in any particular model. Because scale and preference parameters are always confounded, parameters estimated from different data sets should not be directly compared, because it is not clear whether differences are due to preferences, scale, or both. However, if data are available from more than one choice set, then it is possible to recover an estimate of relative scale parameters for the data sets. And, given an estimate of scale, it is then possible to test whether parameter vectors are the same up to a scaling constant.

The scale factor in a MNL model is inversely related to the variance of the equation error (where π is the mathematical constant 3.1416…):

$$\sigma^2 = \pi^2 / 6\mu^2 \qquad 18.13$$

A larger scale is indicative of a smaller variance and, in turn, implies less noise and a better fitting model.[6] If, in a ranking task, respondents become fatigued or confused as they proceed through lower ranks, then the scale parameter would be expected to decrease (variance increases) as ranking depth increases. In an analysis of ranking data, Hausman and Ruud (1987) found a general decrease in scale with ranking depth, although the change in μ was not monotonic. Ben-Akiva et al. (1992) found that scale decreased with rank and recommended that data not be pooled in a single ranking model unless further testing indicated that parameter vectors are equal up to the scaling constants. They suggested a simple graphical method for identifying the scale parameters. This procedure is identical to a method proposed by Swait and Louviere (1993) for comparing multinomial logit models across independent data sets.

Notice in equation 18.10 that the first exploded rank is nominally identical to a choice question where respondents are asked to choose one item from a choice set containing K items. Consequently, not only is it possible to test the stability of preferences across different depths of ranking data, it is also possible to test the cross-validity of choice and ranking models using exploded rank data.

6. VALIDITY TESTS FOR CHOICE MODELS

If data are available from two or more choice sets (either independent sub samples from choice experiments or presumed independent choices in a ranking experiment), relative scale parameters can be recovered, and rescaled parameter vectors can be tested for equality. This can be accomplished by optimally rescaling the set of explanatory variables in one of the data sets.

Consider the case of two data sets X_1 and X_2 with common attributes. Let $\mu_1\beta_1$ represent parameter estimates from X_1, and let $\mu_2\beta_2$ represent parameter estimates from X_2. If both data sets reflect identical preferences ($\beta_1 = \beta_2$) but have different scales ($\mu_1 \neq \mu_2$), casual examination of the estimated coefficients would indicate that tastes were different (because $\mu_1\beta_1 \neq \mu_2\beta_2$). However, it is possible to estimate the relative scale parameter (μ_2/μ_1) and then test the hypothesis that $\beta_1 = \beta_2$ controlling for relative scale.[7]

Swait and Louviere (1993) show how to test the joint hypotheses: H$_1$: $\beta_1 = \beta_2$ and $\mu_1 = \mu_2$ using a two-stage variant of the Chow test. The test proceeds by the following steps:

1. Make an initial estimate of μ_2/μ_1. This can be accomplished by regressing β_1 on β_2 or, more simply, by setting $\mu_2/\mu_1 = 1$.

2. Multiply data points in X_2 by μ_2/μ_1 and pool (vertically concatenate) the data with X_1.
3. Maximize the log likelihood function for the pooled data.
4. Repeat steps 2 and 3 for smaller and larger values of μ_2/μ_1.
5. Plot the values obtained from steps 2 through 4 until a peak is found for maximum likelihood as a function of μ_2/μ_1.
6. At the peak value for μ_2/μ_1, the data have been optimally rescaled.

The hypothesis $H_A : \beta_1 = \beta_2$ can then be tested using the likelihood ratio test statistic:

$$\lambda_A = -2[L_\mu - (L_1 + L_2)] \qquad 18.14$$

where L_μ is the log likelihood value for the optimally adjusted pooled data model, L_1 is the log likelihood value for the X_1 model, and L_2 is the likelihood value for the X_2 model.[8]

If the hypothesis $\beta_1 = \beta_2$ is rejected after optimally adjusting for the scale parameter, then it is clear that the data do not represent the same preferences. If this hypothesis is not rejected, parameters estimated from the pooled data can be used for analysis and inference. Further, if H_A is not rejected, it is possible to test the hypothesis that $H_B: \mu_1 = \mu_2$. This is simply accomplished by pooling X_1 and X_2 (unadjusted) and using the likelihood ratio test statistic:

$$\lambda_B = -2[L_p - L_\mu] \qquad 18.15$$

where L_p is the log likelihood value for the (unadjusted) pooled data.

7. THE FOREST MANAGEMENT EXPERIMENT

Our forest management experiment is based on data collected in a mail survey of Maine residents regarding their preferences for alternative timber harvesting practices. As described in the introduction, forest management in Maine is a controversial subject. After holding discussions with forest management experts in the State, and after focus groups conducted with randomly sampled citizens, we chose seven forest management attributes to include in the experiment (table 18.1). The number of attributes and levels we used resulted in a larger design space and more complex choice problems than those of previous forest valuation studies.

Table 18.1. Forest management attributes, levels, and names

Attributes	Levels	Variable Names
Forest road density	One road every mile	ROADS_1
	One road every ½ mile	ROADS_½
Live trees after harvest	No trees > 6-in. diam./ acre	LIVE_0
	153 trees > 6-in. diam./ acre	LIVE_153
	459 trees > 6-in. diam./ acre	LIVE_459
Dead trees after harvest	**Remove all**	DEAD_0
	5 trees/acre	DEAD_5
	10 trees/acre	DEAD_10
Max. size of harvest area	5 acres	HAREA_5
	35 acres	HAREA_35
	125 acres	HAREA_125
Available for harvesting	**80%**	HVST_80
	50%	HVST_50
	20%	HVST_20
Width of riparian buffers	500 ft. min.	H2O_500
	250 ft. min.	H2O_250
Slash disposal	Leave it where it falls	SLASH_LV
	Distribute along skid trails	SLASH_DST
	Remove all	SLASH_NO

The management practices representing base level (most common) are shown in bold.

As can be seen, most of the levels included for forest management attributes represent more environmentally benign practices relative to the base level. Only two attributes (number of live trees remaining after harvest and maximum size of harvest area) include levels with greater and lesser environmental impact than the base level. Attributes were coded using effects codes with the base level of the attribute the omitted level.[9]

Descriptive information regarding the pros and cons of alternative management practices, as well as a description of the most common practice, was presented by enclosing an information booklet with the questionnaire. Line drawings were used to represent two levels of each management attribute to help respondents conceptualize the management activity being addressed. The first questions included in the questionnaire booklet were quiz questions to help us gauge how well respondents' understood the background information.

The context for evaluating management activities described the State purchasing a 23,000-acre parcel of forest land from a large forest land management company. Respondents were given a description of the parcel and provided with a map showing its approximate location. They were then presented with four management plans to consider for the parcel. Each management plan was composed of randomly assigned levels of each management practice. In addition, a monetary attribute was included in the design, which was a one-time increase in State income taxes to pay for the forest land purchase.[10]

Stated Preference Methods for Valuation of Forest Attributes

Alternative forest management plans were constructed using a completely randomized design across individuals. That is, attribute levels were randomly sampled from the entire design space and placed in potentially unique alternatives for each individual in the sample. Respondents were randomly assigned into sub samples for ranking and choice questions. For the ranking questions, respondents were asked to rank four management plans from most preferred to least preferred. For the choice question, respondents were asked to circle the letter of their most preferred management plan. An example of alternative forest management plans is shown below (table 18.2).[11]

Table 18.2. Sample forest management plans for the choice and ranking experiments

Attributes	Plan A	Plan B	Plan C	Plan D
Forest road density	1 every ½ mile	1 every mile	1 every mile	1 every ½ mile
Dead trees after harvest	5 trees/acre	5 trees/acre	Remove all	10 trees/acre
Live trees after harvest	459 trees/acre	153 trees/acre	459 trees/acre	No trees
Maximum size harvest opening	125 acres	125 acres	35 acres	5 acres
Proportion cut/ set-aside	20% cut/ 80% set-aside	50% cut/ 50% set-aside	50% cut/ 50% set-aside	20% cut/ 80% set-aside
Watershed protection	At least 250-ft. buffer zone	At least 500-ft. buffer zone	At least 250-ft. buffer zone	At least 500-ft. buffer zone
Slash disposal	Distribute along skid trails	Remove all	Leave it where it falls	Remove all
One-time tax increase	$400	$60	$140	$10

Preference parameters for the forest management attributes were estimated using MNL models for full ranks, exploded ranks 1 (choose one of four), exploded ranks 2 (choose one of three) and exploded ranks 3 (choose one of two), as well as an MNL model for responses to the choice question (table 18.3). Even a simple eyeball examination of the results provides valuable information. First, a comparison of the full ranks model with the exploded ranks models shows that the set of salient (statistically different than zero) attributes varied across the different specifications of the ranking model. It appears as though respondents searched for salient attributes in the management plans, and their focus shifted as they progressed through the ranking exercise. This may have resulted from the complexity associated with having to consider seven management attributes plus a tax price. We also note that the number of salient attributes decreased as ranking depth increased, and McFadden's R^2 for lower ranks were less than for exploded rank 1.[12] These indicators suggest that respondents became fatigued as they completed the ranking question. Based on our initial visual observation, the lack of consistent preferences across ranking depths suggests that ranking data should not be pooled to estimate a full ranks model.

A comparison of exploded ranks 1 and choose-one data, which are ostensibly identical response formats, provides substantial insight into preferences for forest management attributes. Here examination shows consistency regarding the saliency of management attributes: the same attributes have a statistically significant impact on respondent choices. The nonsalient attributes were also the same across response formats and included H2OZONE (the width of riparian buffers), ROADS (forest road density), and HAREA (maximum size of harvest area).[13]

The set of attributes that were salient in both response formats were tax price (TAX), the number of live trees remaining after harvest (LIVE), the number of dead trees remaining after harvest (DEAD), the proportion of the forest available for harvest versus set-aside (HVST), and the disposal of slash created by the harvesting operation (SLASH). Focusing first on the number of live trees remaining after harvest, the parameter estimate on LIVE_0 (no live trees > 6-in. diameter after harvest, or clear cutting) was negative and larger in magnitude than any other management attribute. Including a clear cutting alternative in the contingent management plan had a large negative impact on the conditional indirect utility of respondents, even though the word clear cut was not used in the survey. The parameter estimate for the omitted base level (153 trees > 6-in. diameter/acre left after harvest) was computed to be 0.311 in the exploded ranks 1 model and 0.315 in the choose-one model.[14] In both models, then, we identified a quadratic valuation function, where utility was maximized at the base timber harvest level. Utility decreased rapidly from the moderate harvest intensity level to the clearcut harvest level. Utility decreased less rapidly as harvest intensity decreased from moderate to light.

The number of dead trees remaining after harvest was a salient attribute in both the exploded ranks 1 and choose-one models, but the most preferred level was different between the two models (DEAD_5 versus DEAD_10, respectively). This pattern was also identified for the attribute representing the percent of the forest area available for harvest (HVST_20 versus HVST_50, respectively). Why this shift occurred between models is not clear. What is clear, however, is that respondents preferred more dead trees left after harvesting (which mimics one aspect of old-growth forest structure) and a greater proportion of forest area set-aside for conservation, relative to the base level. However, there was lack of convergence across response formats regarding the optimal level of these attributes.

Stated Preference Methods for Valuation of Forest Attributes 335

Table 18.3. Parameters for MNL models estimated using ranking and choice data

Variable	Full Ranks (Std. Err.)	Exploded Ranks 1 (Std. Err.)	Exploded Ranks 2 (Std. Err.)	Exploded Ranks 3 (Std. Err.)	Choice (Std. Err.)
ROAD_1	0.108**	0.061	0.300***	-0.514	0.035
	(0.053)	(0.085)	(0.0093)	(0.109)	(0.076)
LIVE_0	-0.318***	-0.585***	-0.324***	0.016	-0.497***
	(0.074)	(0.133)	(0.126)	(0.145)	(0.114)
LIVE_459	0.115	0.274**	0.106	-0.089	0.182*
	(0.075)	(0.120)	(0.128)	(0.157)	(0.105)
DEAD_5	0.130*	0.324***	0.153	-0.195	0.123
	(0.074)	(0.125)	(0.132)	(0.162)	(0.103)
DEAD_10	0.147**	0.141	0.00004	0.455***	0.331***
	(0.073)	(0.125)	(0.132)	(0.165)	(0.101)
H2OZONE	0.030	0.072	-0.056	0.098	0.017
	(0.052)	(0.087)	(0.089)	(0.107)	(0.074)
HVST_20	0.214***	0.374***	0.075	0.170	-0.103
	(0.075)	(0.117)	(0.132)	(0.166)	(0.108)
HVST_50	0.024	0.010	0.030	0.070	0.357***
	(0.074)	(0.122)	(0.126)	(0.147)	(0.102)
HAREA_125	-0.008	0.178	-0.316**	0.072	0.021
	(0.074)	(0.118)	(0.135)	(0.162)	(0.108)
HAREA_5	-0.075	-0.109	0.145	-0.292*	-0.004
	(0.074)	(0.122)	(0.124)	(0.155)	(0.107)
SLASH_LV	0.123*	0.231*	0.168	-0.048	0.179*
	(0.073)	(0.121)	(0.131)	(0.159)	(0.106)
SLASH_DST	0.116	0.385***	-0.119	0.068	0.023
	(0.075)	(0.126)	(0.140)	(0.152)	(0.105)
TAX	-0.00090***	-0.00083***	-0.00096***	-0.00081***	-0.00152***
	(0.00011)	(0.00025)	(0.00025)	(0.00024)	(0.00027)
L(0)	—	-295.2807	-232.9058	-146.9472	-385.3898
L(β)	—	-251.3436	-210.4798	-132.0054	-332.1598
1-L(β)/L(0)	—	0.1488	0.0963	0.1017	0.1381
N	212	212	212	212	278

*** = significant at 1% level, ** = significant at 5% level, * = significant at 10% level.

The disposition of slash created by the harvest operation was also a salient attribute in the exploded ranks 1 and choose-one models. Again we identified a similar pattern: respondents preferred more environmentally benign practices for slash disposal relative to the base level (remove all slash), but there was lack of convergence across models regarding the optimal level for this attribute.

Results from the Swait and Louviere (1993) procedure for testing hypotheses regarding the equality of parameter estimates in MNL models confirm the results from the eyeball comparisons (table 18.4). The hypothesis that parameter estimates for choose-one (C1) and exploded ranks 1 (ER1) are no different was rejected at the 95% confidence level.[15] We also found that parameter estimates for the exploded ranks data were not equal over all ranking depths. Although we rejected the hypothesis (at the 95% confidence level) that preference parameters for ER1 and exploded ranks 2 (ER2) are the same, we could not reject the hypothesis that preference parameters for ER1 and exploded ranks 3 (ER3) are the same. This is likely due to the relatively large standard errors associated with the ER3 model. Further, we found that model variance increased along with ranking depth. This result is consistent with the idea that respondents become fatigued as they complete a ranking task, reflecting findings reported by Hausman and Ruud (1987) and Ben-Akiva et al. (1992). These results lead us to formally conclude that exploded ranks data should not be pooled to estimate a full ranks model in our case.

Table 18.4. Results for hypothesis tests regarding parameter equality in MNL models: H_A ($\beta_1 = \beta_2$); H_B ($\mu_1 = \mu_2$)

Test	μ_1/μ_2	L_1	L_2	L_μ	λ_A	Reject H_A?	L_p	λ_B	Reject H_B?
ER1: C1	0.97	-251.34	-332.16	-596.33	25.66	Yes[a]	—	—	—
ER1: ER2	1.82	-251.34	-210.48	-476.26	28.87	Yes[a]	—	—	—
ER1: ER3	2.22	-251.34	-132.01	-393.89	21.08	No[a]	-397.85	7.93	Yes[b]

[a] χ^2 statistic for 14 d.f. and 95% confidence level = 23.69
[b] χ^2 statistic for 1 d.f. and 95% confidence level = 3.84

Finally, estimates of compensating variation were computed using the parameter values for the choose-one model shown in table 18.2 and the formula shown in equation 18.9. We considered a reduced impact timber harvest alternative for the contingent forest versus a base level timber harvest alternative representing typical current management practices. We specified the reduced impact (*base level*) alternative to have the following attributes and levels: (1) 459 live trees > 6-in. diameter (*153 live trees > 6-in. diameter*) per acre remaining after harvest, (2) 10 dead trees (*no dead trees*) per acre remaining after harvest, (3) harvest permitted on 50% (*80%*) of the forest, and (4) leave slash where it falls (*remove all slash*). The value of the reduced impact timber harvest alternative, relative to the base level, was estimated to be $1,081.58. This is a per household lump sum amount.

8. SUMMARY AND CONCLUSIONS

Attribute-based stated preference methods are relatively new tools for environmental valuation. They can provide detailed information about citizen preferences for incremental changes in a set of environmental attributes under the control of managers and policy makers. ABMs seem eminently suitable for valuation problems in cost/benefit analyses of forest management and protection alternatives.

However, as with other stated preference methods, such as contingent valuation, the application of ABMs to environmental valuation is not trouble free. An important issue is the convergent validity of different response formats. In the case study reported in this chapter, convergent validity was not established for two nominally identical responses, choose-one and first rank. The lack of convergence may be due to differences in cognitive processes used to answer the questions. Future research needs to investigate the effect of decision context and complexity on responses made to attribute-based stated preference questions.

Although we were unable to recover statistically identical preference parameters using our split-sample design and different response formats, highly similar preferences were recovered, allowing some general conclusions to be made. First, the general public in Maine preferred a balance of timber harvest and natural area protection and they were willing to pay for an increase in the amount of forest land set aside from timber production relative to the base level. This result echoes the findings reported in Garrod and Willis (1996). Second, our results showed that clear cut timber felling greatly reduced conditional indirect utility, similar to findings reported in Hanley et al. (1998) and Boxall and MacNab (2000). Third, as an alternative to clear cutting, the public preferred a medium-intensity felling alternative relative to light-intensity harvests. This result may reflect public awareness of the practice of high-grading stands in which the best trees are selected for harvest, leaving genetically inferior trees for regeneration. Fourth, our results showed that the public prefers timber harvesting alternatives that leave standing dead trees after harvest (mimicking one aspect of old-growth forest structure) and that leave harvesting slash in the woods (which benefits soil productivity and provides habitat for small animals and insects).

The general public in Maine, as represented by our survey respondents, was willing to pay a considerable amount for timber harvesting practices that reduced the biological and amenity impacts on forest ecosystems. Willingness to pay for reduced-impact harvesting alternatives likely reflects the public's concern with a variety of goods and services associated with healthy forest ecosystems, including the provision of timber, recreational opportunities, wildlife habitat, and aesthetically pleasing views. We think that carefully conducted citizen surveys, such as those presented here, can help forest managers and policy makers identify management alternatives preferred by the public and that such information can add balance to public debates regarding forest policy.

9. LITERATURE CITED

ADAMOWICZ, W., J. L. LOUVIERE AND J. SWAIT. 1998a. Introduction to attribute-based stated choice methods. Final Report, Resource Valuation Branch, NOAA, U.S. Dept. of Commerce, Washington DC. 44 p.
ADAMOWICZ, W. P. BOXALL, M. WILLIAMS AND J. LOUVIERE. 1998b. Stated preference approaches for measuring passive use values: Choice experiments and contingent valuation. Am. J. Agr. Econ. 80: 64-75.
ADAMOWICZ, W., J. SWAIT, P. BOXALL, J. LOUVIERE AND M. WILLIAMS. 1997. Perceptions versus objective measures of environmental quality in combined revealed and stated preference models of environmental valuation. J. of Env. Econ. and Manage. 32: 65-84.
BEGGS, S., S. CARDELL AND J. HAUSMAN. 1981. Assessing the potential demand for electric cars. J. Econometrics 17: 1-20.
BEN-AKIVA, M., T. MORIKAWA, AND F. SHIROISHI. 1992. Analysis of the reliability of preference ranking data. J. Business Res. 24: 149-164.
BEN-AKIVA, M. AND S.R. LERMAN. 1985. Discrete Choice Analysis: Theory and Application to Travel Demand. MIT Press, Cambridge, MA. 390 p.
BENNETT, J. AND R. BLAMEY (eds.). 2001. The Choice Modelling Approach to Environmental Valuation. Edward Elgar, Northampton, MA. 269 p.
BOXALL, P.C. AND B. MACNAB. 2000. Exploring the preferences of wildlife recreationists for features of boreal forest management: A choice experiment approach. Can. J. For. Res. 30: 1931-1941.
BOYLE, K.J., T.P. HOLMES, M.F. TEISL, AND B. ROE. 2001. A comparison of conjoint analysis response formats. Am. J. Agr. Econ. 83: 441-454.
GARROD, G.D., AND K.G. WILLIS. 1997. The non-use benefits of enhancing forest biodiversity: a contingent rank study. Ecol. Econ. 21: 45-61.
HAEFELE, M.A., AND J.B. LOOMIS. 2001a. Improving statistical efficiency and testing robustness of conjoint marginal valuations. Am. J. Agr. Econ. 83: 1321-1327.
HAEFELE, M.A., AND J.B. LOOMIS. 2001b. Using the conjoint analysis technique for the estimation of passive use values of forest health. J. Forest Econ. 7: 9-28.
HANEMANN, W.M. 1999. Welfare analysis with discrete choice models. P.33-64 in Valuing Recreation and the Environment: Revealed Preference Methods in Theory and Practice, Herriges, J.A., and C. L. Kling (eds.). Edward Elgar, Northampton, MA. 290 p.
HANLEY, N., R. WRIGHT AND W. ADAMOWICZ. 1998. Using choice experiments to value the environment: Design issues, current experience and future prospects. Environ. & Res. Econom. 11:413-428.
HAUSMAN, J.A. AND P.A. RUUD. 1987. Specifying and testing econometric models for rank-ordered data. J. Econometrics 34: 83-104.
HAUSMAN, J. AND D. MCFADDEN. 1984. Specification tests for the multinomial logit model. Econometrica 52:1219-1240.
HOLMES, T.P., AND W. ADAMOWICZ. Forthcoming. Attribute-based methods. In A Primer on Non-Market Valuation. Champ, P., T. Brown, and K.J. Boyle (eds.). Kluwer Academic Publishers, Dordrecht, The Netherlands.
HOLMES, T.P., K. ALGER, C. ZINKHAN, AND E. MERCER. 1998. The effect of response time on conjoint analysis estimates of rainforest protection values. J. Forest Econom. 4: 7-28.
LOUVIERE, J.L. 1988. Conjoint analysis modeling of stated preferences: A review of theory, methods, recent developments and external validity. J. Transport Econom. and Policy 20: 93-119.
LOUVIERE, J.L., D.A. HENSHER, AND J.D. SWAIT. 2000. Stated Choice Methods: Analysis and Application. Cambridge University Press, Cambridge, UK. 402 p.

MARSCHAK, J. 1960. Binary choice constraints on random utility indicators. P. 312-329 *in* Stanford Symposium on Mathematical Methods in the Social Sciences, K. Arrow (ed.). Stanford University Press, Stanford, CA.

MCFADDEN, D. 1986. The choice theory approach to market research. Marketing Sci. 5: 275-297.

MCFADDEN, D. 1973. Conditional logit analysis of qualitative choice behavior. P. 105-42 *in* Zarembka, P. (ed.). Frontiers in Econometrics, Academic Press, New York.

MOREY, E.R. 1999. Two rums uncloaked: Nested-logit models of site choice and nested-logit models of participation and site choice. P. 65-120 *in* Herriges, J.A., and C.L. Kling (eds.). Valuing Recreation and the Environment: Revealed Preference Methods in Theory and Practice. Edward Elgar, Northampton, MA. 290 p.

SCHABERG, R.H., T.P. HOLMES, K.J. LEE, AND R.C. ABT. 1999. Ascribing value to ecological processes: An economic view of environmental change. For. Ecol. Manage. 114: 329-338.

SWAIT, J., AND J. LOUVIERE. 1993. The role of the scale parameter in the estimation and comparison of multinomial logit models. J. Marketing Res. 30: 305-14.

---

[1] In 1996, a Ban Clearcutting Referendum was placed on the ballot with a more moderate Forest Compact developed by the Governor. The Ban Clearcutting initiative and the Forest Compact were rejected by voters. The Forest Compact was again placed on the ballot in 1997 and was again defeated. Subsequently, the conservation community worked out a 4-point plan that would have (1) placed strict limits on the amount and size of clearcuts, (2) set science-based post-harvest stocking standards, (3) ensured that cutting does not exceed growth, and (4) imposed mandatory audits to ensure the protection of ecosystem integrity. The Maine legislature subsequently voted down the 4-point plan.

[2] For a good review of rating, ranking, and choice methods, see Louviere (1988).

[3] Randomness in an individual's utility function is attributable to variation in preference unobserved by the researcher as well as errors in perception, discrimination, and optimization by the consumer (McFadden 1986).

[4] The cumulative distribution of the EV1 is: $F(\varepsilon) = \exp[-e^{-\mu(\varepsilon-\eta)}]$ (where η is a location parameter and μ is a positive scale parameter).

[5] The IIA property can be tested using the standard Hausman-McFadden test (1984). If the IIA property is violated, other modeling approaches are available, such as the nested form of MNL.

[6] Louviere et al. (2000, pp. 235-236) show that as variance approaches infinity, scale approaches zero, and the MNL model predicts equal choice probability for all alternatives due to a lack of discrimination between alternatives. Conversely, as variance approaches zero and scale approaches infinity, the MNL model perfectly discriminates between alternatives, and the logit function behaves as a step function.

[7] The scale parameter can also be estimated using a nested logit model (see Louviere et al. 2000). The main advantage of the full information maximum likelihood method is that a standard error for the scale parameter is estimated. Scale can also be parameterized with individual or design characteristics.

[8] As noted by Ben-Akiva et al. (1992), steps 1 through 6 can also be used to test the stability of parameter estimates from exploded ranks.

[9] For a description of effects coding, see Holmes and Adamowicz (2003) or Louviere et al. (2000).

[10] Tax prices used were $1, $10, $20, $40, $80, $120, $140, $160, $180, $200, $400, $800, and $1600.

[11] The option of not choosing (or not providing ranks for) any of the alternatives was included in a later question where people were asked whether or not they would vote for each of the alternatives if they were presented in a referendum.

[12] McFadden's R^2 is computed as $(1 - L(\beta)/L(0))$, where $L(\beta)$ is the likelihood value computed using the full set of parameter estimates, and $L(0)$ is computed using an intercept only.

[13] The lack of significance of riparian buffers may reflect a relatively high standard for the base level (250-ft. buffer) and indifference between the base level and a more stringent standard. Lack of significance for road density may indicate ambivalence across the sample between the gain in access due to greater road density and the loss of ecosystem services.

[14] The parameter value for the omitted attribute level can be computed for effects coded variables. The value of the parameter for the Lth level of an attribute is the sum $b_1(-1) + b_2(-1) + \ldots + b_{L-1}(-1)$ where b_n is the parameter estimate on the nth level ($n \neq L$) of an effects coded variable.

[15] This result is consistent with results reported in Boyle et al. (2001).

Chapter 19

Estimating Forest Recreation Demand Using Count Data Models

Jeffrey E. Englin, Thomas P. Holmes, and Erin O. Sills
University of Nevada - Reno, USDA Forest Service and North Carolina State University

Forests, along with related natural areas such as mountains, lakes, and rivers, provide opportunities for a wide variety of recreational activities. Although the recreational services supplied by forested areas produce value for the consumers of those services, the measurement of recreational value is complicated by the fact that access to most natural areas is non-priced. Because outdoor recreation often competes with commodity uses of forests, such as timber harvesting or mineral extraction, failure to account for the recreational use of forest land makes it impossible to determine the efficient use of forest resources.

A key insight attributed to Harold Hotelling is that the price of recreational access can be inferred from information on travel costs. Subsequent development of this idea was undertaken by Marion Clawson (1959) and, a few years later, articulated in a general work on the economics of outdoor recreation (Clawson and Knetsch 1966). The basic Hotelling-Clawson-Knetsch (HCK) approach to estimating recreation demand is to statistically regress the number of trips taken to a recreational site on the round-trip cost of travel between trip origins and the site. A set of demand shift variables are also typically included in the specification to control for socio-economic characteristics of visitors, indicators of site quality, and costs associated with visiting substitute sites. Once a travel cost demand curve is estimated, the value of a recreational site can be computed by integrating the area under the demand curve.

Two types of data can be used to estimate travel cost models (see, for example, Bockstael et al. 1991 and Freeman 1993). The early studies

Sills and Abt (eds.), Forests in a Market Economy, 341–359. ©Kluwer Academic Publishers. Printed in The Netherlands.

typically used aggregate data on origin zones; these are often referred to as zonal travel cost models. Per capita visitation rates for each origin zone (often counties, but also distance zones) were computed, and distances were translated into travel costs using cost per mile multipliers. Socio-economic variables for origin zones were proxy variables for the representative visitor, and prices based on travel costs to substitute sites were included in the specification.

The second type of data that can be used to estimate travel cost models is based on individual observations of visitation rates and socio-economic variables (referred to as individual travel cost models). Individual data do not rely on the representative visitor assumption. The added precision in describing individual characteristics and trip decisions has led to the development of a rich array of empirical methods and, in particular, models based on random utility maximization (RUM).

The RUM approach models the choice of a recreation site from among a set of alternative sites as a utility-maximizing decision, where utility includes a stochastic component. RUM models emphasize the impact of site quality on recreation demand and are estimated using either multinomial or nested logit models. Forestry examples include Englin et al. (1996) and Pendleton and Shonkwiler (2001).

Another approach that focuses attention on site quality is the hedonic travel cost (HTC) method. The HTC method is used to estimate the demand for site characteristics using a two-step procedure (Brown and Mendelsohn 1984). In the first stage, marginal values (implicit prices) of the site characteristics are estimated for each origin zone. Then, demand functions for characteristics are estimated in the second stage across all origins. Applications to forestry include Englin and Mendelsohn (1991), Holmes et al. (1997), and Pendleton et al. (1998).

During the past decade, there has been an explosion of interest in the application of count data models based on the Poisson distribution to estimation of HCK models of recreation demand. In this chapter, we provide an overview of the major developments in count data travel cost modeling and show how they can be applied to forest-based recreation.

In contrast to earlier HCK modeling that used ordinary least-squares (OLS) regression, count data models emphasize the non-negative, integer nature of data on the number of trips taken and are most useful when the counts (per person) are small. This is often the case with forest recreation, such as backcountry trips or adventure activities, which most people participate in only a few times a year. Although the normal distribution is a good approximation of the Poisson distribution (which is sometimes called the "law of rare events") if the mean of the distribution is large, the normal distribution provides a poor approximation of the Poisson for small mean values. This is due to the skewness of the Poisson distribution (Kalbfleisch 1985). Count data estimators place positive probability only on possible, discrete events. OLS estimators can place positive probability on fractional and negative events (Creel and Loomis 1990). Thus, for small counts of recreational trips, a count data distribution is more likely to represent the true data-generating process than is a normal distribution.

The remainder of the chapter is divided into four sections. The next section provides a brief history of the development of the count data travel cost model. The second section outlines the basic theoretical and empirical issues that must be considered in designing a count data analysis of recreation demand. In section 3, we use count data models to estimate the value of rain forest protection in Brazil. In the final section, we summarize the chapter and discuss the needs for future research in count data recreation demand modeling.

1. A BRIEF HISTORY OF COUNT MODELS IN TRAVEL COST ANALYSIS

Over the past two decades, applied econometricians have paid increasing attention to estimation and testing of count data models. Early econometric analyses of count data models include the effects of research and development on patents issued (Hausman et al. 1984) and the relationship between urban air quality and respiratory illness (Portney and Mullahy 1986).

Not long after these foundational studies, it was realized that data collected for the HCK class of travel cost models were amenable to analysis using count data models. To the authors' knowledge, Shaw (1988) was the first to apply count data models to recreation demand. Shaw recognized that recreation data collected on site are truncated and may suffer from endogenous stratification (people who frequently visit a site are more likely to be sampled than people who rarely visit), and that failure to correct for these problems leads to biased estimates of population parameters. Shaw's estimator is presented in section 2.4.

Economists are always cognizant of the need for empirical models to be consistent with an underlying theoretical foundation. In 1993, Hellerstein and Mendelsohn provided such a foundation for count data travel cost models. They realized that on any choice occasion, the decision of whether or not to take a trip to a specific site can be modeled using a binomial distribution and, as the number of choice occasions increases throughout a recreational season, the binomial distribution asymptotically converges to a Poisson distribution.

During the 1990s, Poisson and negative binomial count data models (presented in sections 2.2 and 2.4) were estimated for a variety of recreational resources. Hellerstein (1991) estimated count data models for trips to the Boundary Waters Canoe Area and showed how models could be estimated using aggregate (zonal) data. Creel and Loomis (1990) tested a variety of Poisson and negative binomial estimators and found that count data models were more appropriate for estimating and predicting the demand for deer hunting in California than were OLS and nonlinear least-squares estimators. Yen and Adamowicz (1993) evaluated the statistical properties of welfare measures computed using count data models of the demand for hunting bighorn sheep and suggested caution when evaluating consumer surplus measures derived from truncated estimators (which are used for analysing on-site data). Englin and Shonkwiler (1995) developed a truncated, endogenously stratified negative binomial model and used it to estimate long-run demand for overnight hikes in the Cascade Mountains. Ovaskainen et al. (2001) used the Englin-Shonkwiler estimator to model the demand for forest recreation trips in Finland.

A recent development in count data modeling is based on the realization that observations of zero trips may be generated either by people who are not in the market (they would not take a trip at any positive price) or by people who are in the market but did not take a trip during the observation period (the price faced in the observation period was too high). Zero-inflated models (also referred to as augmented count or double-hurdle models) have not been as frequently applied as other count data models, probably because they require samples of the entire population. Shonkwiler and Shaw (1996) clarified the nature and interpretation of hurdle count data models, including single-hurdle selection-type models and zero-inflated models. Haab and McConnell (1996) presented zero-inflated models for beach trips; Shaw and Jakus (1996) showed how to estimate a zero-inflated model of the demand for rock climbing; and Gurmu and Trivedi (1996) estimated the demand for lake recreation using zero-inflated models. This approach is discussed in section 2.5.

2. THEORY AND EMPIRICAL ANALYSIS

2.1 Linear Exponential Demand and Welfare Estimates

Unlike demand functions based on the normal distribution, expected values in demand functions based on count data models are restricted to be non-negative. A functional form that guarantees positive mean values is the linear exponential (semi log) demand function:

$$E[Q_i] = e^{X_i\beta} \qquad 19.1$$

where $E[Q_i]$ is the expected number of visits to a site by individual i, X_i is a vector of observations on independent variables associated with individual i (including the travel cost), and the βs are parameters to be estimated. The specification in equation 19.1 makes a clear distinction between the functional form of the demand curve and the distributional assumptions used to obtain estimates of the demand function parameters. For count data models, the demand function is linked with a count data distribution by the relationship:

$$\lambda_i = E[Q_i] = e^{X_i\beta} \qquad 19.2$$

where λ_i is the mean of the count data distribution (for individual i).

Equation 19.1 represents expected, not actual, demand, because the equation errors associated with individual heterogeneity do not enter the expression. Therefore, Marshallian consumer surplus in a count model is computed for the typical consumer.

As usual, consumer surplus is found by integrating the area under the demand curve from a lower price p_0 to an upper price p_1. For the linear exponential demand function, this integration yields the expression:

$$CS_i = \int_{p_0}^{p_1} \lambda_i dp = \frac{\lambda_i(X_i^{p_1}) - \lambda_i(X_i^{p_0})}{\beta_{tc}} \qquad 19.3$$

where β_{tc} is the parameter estimate on travel cost, $X_i^{p_1}$ is the X_i vector substituting p_1, and $X_i^{p_0}$ is the X_i vector substituting p_0. If p_1 is the choke price (the price at which the expected number of trips equals zero), then consumer surplus is written as:

$$CS_i = \int_{p_0}^{p_1} \lambda_i dp = \frac{-\lambda_i(X_i^{p_0})}{\beta_{tc}} \qquad 19.4$$

because $\lambda_i(X_i^{p_1}) = 0$. Equation 19.3 would be useful if the analyst wanted to estimate the change in surplus associated with a marginal increase in price, such as an increase in user fees. Equation 19.4 provides an estimate of total consumer surplus associated with the site. By dividing total consumer surplus by the number of trips (λ_i), it is easily seen that Marshallian consumer surplus per trip is $-1/\beta_{tc}$.

It is important to recognize that welfare estimates should only be applied to the sample frame from which the X_is are drawn. If the estimator used to obtain the βs recovers the population parameters, then consumer surplus and latent demand (the desired number of trips) are found by simulating the demand equation using population means for the independent variables. This procedure obtains a consumer surplus estimate for the typical member of the population only if the population parameters were accurately recovered from the available sample.

To test hypotheses about consumer surplus estimates (such as the hypothesis that consumer surplus associated with a project exceeds project cost), it is necessary to obtain estimates of variance. Englin and Shonkwiler (1995) showed that the second-order Taylor series approximation of the variance of consumer surplus associated with linear exponential demand is:

$$Var\left(\frac{1}{\beta_{tc}}\right) = \frac{V}{\beta_{tc}^4} + 2\frac{V^2}{\beta_{tc}^6} \qquad 19.5$$

where V is the variance of β_{tc}. Of course, if estimates of exact changes in welfare are desired, then Hicksian measures are required. Bockstael et al. (undated) determined the compensating variation (CV) and equivalent variation (EV) formulas to be:

$$CV = \frac{1}{\beta_y}\ln(1+\frac{\lambda\beta_y}{\beta_{tc}}) \qquad 19.6$$

and

$$EV = -\frac{1}{\beta_y}\ln(1-\frac{\lambda}{\beta_{tc}}) \qquad 19.7$$

where β_y is the coefficient on the income variable, and β_{tc} is the coefficient on the travel cost variable. Englin and Shonkwiler (1995) provide a method for calculating variances around these Hicksian welfare measures.

2.2 Econometric Analysis of Single-Site Count Models

Two versions of single-site count models have been developed. The first version applies when data are available only for a specific site. These models are based on either data for individual visitors or zonal data. The variation in prices needed to estimate the demand curve is obtained by pooling different individuals who face different travel costs. Other explanatory variables are included in the specification to control for variation in socio-economic characteristics (such as income).

Studies based on a specific site are unusual. In the second version of single-site models, data are pooled across individuals and sites. This approach imposes the restriction that parameters are the same across all of the pooled sites. If this restriction is accepted and individuals and sites are both pooled, then the independent variables include the characteristics of the sites as well as the individuals. Examples of this approach include Creel and Loomis (1990, 1992), Englin and Shonkwiler (1995), and Ovaskainen et al. (2001).

Single-site count data models are usually estimated using either the Poisson or the negative binomial distribution. The probability density function (*pdf*) for the Poisson is a one-parameter distribution (the mean equals the variance) and is written as:

$$\Pr(Q_i = q_i) = \frac{e^{-\lambda_i} \lambda_i^{q_i}}{q_i!} \qquad 19.8$$

where q_i is a non-negative integer, and the log-likelihood function for a sample of size n is given by:

$$\ln L = \sum_{i=1}^{n} [-e^{X_i \beta} + q_i X_i \beta - \ln q_i!] \qquad 19.9$$

(recall that $\lambda_i = exp(X_i\beta)$). Parameter estimates are obtained by maximizing the log-likelihood function.

The mean-variance equality restriction of the Poisson model has been viewed as its major limitation. One way to account for over-dispersion (variance > mean) in count data is to include a stochastic variable (ε) in equation 19.2 that accounts for heterogeneity across people and allows $\lambda_i = exp(X_i\beta + \varepsilon_i)$ to vary according to a specific probability law. Cameron and Trivedi (1986) show that if $exp(\varepsilon)$ follows a gamma (Γ) distribution, then the compound count data generation process follows a negative binomial distribution. The *pdf* for the negative binomial distribution is:

$$\Pr(Q_i = q_i) = \frac{\Gamma\left(q_i + \frac{1}{\alpha}\right)}{\Gamma(q_i + 1)\Gamma\left(\frac{1}{\alpha}\right)} (\alpha\lambda)^{q_i} (1 + \alpha\lambda)^{-(q_i + \frac{1}{\alpha})} \qquad 19.10$$

where $1/\alpha$ is a dispersion parameter. Other forms of the dispersion parameter are possible (Cameron and Trivedi 1986). The distribution in equation 19.10 has conditional mean λ_i and conditional variance $\lambda_i (1 + \alpha\lambda_i)$. Since $\lambda_i > 0$ and $\alpha > 0$, it is clear that the variance is greater than the mean. If the dispersion parameter is not different from zero, the negative binomial model reduces to the Poisson. Software is available for estimating the parameters of both the Poisson and the negative binomial models.

2.3 Practical Issues in Model Specification

Until now this chapter has focused on what may be called the science of recreation count modeling. These tools, while powerful, must be implemented in a manner appropriate to a given context. A number of practical issues arise related to model specification.

To see the question of model specification and welfare estimation, return to the formula for consumer surplus in equations 19.3 and 19.4. Two methods for incorporating trip attributes in the welfare estimates are possible. First, changes in X affect total welfare, and a shift $(X_1 - X_2)$ provides a measure of the associated change in consumer surplus. Englin and Shonkwiler (1995) used this approach to derive the long-run shift in hiking values from demographic shifts over a four-decade period. This method could also be used to derive impacts resulting from changes in site characteristics if that information were included in the Xs. Notice, however, that the change in consumer surplus is entirely driven by changes in visitation.

In some situations, this simple approach may be unsatisfactory. Suppose one were interested in estimating the impact of clearcuts on the economic welfare of hikers. Both the number of trips and the consumer surplus per trip would be affected. A practical remedy is the varying parameter model where the slope of the demand curve is a function of the level of characteristics (Vaughan and Russell 1982). This is accomplished by adding a term that includes the travel cost interacted with the characteristic level. Total consumer surplus in this model becomes:

$$CS_i = \frac{-\lambda_i(X_i^{P_0})}{\beta_{tc}(X_i)} \qquad 19.11$$

where $\beta_{tc}(X_i)$ indicates that the parameter estimate on travel cost is a function of site quality. In general, one simply interacts travel cost and the characteristics linearly, but other specifications are possible.

A second related issue is the measurement of forest ecosystem attributes in a pooled-site model. Given that GIS data are available in many areas, the researcher has several options. Consider the evaluation of hiking trails that pass through different ecotypes where trails cover major changes in elevation. The simplest alternative is to measure the ecotype as present or absent along the trail (Englin and Shonkwiler 1995). A second approach is to use the total area of forest type that a trail goes through, or the total length of the trail that passes through a given ecotype (Pendleton et al. 1998). A third alternative is to use a latent characteristics model to construct bundles of attributes that represent the holistic quality of ecotypes (Pendleton and Shonkwiler 2001). The approach taken will depend, among other things, on the quality of the data available and the variation in the relevant ecosystems. Focus groups and other scoping methods can indicate which of these measures is most relevant to recreationist decision-making.

2.4 On-Site Survey Count Data Estimators

Recreation demand models are often estimated using survey responses collected from on-site samples of visitors. This is because it is generally less expensive to collect data on site than to collect data from the general population. However, it is useful to be able to estimate population parameters from on-site (truncated) samples so that total demand and value can be computed. This is accomplished by adjusting the untruncated models (Grogger and Carson 1991).

The *pdf* of a variable truncated at zero (i.e., zeros are not observed) is simply the untruncated *pdf* $f(q_i)$ divided by the area under $f(q_i)$ where $q_i > 0$. This guarantees that the area under the truncated *pdf* equals 1. For the Poisson distribution, the probability that q_i exceeds zero is $(1 - exp(-\lambda_i))$, and, dividing the expression in equation 19.8 by this probability, the conditional probability for a zero-truncated model is:

$$\Pr(Q_i = q_i) = \frac{e^{-\lambda_i}(\lambda_i)^{q_i}}{q_i! \left[1 - e^{-\lambda_i}\right]} \qquad 19.12$$

This procedure can also be used to obtain the conditional probability for the negative binomial distribution (for example, see Creel and Loomis 1990)

Shaw (1988) recognized that, in addition to truncation at zero, on-site samples are endogenously stratified. That is, people who visit a site often are more likely to be sampled than are people who visit infrequently. He showed that the on-site *pdf* of the ith person in the population is the product of the untruncated *pdf* and the variable q_i/λ_i, which is the ratio of actual trips to expected number of trips for the representative individual with characteristics X_i:

$$\Pr(Q_i = q_i) = \frac{e^{-\lambda_i}(\lambda_i)^{q_i-1}}{(q_i - 1)!} \qquad 19.13$$

A comparison of equations 19.13 and 19.8 shows that population parameters for the Poisson model, controlling for truncation and endogenous stratification, can be estimated from an on-site data sample by replacing q_i with $(q_i - 1)$. Unfortunately, this convenient result does not hold for the negative binomial model. As shown by Englin and Shonkwiler (1995), the on-site sample's negative binomial density function (found by multiplying equation 19.10 by the ratio q_i/λ_i) is:

$$\Pr(Q_i = q_i) = \frac{q_i \Gamma(q_i + \frac{1}{\alpha})\alpha_i^{q_i}\lambda_i^{q_i-1}[1 + \alpha_i\lambda_i]^{-\left(q_i + \frac{1}{\alpha}\right)}}{\Gamma(q_i + 1)\Gamma\left(\frac{1}{\alpha}\right)}. \qquad 19.14$$

While equation 19.13 can be estimated in standard packages, the likelihood function associated with equation 19.14 must be programmed.

2.5 Population Samples, the Participation Decision, and Zero-Inflated Models

Prior research has shown that truncated count data models do not always provide good estimates of population parameters, and substantial benefits may be gained by collecting information on nonparticipants (Yen and Adamowicz 1993). It is logical that the most direct way to estimate population parameters is to collect trip data from a sample of the population. These data would include information on people who did not take recreation trips to the site(s) of interest. Recent research has shown that modeling the participation decision can increase the efficiency of parameter estimates and provide planners with information about the segmentation of recreation markets (Haab and McConnell 1996).

The zero-inflated Poisson (ZIP) and zero-inflated negative binomial (ZINB) models are more general than the Poisson and negative binomial models in that they relax the restriction that an identical process generates both the zeros and the positive integers. Recreation data collected from users and nonusers of a resource provide two kinds of information: (1) whether or not to participate, and (2) the quantity demanded conditional on the participation decision. Poisson and negative binomial count data models do not extract information about the participation decision from the zeros in the data but treat the zeros as being generated by the same process that generates positive observations.

Instead of a single data-generation process, the ZIP/ZINB models consider that (1) $q_i \sim 0$ with probability p_i, and (2) $q_i \sim$ Poisson or negative binomial with probability $1 - p_i$. For the Poisson model, this implies that:

$$q_i = 0 \quad \text{with probability } p_i + (1 - p_i)e^{-\lambda_i}$$

$$q_i = k \quad \text{with probability } (1 - p_i)\frac{e^{-\lambda_i}\lambda_i^k}{k!} \qquad 19.15$$

where $k = 1,2,3,...$ are positive integers, and $exp(-\lambda_i)$ is the Poisson probability of taking zero trips. Note that in equation 19.15, zero trips can be generated by both a binomial process (for people not in the market) and a Poisson process (for people in the market who took zero trips). This later expression, $(1 - p_i)exp(-\lambda_i)$, represents the probability of a corner solution by potential users.

The (binary) recreation participation decision can be specified as a logistic model:

$$\log\left(\frac{1 - p_i}{p_i}\right) = Z_i\gamma \qquad 19.16$$

where γ is a vector of participation-decision parameters, and Z_i is a vector of explanatory variables that may or may not share variables with X_i. Expected consumer surplus per year is estimated by:

$$CS_i = (1 - p_i)(\frac{-\lambda_i(X_i^{P_0})}{\beta_{tc}}) \qquad 19.17$$

where β_{tc} is the parameter estimate on the travel cost variable. Expected consumer surplus per trip remains $(-1/\beta_{tc})$.

2.6   Demand System Analysis

Demand system analysis derives from the realization that there may be several sites that have related demand functions. If so, partial equilibrium analysis must account for multiple sites. Systems of count data demand functions can be motivated by recognizing that the limiting distribution of a multinomial distribution (where an individual chooses where to recreate from a choice set containing multiple sites) is a system of independent Poisson distributions (von Haefen and Phaneuf 2002).

In a demand system, the prices for sites and their substitutes change simultaneously. In the case of n sites being in the partial equilibrium, the demand system can be written as:

$$\ln(q_{ij}) = \alpha_j + \sum_{j=1}^{n} \beta_j p_{ij} + \gamma_j m_i \qquad 19.18$$

where q_{ij} is the number of trips taken by individual i to site j, p_{ij} are travel costs facing individual i for trips to site j, m_i is individual i's income, and α_j, β_j, and γ_j are parameters to be estimated.

Given a set of demand functions of the form shown in equation 19.18, an important question is whether they can be integrated back to the expenditure function (for example, see Bockstael et al. 1991) or, through inversion, the indirect utility function. LaFrance (1990) showed that if the linear exponential demand functions are treated as an incomplete demand system, the associated partial utility function could be recovered. Assuming no income effects, the partial indirect utility function consistent with equation 19.18 is:

$$m - \left(\frac{\alpha_1}{\beta_{11}}\right) e^{\sum_{i=1}^{k} \beta_{ii} p_i} - \sum_{i=k+1}^{n} \left(\frac{\alpha_i}{\beta_{ii}}\right) e^{\beta_{ii} p_i} \qquad 19.9$$

The conditions that a system of semi logarithmic demand functions must fulfill to form an integrable demand system have been well documented. Empirically, the conditions are simply restrictions on the relationships between the intercept, cross-price effects, and the income effect in the model. The intercept restriction is:

$$\alpha_j \geq 0 \qquad 19.20$$

or non-negativity, where α_j is the intercept for the jth site. A second restriction is that the income effect (β) is restricted to be the same across equations. A final restriction is that the Marshallian cross-price effects are all zero.

For a Poisson count demand system, the likelihood function is simply the product of the single-site demands shown in equation 19.8. The joint likelihood function is:

$$\prod_{i=1}^{n} \frac{e^{-\lambda_i} \lambda_i^{q_i}}{q_i!} \qquad 19.21$$

where n is the number of sites and the latent quantity demanded for a given site i, λ_i, is $exp(X_i\beta)$. Equation 19.21 is maximized subject to the restrictions described above.

A good example of the application of a Poisson demand system to forest recreation is the study by Englin et al. (1998). They focused on the impact of exchange rates on demand for backcountry canoeing in four wilderness parks in Canada. Because visitors from the United States use these parks, a shift in the exchange rate not only changes the prices of all parks but also changes the relative prices of parks within the system to different consumers. This is because the proportion of the travel costs that occur within the United States versus Canada differs across individuals. For Canadian visitors, the Canadian costs comprise total travel cost. For some visitors, such as visitors from the north central United States, the Canadian costs comprise a large proportion of total travel costs. However, for other visitors, such as from the southern United States, Canadian costs are a tiny proportion of total costs.

Measuring welfare in a setting where relative price shifts are simultaneously distributed across several substitute sites requires a Hicksian framework for analysis. As pointed out by Englin et al. (1998), the restriction that Marshallian cross-price effects must equal zero does not mean that the Hicksian cross-price effects are zero. The Hicksian cross-price effects can be calculated as:

$$s_{ijk} = q_{ij} \frac{\partial q_{ik}}{\partial m_i} = \gamma q_{ik} q_{ij} \qquad 19.22$$

where s_{ijk} is the Hicksian substitution effect between sites j and k for individual i, and the q's are quantities of trips to the sites in the system by individual i. This is simply an application of the Slutsky formula. The Hicksian cross-price effects are symmetric (i.e., $s_{ik} = s_{kj}$) for any individual i. If no individual takes trips to specific pairs of sites, then some cross-price effects are zero. For example, Englin et al. (1998) found that the cross-price effects for the most remote and most developed parks were zero, suggesting that subsets of parks provide opportunities for different types of recreationists.

3. APPLICATION: RAIN FOREST VALUATION

To demonstrate how count data methods can be applied to real world forestry problems, we present a case study of tropical rain forest protection in southern Brazil. There are relatively few studies of the *in situ* value of tropical rain forests. This example provides estimates of the recreational value of an area designated by UNESCO (United Nations Educational, Scientific and Cultural Organization) as the Lagamar Biosphere Reserve and, to our knowledge, represents the first application of a count data model to recreation in Brazilian protected areas.

3.1 Background

The Atlantic Coastal Forest stretches for more than 3,000 kilometres along the coast of Brazil. This ecosystem is rich in biological diversity and endemic species and is considered to be one of the most endangered ecosystems in the world. The largest remaining contiguous area of this forest type occurs in southern Brazil in the Environmental Protection Area (Área de Proteção Ambiental, or APA) of Guaraqueçaba.

The forests within and surrounding the APA have been protected in large part due to the area's isolation, which has limited tourism and other forms of economic development. Beyond the mountains and the bay that form the boundary of the region lie some of Brazil's largest and most economically developed cities, which are potential sources both of deforestation pressure and of tourists.

At the time of the study, recreation of any type was very limited in the APA due to difficult accessibility. The APA could be reached by following a dirt road, much of it in poor repair, for more than 60 kilometres. Adventure tourism in the immediately surrounding area included mountaineering, hiking, and camping in the protected Marumbi Area (Serra do Mar mountains) and primitive beach recreation on the protected Ilha do Mel Ecological Station. Immediately outside of these primitive areas, mass tourism was occurring at heavily developed beaches along the coast.

3.2 Sampling Methods and Data

Data for this case study were collected from Brazilian tourists at adventure tourism sites and other locations. On-site interviews were conducted at six adventure tourism sites in the study area. The off-site sample was drawn from tourists at 25 popular locations outside of the adventure tourism areas. All respondents were asked to indicate how many trips they had taken to each site during the past year. Respondents were also asked where they lived, what recreational activities they participated in, what investments the government should make (if any) in public recreation, and some socio-economic questions. Complete records were obtained for 143 people on-site and 337 people off-site. Following Creel and Loomis (1990), trips to adventure tourism sites (the APA, Marumbi, and Ilha do Mel) were pooled and treated as a single site. This approach imposes the assumption that parameters are the same across the pooled sites.

Distances were computed from origin-destination data using the official Brazilian road atlas. Travel cost (COST) was estimated by multiplying $0.15/mile times the round-trip distance. Income was defined as monthly household income and was included in the model in logarithmic form (LINC). Numerical scales were created identifying the number of adventure activities (such as mountaineering and hiking) and passive activities (such as sightseeing and picnicking) people participated in (ACTIVE, PASSIVE). Socio-economic variables in the model are respondent age (AGE) and gender (SEX). Finally, a dummy variable indicates the importance of paving the access road into the region (ACCESS). While paving the road would reduce the time and effort required to access the recreation sites, it would also change the character of the region by promoting economic development.

3.3 Results

We estimated a variety of models including Shaw's Poisson and the Englin-Shonkwiler negative binomial (using on-site data) and the ZIP model (using off-site data). Because the dispersion parameter was not significant in the zero-inflated negative binomial model using off-site data, estimates for this model are not reported here.

As shown in table 19.1, each model had the expected negative parameter estimate on the travel cost variable, which is consistent with a downward-sloping demand curve. Also consistent across the models, men and those who participated in more adventure activities took more trips. The dispersion parameter was marginally significant in the Englin-Shonkwiler negative binomial model, suggesting that it was more suitable than Shaw's Poisson for these data.

The ZIP model reveals information about the decision of whether or not to participate in adventure tourism. As can be seen in the bottom panel of table 19.1, those who participated in more adventure-related activities, those who did not think that access to the adventure sites should be improved, and those with lower incomes were more likely to visit the adventure tourism sites. Using equations 19.15 and 19.16 and the parameter estimates from the ZIP model, we estimated that 55% of the population would not take a trip to the adventure sites. Of the estimated 45% of the population who were in the market, only one-third (or 15% of the population) were likely to have taken a trip in the survey period. An estimated two-thirds of the people in the market did not take a trip but were located at the demand corner point. This result suggests that there is a large potential demand for primitive recreation sites in this region of Brazil.

Table 19.1. Parameter estimates for on-site and off-site count data models in southern Brazil

Variable	Shaw's Poisson (St. Err.)	Englin-Shonkwiler Negative Binomial (St. Err.)	Zero-inflated Poisson (St. Err.)
Trips equation	- On-site -		- Off-site -
Constant	1.512**	1.50	-0.761
	(0.766)	(1.064)	(0.737)
COST	-0.016***	-0.022*	-0.007***
	(0.004)	(0.013)	(0.002)
ACTIVE	0.360***	0.402***	0.190**
	(0.049)	(0.063)	(0.080)
PASSIVE	0.254***	0.308***	-0.026
	(0.096)	(0.119)	(0.134)
AGE	-0.022**	-0.025*	-0.005
	(0.011)	(0.015)	(0.006)
SEX	-0.426**	-0.570**	-0.611**
	(0.176)	(0.228)	(0.287)
LINC	-0.089	-0.063	0.193*
	(0.098)	(0.129)	(0.106)
σ (overdispersion)	—	0.151*	—
		(0.084)	
Participation equation			
Constant	—	—	3.121
			(1.920)
ACTIVE	—	—	0.516**
			(0.216)
LINC	—	—	-0.547**
			(0.284)
ACCESS	—	—	-1.26**
			(0.542)
N	143	143	337
Predicted trips/year	1.10	1.45	0.39
CS/trip [a]	$62.50	$45.45	$142.86
CS/year [a]	$58.13	$65.91	$55.71

[a] CS = Consumer Surplus
*** = significant at 1% level, ** = significant at 5% level, * = significant at 10% level

In comparison, results from the on-site sample indicated that from 67% (using parameters from Shaw's Poisson model) to 77% (using parameters from the Englin-Shonkwiler negative binomial model) of the population were in the market. The discrepancy between market shares estimated using on-site and off-site models is due to the ability of the off-site model to locate the corner point of the recreation demand curve.

Tobias and Mendelsohn (1991), using the same travel cost per kilometer as in our study, estimated consumer surplus per trip to be $35 for domestic tourists visiting a rain forest reserve in Costa Rica. Our estimates, using on-site and off-site data, are higher than reported for Costa Rica. This is not surprising given the proximity of relatively affluent urban areas (such as São Paulo and Curitiba) to our study sites. Taken together, these studies indicate that protection of tropical rain forests can provide significant recreational benefits to citizens in developing countries.

4. CONCLUSIONS

Estimates of the value of outdoor recreation provide policy makers with information that is essential to planning multiple-use management of forests. Count data models are a relatively new addition to recreation demand modeling and focus attention on the nature of the underlying processes that generate data on recreational trips. If the number of recreational trips taken by visitors is small, the true data-generation process cannot be normally distributed. In these cases, OLS models are inappropriate for estimating travel cost demand curves. Count data models that are consistent with the non-negative, integer nature of trip data are required. While a Poisson model is often the starting point for estimation, the potential for over-dispersion should always be evaluated.

Researchers often collect data by sampling recreationists on-site in order to save money on survey costs. If data are collected on-site, adjustments for truncation and endogenous stratification must be made in order to estimate population parameters. However, these estimators are founded on the (generally untested) assumption that identical processes generate positive trips and zero trips in the population.

Although off-site data may be more expensive to collect than on-site data, the gain in information may exceed the incremental cost. That gain includes the ability to jointly model the participation decision along with the number of trips taken. Such zero-inflated models permit the researcher to isolate three types of people in the population: (1) those who would never take a trip, (2) those who would take a trip if the price were low enough, and (3) those who are trip-takers. In cases where there exists a large potential recreation demand, such as our case study in Brazil, this representation should result in more accurate estimates of the recreational value of forests.

5. LITERATURE CITED

BOCKSTAEL, N., W.M. HANEMANN, AND I. STRAND (eds.). Benefit analysis using indirect or imputed market methods, Vol. II. Department of Agricultural and Resource Economics, University of Maryland, College Park (undated).

BOCKSTAEL, N.E., K.E. MCCONNELL, AND I. STRAND. 1991. Recreation. P. 227-270 in Measuring the Demand for Environmental Quality, Braden, J.B., and C.D. Kolstad (eds.). Elsevier, New York. 370 p.

BROWN, G., JR., AND R. MENDELSOHN. 1984. The hedonic travel cost method. Rev. Econ. Stat. 66:427-433.

CAMERON, A.C,. AND P. TRIVEDI. 1986. Econometric models based on count data: Comparisons and applications of some estimators and tests. J. Applied Econ. 1:29-53.

CLAWSON, M. 1959. Methods of measuring the demand for and value of outdoor recreation. Resources for the Future Reprint No. 10, Washington DC.

CLAWSON, M., AND J.L. KNETSCH. 1966. Economics of outdoor recreation. Johns Hopkins University Press (for Resources for the Future), Baltimore, MD. 328 p.

CREEL, M., AND J. LOOMIS. 1990. Theoretical and empirical advantages of truncated count data estimators for analysis of deer hunting in California. Am. J. Agric. Econ. 72:434-441

CREEL, M., AND J. LOOMIS. 1992. Modeling hunting demand in the presence of a bag limit, with tests of alternative specifications. J. Environ. Econ. Manage. 22: 99-113.

ENGLIN, J., AND R. MENDELSOHN. 1991. A hedonic travel cost analysis for valuation of multiple components of site quality: The recreation value of forest management. J. Environ. Econ. Manage. 21: 275-290.

ENGLIN, J., AND J.S. SHONKWILER. 1995. Estimating social welfare using count data models: An application to long-run recreation demand under conditions of endogenous stratification and truncation. Rev. Econ. Stat. 77: 104-112.

ENGLIN, J., P.C. BOXALL, K. CHAKRABOTY, AND D.O. WATSON. 1996. Valuing the impacts of forest fires on backcountry forest recreation. For. Sci. 42: 450-455.

ENGLIN, J., P. BOXALL, AND D. WATSON. 1998. Modeling recreation demand in a Poisson system of equations: An analysis of the impact of international exchange rates. Am. J. Agric. Econ. 80: 255-263.

FREEMAN, A.M. III. 1993. The measurement of environmental and resource values: Theory and methods. Resources for the Future, Washington DC. 516 p.

GROGGER, J., AND R. CARSON. 1991. Models for truncated counts. J. Applied Econometrics. 6:225-238.

GURMU, S., AND P.K. TRIVEDI. 1996. Excess zeros in count models for recreational trips. J. Bus. Econ. Stat. 14: 469-477.

HAAB, T.C., AND K.E. MCCONNELL. 1996. Count data models and the problem of zeros in recreation demand analysis. Am. J. Agric. Econ. 78: 89-102.

HAUSMAN, J.A., B.H. HALL, AND Z. GRILICHES. 1984. Econometric models for count data with an application to the patents–R&D relationship. Econometrica 52: 909-938.

HELLERSTEIN, D. 1991. Using count data models in travel cost analysis with aggregate data. Am. J. Agric. Econ. 73: 860-867.

HELLERSTEIN, D., AND R. MENDELSOHN. 1993. A theoretical foundation for count data models, with an application to a travel cost model. Am. J. Agric. Econ. 75: 604-611.

HOLMES, T., B. SOHNGEN, L. PENDLETON, AND R. MENDELSOHN. 1997. Economic value of ecosystem attributes in the Southern Appalachian Highlands. P. 187-190 in Proceedings of the Conference on Integrating Social Sciences and Ecosystem Management, H.K. Cordell (ed.). General Technical Report SRS-17, USDA Forest Service, Asheville, NC. 230 p.

KALBFLEISCH, J.G. 1985. Probability and statistical inference. Springer-Verlag, New York. 703 p.

LAFRANCE, J. 1990. Incomplete demand systems and semi-logarithmic demand models. Australian J. of Agric. Econ. 34: 118-131.

OVASKAINEN, V., J. MIKKOLA, AND E. POUTA. 2001. Estimating recreation demand with on-site data: An application of truncated and endogenously stratified count data models. J. Forest Econ. 7: 125-144.

PENDLETON, L.H., B. SOHNGEN, R. MENDELSOHN, AND T. P. HOLMES. 1998. Measuring environmental quality in the Southern Appalachian Mountains. For. Sci. 44: 603-609.

PENDLETON, L.H., AND J.S. SHONKWILER. 2001. Valuing bundled attributes: a latent characteristics approach. Land Econ. 77: 118-129.

PORTNEY, P., AND J. MULLAHY. 1986. Urban air quality and acute respiratory illness. J. Urban Econ. 20: 21-38.

SHAW, DAIGEE. 1988. On-site samples' regression: Problems of non-negative integers, truncation and endogenous stratification. J. Econometrics 37: 211-223.

SHAW, W.D., AND P. JAKUS. 1996. Travel cost models of the demand for rock climbing. Ag. Res. Econ. Rev. 25: 133-142.

SHONKWILER, J.S., AND W.D. SHAW. 1996. Hurdle count-data models in recreation demand analysis. J. Agric. Res. Econ. 21: 210-219.

TOBIAS, D., AND R. MENDELSOHN. 1991. Valuing ecotourism in a tropical rain-forest reserve. Ambio 20(2): 91-93.

VAUGHAN, W.J., AND C.S. RUSSELL. 1982. Valuing a fishing day: An application of a systematic varying parameter model. Land Econ. 58: 451-463.

VON HAEFEN, R.H., AND D.J. PHANEUF. 2002. A note on estimating nested CES preferences for outdoor recreation. Unpublished Working Paper. Department of Agricultural and Resource Economics, North Carolina State University, Raleigh. 18 p.

YEN, S.T., AND W.L. ADAMOWICZ. 1993. Statistical properties of welfare measures from count-data models of recreation demand. Rev. Agric. Econ. 15: 203-215.

Chapter 20

Forest Ecosystem Services As Production Inputs

Subhrendu K. Pattanayak and David T. Butry
Research Triangle Institute and USDA Forest Service

Are we cutting down tropical forests too rapidly and too extensively? If so, why? Answers to both questions are obscured in some ways by insufficient and unreliable data on the economic worth of forest ecosystem services. It is clear, however, that rapid, excessive cutting of forests can irreversibly and substantively impair ecosystem functions, thereby endangering the flow of several socially valuable goods and services from standing forests. One reason for such excessive deforestation is failure to consider the full range of goods and services provided by the forests, particularly latent and complex ecosystem services.

Forests provide ecosystem services by sequestering carbon, maintaining habitat and biodiversity, stabilizing hydrological flows, mitigating soil erosion, and improving microclimates. Public protection of tropical forests is necessary because the market mechanism cannot provide the optimal level of ecosystem services. The level of public support for forest protection depends on the net benefits of providing these services. Recent surveys of valuation studies reveal that economic benefits of forest ecosystem services are not well understood and are rarely quantified (WRI 2000). Two challenges are posed in the literature. First, recent reviews show that (1) ecosystem valuation studies have framed the valuation question incorrectly and have applied inappropriate methods (Bockstael et al. 2000), and (2) valuation studies have overlooked livelihood values of natural resources in developing countries, focusing largely on amenity values in developed countries (Deacon et al. 1998). Second, valuation of ecological services that are inputs into production processes have typically relied on data-intensive approaches, such as the measurement of full profit functions, instead of focusing on

Sills and Abt (eds.), Forests in a Market Economy, 361–378. ©Kluwer Academic Publishers. Printed in The Netherlands.

demand for a weak complement, which substantially reduces the data requirements (Huang and Smith 1998). This chapter addresses these issues with a case study from Indonesia in which forest protection policies in upstream watersheds stabilize hydrological flows in downstream farms.

1. DEFINING AND VALUING FOREST ECOSYSTEM SERVICES

1.1 What Are Forest Ecosystem Services?

Adapting a definition by Daily (1997), forest ecosystem services are the conditions and processes through which forest ecosystems, and the species that make them up, sustain and fulfill human life. Forests maintain biodiversity and the production of ecosystem goods, such as timber and pharmaceutical precursors, and ecosystem services that are actual life-support functions, such as microclimate regulation and watershed services. Forest ecosystems also confer many intangible aesthetic and cultural benefits. Below we catalog a longer list of goods and services from forest ecosystems. Our list of potential goods and services focuses on direct and indirect benefits to human beings because valuation, as described here, is mostly for people (Freeman 1996).

The World Resources Institute (WRI 2000) categorizes forest ecosystem services into two basic groups. Goods include timber, fuelwood, drinking and irrigation water, fodder, nontimber forest products (such as vines, bamboo, and leaves, as described in chapter 15), food (honey, mushrooms, and fruits), and genetic resources. Services include removing air pollution, emitting oxygen, cycling nutrients, maintaining an array of watershed functions, maintaining biodiversity, sequestering carbon (further discussed in chapter 13), moderating weather extremes, generating soil, providing employment, providing human and wildlife habitat, contributing aesthetic beauty, and providing recreation (further discussed in chapter 19).[1]

1.2 Taking Stock of Forest Ecosystems Services

WRI (2000) provides an excellent evaluation of the current state of forest ecosystems, discussing timber, fuelwood, watershed services, biodiversity and carbon, noting that forest cover, now accounting for about 25% of the world's land surface, has been reduced by 20% to 50% since preagricultural times. Most developing countries rely on timber exports, while in most industrialized countries, the majority of timber comes from production forests. Fuelwood accounts for about 15% to 80% of the primary energy

Forest Ecosystem Services as Production Inputs

supply in developing countries, with use concentrated among the poor. Forests harbor about 66% of the known terrestrial species and have the highest species diversity, including threatened species, and endemism of any ecosystem. Forest vegetation and soils sequester nearly 40% of all terrestrial stored carbon.

Nearly 30% of the world's major watersheds—particularly in tropical montane forest regions—have lost more than 75% of their original forest cover. The greatest threats to forest extent and condition today are conversion to other land uses and fragmentation by agriculture, logging, and roads. Although 66% of all fuelwood comes from roadsides, community woodlots, and wood industry residues, fuelwood collection causes local deforestation in parts of Asia, Africa, and Latin America (WRI, 2000).

1.3 Valuing Ecosystem Services

Arrow et al. (2000) call for ecosystem valuation because ecosystem management requires detailed bookkeeping of costs and benefits and evaluation of tradeoffs. We approach this issue by addressing three broad questions: Why value? What can be valued? How to value?

Valuation, described as the search for an integrative metric, is conducted for one of three reasons (Pritchard et al. 2000): (1) to show that natural systems are indisputably linked to human welfare and are represented in the decision-making process, (2) to describe the relative importance of various ecosystem types, or (3) to justify or critique particular decisions in particular places, e.g., cost/benefit analyses. Consequently, valuation appeals to diverse constituencies ranging from free-market advocates who believe it will improve economic efficiency, to managers in search of integrative metrics to guide decision making, to environmentalists who believe that the standing of neglected natural resources will be enhanced by the recognition of their value (Carpenter and Turner 2000). In general, while economic valuation of ecosystem services is neither necessary nor sufficient for conservation (Heal 2000), it can guide public decisions and ecosystem management by providing estimates of the incremental value or cost of changes in ecosystem conditions (Bockstael et al. 2000).

Economic valuation of forest ecosystem services can only address services that are directly or indirectly useful to human beings, including nonconsumptive uses that provide some psychological benefit. A serious discussion of what can be valued was triggered by attempts to value all of nature's services (e.g., Costanza et al. 1997). While consensus has not been reached, a critical review of such global valuation studies (Bockstael et al. 2000) provides three vital considerations for ecosystem valuation. First, analysts should study possible changes to specific forest ecosystem conditions. All-or-nothing changes are irrelevant for policy analysis and uninteresting, perhaps even trivial, from an academic perspective. Second, we should not scale up small changes in specific and localized components of individual forest ecosystems to generate aggregate forest ecosystem values because ecosystem valuation fails simple additivity tests.[2] Finally, the analysis must satisfy the most fundamental economic valuation criterion, namely, that ecosystem values do not exceed ability to pay.

Ecosystem valuation is complicated by the fact that ecosystem services, as quasi-public goods and externalities, are not well accounted for in market mechanisms (Arrow et al. 2000). The key to valuing a change in an ecosystem function lies in establishing the link between that function and some service flow valued by people. The analysis must reflect the intricate web of physical relationships between processes and conditions that link causes and effects in different parts of the ecosystem. If that link can be established, then the economist's concept of derived demand can be applied (Freeman 1996). Some ecosystem functions are related to useful ecosystem services, such as photosynthesis producing useful plant material. Other examples are indirect, subtle, and latent, such as photosynthesis generating wildflowers that support bees, which pollinate commercial fruits. Once we establish how a policy will change photosynthesis capacity and therefore plant and fruit production, we can analyze the demand for the plant material or the commercial fruits to derive a measure of willingness to pay for (or willingness to accept) changes in policy-induced photosynthetic services. These money measures are based on consumer sovereignty as opposed to some external prescription of how consumers should make choices (Bockstael et al. 2000). Travel cost (chapter 19), contingent valuation (chapter 17), hedonic property and wage, and productivity analysis are among the typical valuation methods that apply derived demand theory (see Freeman 1993).

1.3.1 What Forest Ecosystem Services Have Been Valued?

The study by Costanza et al. (1997), which included some forest ecosystem services, failed to satisfy the basic tenets of valuation. Other attempts to link forest ecosystem functions to economically valuable ecosystem services are rare (Freeman 1996). Researchers have measured values for specific attributes of forests (particularly for recreation uses), though the analysis has typically not included ecological models that link the attributes to specific forest ecosystem functions. We observe similar gaps in studies of two other ecosystems, wetlands and atmosphere, which are more often the subject of economic analysis. In both cases, some proxy for the ecosystem service (e.g., saline concentrations in estuarine wetlands or atmospheric ozone concentrations in farming counties) is related to a production activity (e.g., shrimp or corn), but the link between the ecosystem functioning and the service has not been spelled out.

Economic analyses of watershed services have typically concentrated on soil erosion effects (Pattanayak [forthcoming]). Empirical economic analyses of soil erosion have used resource accounting approaches, econometric production functions methods, or mathematical programming models. In the econometric approach, the production functions are usually either aggregative (nation or statewide), thereby losing site-specific details, or simple, with just two or three arguments.[3] In all cases the value of soil erosion is estimated in terms of its effect on economic productivity. By proposing the use of econometric methods to analyze ecosystem services as production inputs, the approach employed in this chapter most resembles soil valuation studies that use production functions. The critical distinction between these studies and our proposal relates to the link between ecosystem functions and the resulting services. The erosion studies typically focus on managed agronomic systems and on-farm economic productivity losses. They do not discuss or rigorously analyze off-site consequences of or the linkages with upstream ecological phenomena.

1.3.2 Analytical Framework

The economic principles for valuation are straightforward, and the economic value of ecosystem services can be viewed as the outcome of three sets of functional relationships (Freeman 1993; subsequently adapted for ecosystem services by Kramer et al. 1997 and Pattanayak [forthcoming]). Public policies combined with private decisions affect forested watersheds, change watershed flows, and, thereby, generate changes in ecosystem services. These services affect private production activities of economic agents and consequently their economic welfare. The change in welfare, evaluated in terms of market prices of private commodities, is the use value of ecosystem services.

1.4 Freeman's Three-Stage Approach

The first stage of analysis relates an index of ecosystem service (e.g., quantity or rates of runoff, streamflow, erosion, and sediment) to public and private land use decisions (e.g., national parks or on-farm agroforestry) that are amenable to public policy (figure 20.1). The structural relationship is conditioned by time lags and various environmental characteristics, including geologic substrate, topography, and climate. Direct and indirect public policies thus cause changes in ecosystem services. Variations between policies and their impact, within the available data set, enable the analyst to use the associated variations in ecosystem services to compute economic values. Exclusive private provision, without any public support (such as subsidies, taxes, provision of information, and technical expertise and credit) is not typical because of the inherent 'public good' characteristics of the provision process and of the ecosystem services themselves.

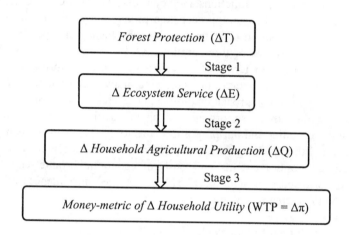

Figure 20.1. Freeman three-stage framework for valuation

The second stage quantifies human use of the ecosystem service. Households use their labor and other inputs, conditional on the nonmarket ecosystem service and other fixed inputs, to produce a vector of commodities for the market and domestic consumption. Ecosystem services can thus be considered a fixed input in either home production of final services, which yield utility (household production theory), or agricultural production (production theory).[4]

In the third stage, the economic value of an ecosystem service or willingness to pay (WTP) for an ecosystem service is determined in terms of the market value of commodities related to that ecosystem service. Models from welfare economics are used to express the money-metric of utility changes or WTP in terms of expenditures changes that depend on the utility level and therefore consumption choices (Freeman 1993). These choices are directly or indirectly driven by market prices of all outputs and inputs, levels of ecosystem service, other fixed inputs, and exogenous income. Thus we can describe WTP as a function of all these exogenous variables listed and measure it by estimating expenditure function or indirect utility functions (Freeman 1993). Alternatively, WTP can be measured as increases in producer surplus, $\Delta \pi$, if markets are complete (Pattanayak and Kramer

2001).[5] Ecosystem services are valued because they are expected to increase utility (and profits). Below we describe an approach proposed by Huang and Smith (1998) to estimate producer surplus changes with input demand functions that have special properties.

1.5 Weak Complementarity for Valuation

Much of the environmental valuation literature has focused on weak complements to environmental goods—goods that are nonessential inputs to household consumption (Freeman 1993). Analysts estimate how the demand for the weak complement shifts in response to changes in environmental quality and measure WTP for environmental quality as the change in consumer surplus. Huang and Smith (1998) develop production analogs of the weak complementarity logic to show that input demand can be used to measure the change in producer surplus induced by a change in environmental inputs into production. In this approach, shown in equation 20.1, WTP for ecosystem services ($E_1 - E_0$) is estimated using Hotelling's lemma and the input demand curve, $L(P_L \mid E, \bullet)$ [6]. Profits can be measured by integrating the input demand function from the market price, P_{L0}, to the choke price, $P_{LC}(E)$.

$$\begin{aligned} WTP &= \Delta \pi(P_Q, P_L \mid \Delta E, Z_Y) \\ &= \int_{P_{L0}}^{P_{LC}(E_1)} L(P_L \mid E_1, \bullet) dP_L - \int_{P_{L0}}^{P_{LC}(E_0)} L(P_L \mid E_0, \bullet) dP_L \\ &= \int_{P_{L0}}^{P_{LC}(E_1)} -\frac{\partial \pi(P_L \mid E_1, \bullet)}{\partial P_L} dP_L - \int_{P_{L0}}^{P_{LC}(E_0)} -\frac{\partial \pi(P_L \mid E_0, \bullet)}{\partial P_L} dP_L \end{aligned} \qquad 20.1$$

The choke price, at which labor demand (L) is equal to zero, depends on the ecosystem condition. An improvement in the ecosystem that generates an ecosystem service will expand the demand for the weakly complementary production input, raise the choke price, and increase profits. WTP for the ecosystem service is, therefore, equal to the change in profits calculated from the two input demand curves. This logic is illustrated in figure 20.2. The basic intuition is that increased ecosystem service raises the value of the marginal product of farm labor because it is a complement. Consequently, the value of the ecosystem service or the amount the household will be willing to pay should equal the increased marginal value product of labor, which is equivalent to the profit increase.

There are two important theoretical conditions for application of the weak complementarity logic in ecosystem valuation. First, the production input in question must be nonessential, so that we can define a choke price.

Without this binding price, the compensation or surplus measure would be infinite (see Freeman 1993 for details). Second, at the choke price, the marginal productivity of ecosystem service must be zero, implying that the production input is a necessary complement to using the ecosystem service. If this were not the case, then we could not value the ecosystem service by analyzing only this production input, because the service would be productive irrespective of the demand for this input. In addition, the induced change in labor demand should not be large enough to induce labor price effects. Huang and Smith (1998) suggest that by focusing on the demand function for a weak complement (e.g., labor), researchers could substantially economize on the data demands for valuation by avoiding estimation of full profit functions (Pattanayak and Kramer 2001).[7]

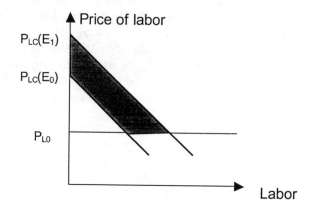

Figure 20.2. WTP for ecosystem service as change in area under demand curve of a weak complement: agricultural labor

2. THE CASE OF DROUGHT MITIGATION FROM RUTENG PARK ON FLORES, INDONESIA

Since the time of Dutch colonial rule, the forests of the Manggarai region on Flores Island have been protected to different degrees across watersheds. In 1993, the government of Indonesia established Ruteng Park on 32,000 ha to prevent further deforestation, initiate reforestation and land conservation, and enhance watershed protection. A recent evaluation of water resources in the region finds that the forests provide drought-mitigation service by protecting streams and rivers (Binnies 1994). Two forest hydrology studies in addition to the Binnies study suggest that in many Manggarai watersheds, forests are net producers of baseflow, the non-episodic residual streamflow that is left over after rain has cycled out of the hydrological system (Swiss Intercooperation 1996, Priyanto 1996). We apply the three-stage framework to this case because the economic value of ecosystem services is unknown even though there is substantial biophysical evidence that Ruteng Park provides drought mitigation to the downstream farmers.

2.1 Applying the Freeman Framework to Ruteng Park

In stage 1, we assume that the establishment of Ruteng Park produces a drought-mitigation service that can be measured as a change in baseflow. The forest hydrology literature posits that extensive tree cover helps maintain baseflow levels in areas with environmental characteristics similar to Ruteng, i.e., steep terrain, intense rainfall, and clayey and compacted soil (Bonell and Balek 1993). The studies by Binnies (1994), Swiss Intercooperation (1996), and Priyanto (1996) suggest that Ruteng forests are net producers of baseflow. The studies do not, however, report precise estimates of enhanced baseflow by watershed. In stage 2, the primary economic role of baseflow is as a fixed input in agricultural production; i.e., it provides soil moisture that enhances farm productivity. In stage 3, improved agricultural production changes the economic welfare of agricultural households downstream of Ruteng Park. This change in welfare is a measure of the value of drought mitigation. As shown above, the value of drought- mitigation services can be measured by computing the incremental producer surplus resulting from the incremental baseflow.

2.2 Ruteng Data

The empirical model presented in the next subsection is based on secondary hydrological and forest statistics and household survey information on the economic activities of the Manggarai people. A water balance model was used to derive baseflow volumes for 37 subwatersheds in the buffer zone of the park, which correspond to current land use (Priyanto 1996). This cross-sectional variation in current baseflow is sufficient to econometrically establish the influence of baseflow on agricultural profits.[8] The household data are drawn from a socioeconomic survey of 500 households that was conducted in the Ruteng area in 1996. Because the hydrological effects of the park dissipate over geographical distance, the survey was restricted to the 47 villages in the buffer zone of Ruteng Park, contiguous to the protected area. The average Ruteng household exhibits a heavy reliance on agriculture, primarily growing coffee and rice and keeping chicken and pigs. Eighty-seven percent of the local people are employed in agriculture. There are a few nonagricultural employment opportunities, including positions with the local government, nongovernmental organizations, kiosks, and logging crews. The statistics on both hiring-in and hiring-out labor, the fact that a large proportion of households report input and output prices, and the proximity of roads and other market infrastructure (e.g., stores and credit facilities) provide some evidence that markets are complete for agricultural products and labor. While the Ruteng region receives on average 2.5 meters of rainfall annually, only about 40% stays in the system as baseflow.

Given that we implement our empirical model by combining socioeconomic survey data with the ecological data related to the forest hydrology model, the precision of our ecological data is important. By merging the two data sets within a geographical information system (GIS), we can potentially improve the general precision of the data set and compute spatially explicit ecological indices. For example, if portions of two streams contribute baseflow to a particular village, we can use GIS to compute the fraction of the total baseflow from any one stream that goes to the particular village by first calculating the fraction of the total stream that passes over the specific village. The contributions of each stream can then be summed. Without GIS, we would calculate a crude weighted average of the baseflow in the two streams, based on eye-balled proportions. We investigate the implications of using data generated with and without GIS (analysis not reported in this chapter).

2.3 Valuation of Drought Mitigation

Applying the Huang and Smith logic and focusing on demand for agricultural labor, we see that agricultural labor can be conceived as a weak complement to baseflow because it satisfies the two necessary conditions for weak complementarity. First, it is possible that labor demand is nonessential, so that at a choke price of P_{LC}, demand for labor is zero. Because nonagricultural sources of income make substantial contributions to household full income (note, not cash income), agricultural labor is a nonessential input to household full income as households switch to other activities when the price of labor is too high. Second, the marginal productivity of baseflow is zero at the choke price, implying that changes in baseflow have no welfare significance unless the effective wage is low enough to make labor demand positive. This follows from the fact that baseflow is useful to the farming households only as a farming input and it is impossible to farm without labor.[9] We estimate the three most common functional forms of labor demand: linear, log-linear, and semilog, described in equations 20.2 to 20.4:

Linear: $L = \alpha + \beta P + \gamma z$ and Welfare Est. $= \dfrac{\hat{L}_1^2 - \hat{L}_0^2}{-2\beta}$ 20.2

Log-linear: $L = e^{\alpha} P^{\beta} Z^{\gamma}$ and Welfare Est. $= \dfrac{P_0 \hat{L}_0 - P_1 \hat{L}_1}{\beta + 1}$ 20.3

Semilog: $L = e^{(\alpha + \beta P + \gamma z)}$ and Welfare Est. $= \dfrac{\hat{L}_1 - \hat{L}_0}{-\beta}$ 20.4

where $\hat{L}_0$ and $\hat{L}_1$ are the predicted baseline labor demand evaluated at mean wage with and without drought mitigation, β is the regression coefficient for wage, and Z is a vector of all other variables including output prices.[10]

Labor demand is hypothesized to be a function of the price of labor; the price of the primary outputs (coffee and rice); and fixed inputs, including baseflow, farm size, soil condition (erosivity), and an irrigation index. Table 20.1 summarizes the expected relationships.

The signs, sizes, and significance of the estimated coefficients are the criteria for evaluating the theoretical performance of our models. We expect labor demand to be negatively correlated with the price of labor. Because prices of rice and coffee reflect returns to labor in farming or the effective Note that the coefficient on erosivity should be negative because it is a negative fixed input. The key parameter in our model is the coefficient on the baseflow variable; its sign and the size will reflect the relative contribution or value of drought mitigation from the forests of Ruteng Park for the farming households. Finally, we include a set of household characteristics, family size, average age, and ratio of ill, adult, and male family members, to test the complete labor market assumption (Pattanayak and Kramer 2001).[11]

Note that the coefficient on erosivity should be negative because it is a negative fixed input. The key parameter in our model is the coefficient on the baseflow variable; its sign and the size will reflect the relative contribution or value of drought mitigation from the forests of Ruteng Park for the farming households. Finally, we include a set of household characteristics, family size, average age, and ratio of ill, adult, and male family members, to test the complete labor market assumption (Pattanayak and Kramer 2001).[12]

Table 20.1. Descriptive statistics and expected signs

Variables	Units	Mean	Expected sign
Labor	Days	115.30	
Price of coffee	$ per kilogram	1.78	(+)
Price of rice	$ per kilogram	0.18	(+)
Price of labor	$ per day	0.90	(−)
Farm size	Hectares	1.2	(+)
Water condition	Baseflow in meters / ha / year	0.4	(+)
Irrigation index	% of farm irrigated	0.1	(+)
Soil condition	Erosivity in tones / ha/ year	2.1	(−)
Family size	Number	4.3	~
Ratio of adults in family	Ratio	0.77	~
Ratio of ill in family	Ratio	0.77	~
Average age	Years	26.7	~
Ratio of males in family	Ratio	0.49	~

2.4 Three Estimated Models

The results of the three labor demand models, for each of the three functional forms (linear, log-linear, and semi-log), are reported in table 20.2.

Table 20.2. Models of labor demand

Variables	Linear		Log-linear		Semi-log	
Constant	89.06	*	5.27		4.35	
Price of coffee	20.86		0.09		0.07	
Price of rice	256.01	***	0.54	***	1.80	**
Price of labor	-119.51	***	-0.88	***	-1.10	***
Farm size	17.03	***	0.15	***	0.17	***
Baseflow	69.54	***	0.23	***	0.67	***
Irrigation	33.93	**	0.06	***	0.52	***
Erosion	-11.53	***	-0.17	**	-0.12	***
Family size	0.24		0.06		0.01	
Adult ratio	43.91		0.08		0.32	
Ill ratio	-4.80		-0.02		-0.02	
Average age	0.36		0.15		0.00	
Male ratio	-0.93		-0.02		-0.01	
Adj. R^2	0.16		0.15		0.15	
F-statistic	8.96	***	8.29	***	8.22	***
Sample size	494		494		494	

***, **, * = significant at 1%, 5%, and 10% level, respectively.

All models are statistically significant. The models explain about 14% to 20% of the variation in labor demand, which is not unusual in a cross-sectional data set. Although the log-linear model has the highest R^2, all models have similar explanatory power. All variables have expected signs and significance in all models, except that price of rice is insignificant and

Forest Ecosystem Services as Production Inputs

weakly related to labor-demand in the log-linear and semi-log models. All five household variables are individually and jointly insignificant in all models—validating the complete labor market assumption. Critically, baseflow has a positive and significant coefficient in all models.

2.5   Economic Value of Baseflow: Elasticity

The positive and significant coefficient of the baseflow variable supports the hypothesis that drought mitigation services enhance agricultural profits. To compare across models with different functional forms, we consider the elasticity of labor demand with respect to baseflow, which is a reflection of the marginal productivity. The precision with which we map the ecological data onto the economic model influences our estimate of the economic contribution of drought mitigation; mapping without GIS (not reported here) tends to overstate the economic contributions of baseflow. The estimated elasticities of 0.21 to 0.26 in Table 20.3 provide a credible approximation of the economic contribution of baseflow to agricultural profitability in Flores, Indonesia.

2.6   Policy Simulation: Valuing 10% and 25 % Increases in Baseflow

We do not have projections of the baseflow levels that will result from forest protection and regeneration in Ruteng Park. Therefore we evaluate two alternative forest hydrology scenarios in which forest protection induces baseflow increases of 10% and 25%. Using the welfare change formulae presented in equations 20.2 to 20.4 and the estimated parameters from table 20.2, we find that a 10% increase in watershed baseflow is estimated to increase profits by $4 to $25 or by 1% to 7% for the typical household. Note, a typical household is one with average household characteristics and profits equal to $350 annually. A 25% increase in baseflow would increase profits by $11 to $65 or by 3% to 19%. The results are similar across functional forms. Collectively they suggest that ecosystem services in the form of increased baseflow can make substantial economic contributions to the farming households in the immediate downstream of the park.

Table 20.3. Elasticities and simulated values

Specification	Elasticity of labor w.r.t baseflow	Drought mitigation benefits of baseflow increase	
		10%	25%
Linear	0.26	$5.39	$13.95
Log-linear	0.23	$25.18	$65.28
Semilog	0.21	$4.12	$10.73

2.7 Policy Implications

Hydrological stabilization, such that downstream drought conditions are mitigated, is one among several ecosystem functions of the forested watersheds within Ruteng Park. The estimated elasticity of 0.2 and the projected profitability increases of 5% to 10% reveal that watershed management that effectively mitigates drought could increase the annual agricultural profit of each household. Pattanayak and Kramer (2001) argue that increased forest cover will mitigate droughts by increasing baseflow only when the Ruteng watersheds have a particular mix of climatic and physiographic features, and therefore policy makers should adopt a selective approach targeting specific watersheds for forest conservation. We wish to emphasize that regardless of the mechanism that effectively mitigates drought, it is clear that increases in baseflow have positive economic value. While it is not our purpose to conduct a comprehensive cost/benefit analysis, the estimates of profit increases reported above can be compared with watershed regreening costs to judge the overall worthiness of investments in Ruteng Park. Finally, we reiterate that the value of drought mitigation constitutes just one element in the calculation of the net present value of the overall integrated conservation and development project for Ruteng Park. Thus, the net impact of reforestation may be positive when all benefits are considered.

3. INSIGHTS FOR FUTURE RESEARCH

This case study shows that hydrological modeling can be combined with microeconometric techniques to value drought mitigation provided by forested watersheds in an agrarian region of Southeast Asia. The literature review and the mechanics of this case study offer insights for ecosystem valuation methods and future research that are described in detail in Pattanayak (forthcoming).

3.1 Insights for Ecosystem Valuation

3.1.1 Conceptual Framework

The three-stage approach described in section 2 (and figure 20.1) organizes the valuation of ecosystem services in terms of changes in producer surplus. It presents a generalizable framework for measuring the economic value of ecosystem services as they contribute to production activities.

3.1.2 Indexing Ecosystem Services

As discussed by Freeman (1996), the key to ecosystem valuation lies in establishing the link between ecosystem function and some service flow valued by people. The Ruteng study offers two ideas regarding index construction, which is central to operationalizing Freeman's idea. First, the Ruteng study illustrates that cross-sectional variation in current levels of the ecosystem service, i.e., annual baseflow, enables analysts to generate useful policy information even without predictions of the changes in the ecosystem service that will result from policy changes and human behavior. Second, value estimates are significantly influenced by the degree of precision offered by GIS in measuring ecological variables such as baseflow.

3.1.3 Applying Weak Complementarity

The major advantage of weak complementarity, as opposed to estimating the full profit systems (Pattanayak and Kramer 2001), is data efficiency. In comparing ecosystem values for commensurable baseflow measures (non-GIS), we find that the welfare estimates are close—on the order of only a few dollars. The similarity of the two results suggests that the weak complementarity logic presents significant methodological efficiencies by using considerably fewer data. Estimates based on demand for a weak complement may be a lower bound of ecosystem loss when there is more than one such complement. As shown by Bockstael and Kling (1988), the weak complementarity logic can be applied to multiple market complements. In application the trick will be to find the most relevant or substantive complement. Labor productivity is the primary economic contribution of hydrological stabilization in our study area.[13]

3.2 Future Research

The method described in this chapter generates at best an approximate value of complex ecosystem services. For this and other reasons, Pattanayak (forthcoming) calls for the use of other methods to value ecosystem services, in addition to the profit/producer surplus-based approach, to judge the robustness of the empirical estimates. Although household and sample level values of ecosystem services are desirable policy information, their usefulness is limited to the socioeconomic and geographic context in which the values are derived. Given the costs of conducting new research for site-specific environmental resources, it is important to develop methods for transferring benefit estimates from one site to similar sites in a theoretically correct manner. Future research could focus on adapting the calibration strategy described and illustrated by Smith et al. (2002) and on applying a meta-analytic approach for combining estimates from several existing valuation studies to develop a value function of ecosystem services.

4. CONCLUSION

This chapter offers a conceptual and empirical framework for valuing ecosystem services and some suggestions for future research. Although forest protection is professed to generate several ecosystem benefits, recent surveys of valuation studies reveal that economic benefits of forest ecosystem services are not well understood and are rarely quantified (WRI 2000). We discuss issues surrounding valuation of forest ecosystem services and illustrate a method for valuing watershed services. We focus on estimating livelihood values to poor farming communities from protected tropical watersheds by estimating demand for a weak complement of ecological services—agricultural labor. We address these research issues with a case study from Indonesia in which forest protection policies in upstream watersheds in Flores stabilize hydrological flows in downstream farms.

5. LITERATURE CITED

ADAMOWICZ, W., J. FLETCHER, AND T. GRAHAM-TOMASI, 1989. Functional Forms and Statistical Properties of Welfare Measures. Am. J. of Agri. Econ 71(2): 414-421.

ARROW, K., G. DAILY, P. DASGUPTA, S. LEVIN, K. MALER, E. MASKIN, D. STARRETT, T. STERNER, AND T. TIETENBER, 2000. Managing ecosystem resources. Environ. Sci. Tech. 34: 1401-1406.

BINNIES AND PARTNERS. 1994. Master plan report: Integrated water resources in Flores Island, Indonesia. Volume 2. Directorate General of Water Resources Development, Ministry of Public Works, Indonesia.

BOCKSTAEL, N., M. FREEMAN, R. KOPP, P. PORTNEY AND V. K. SMITH, 2000. On measuring economic values for nature. Environ. Sci. Tech. 34 (8): 1384-1389.

BOCKSTAEL, N., AND C. KLING, 1988. Valuing Environmental Quality: Weak Complementarity with Sets of Goods. Am. J. Agr. Econ 70(3): 654-662.

BONELL, M., AND J. BALEK. 1993. Recent scientific developments and research needs in hydrological processes of the humid tropics. P. 167-260 *in* Hydrology and Water Management in the Humid Tropics, Bonnell, M., M. Hufschmidt, and J. Gladwell (eds.). UNESCO / Cambridge University Press, New York, NY.

CARPENTER, S., AND M. TURNER. 2000. Opening the black boxes: Ecosystem science and economic valuation. Ecosystems. 3(1): 1-3.

CHAMBERS, R. 1988. Applied production analysis: A dual approach. Cambridge University Press. New York, NY. 331 p.

COSTANZA, R., R. D'ARGE, R. DE GROOT, S. FARBER, M. GRASSO, B. HANNON, K. LIMBURG, S. NAEEM, R. O'NEILL, J. PARUELO, R. RASKIN, P. SUTTON, AND M. VAN DEN BELT. 1997. The value of the world's ecosystem services and natural capital. Nature. 387:253-260.
DAILY, G., 1997. What are ecosystem services? P. 1-10 *in* Nature's Services: Societal Dependence on Natural Ecosystems, G. Daily (ed.). Island Press, Washington DC.
DEACON, R., D. BROOKSHIRE, A. FISHER, A. KNEESE, C. KOLSTAD, D. SCROGIN, V. SMITH, M. WARD AND J. WILEN. 1998. Research trends and opportunities in environmental and natural resource economics. Environ. Res. Econ. 11(3-4):383-397.
FREEMAN, A. 1993. The measurement of environmental and resource values: Theory and methods. Resources for the Future, Washington DC. 516 p.
FREEMAN, A. 1996. On Valuing the Services and Functions of Ecosystems. P. 241-254 *in* Human Activity and Ecosystem Function: Reconciling Economics and Ecology, Simpson, R. and N. Christensen (eds.). Chapman and Hall, New York.
HEAL, G. 2000. Valuing ecosystem services. Ecosystems 3(1): 24-30.
HUANG, J. and V. K. SMITH. 1998. Weak complementarity and production. Econ. Letters 60: 329-333.
KRAMER, R., D. RICHTER, S. PATTANAYAK, AND N. SHARMA. 1997. Economic and ecological analysis of watershed protection in Eastern Madagascar. J. Environ. Manage. 49: 277-295.
PATTANAYAK, S. K., AND R. KRAMER. 2001. Worth of watersheds: A producer surplus approach for valuing drought control in Eastern Indonesia. Environ. Dev. Econ. 6: 123-145.
PATTANAYAK, S.K. Forthcoming. Valuing watershed services: Concepts and empirics from Southeast Asia. Agr., Ecosystems, Environ.
PRITCHARD, L., C. FOLKE, AND L. GUNDERSON, 2000. Valuation of ecosystem services in institutional context. Ecosystems. 3(1): 36-40.
PRIYANTO, A. 1996. Hydrology specialist report on Ruteng. Directorate General of forest protection and nature conservation. Ministry of Forestry. Jakarta, Indonesia. 25 p.
SMITH, V.K., G.L. VAN HOUTVEN, AND S.K. PATTANAYAK, 2002. Benefit transfer via preference calibration: Prudential algebra for policy. Land Econ. 78(1):132-152.
SWISS INTERCOOPERATION. 1996. Performance of springs in Manggarai. Summary of Study by PDAM Ruteng. Ruteng, Flores, Indonesia. 18 p.
WRI 2000. Taking stock of forest ecosystems. P. 87-102 *in* People and Ecosystems: The Fraying Web of Life. World Resources 2000-2001. World Resources Institute, Washington DC

---

[1] Pattanayak (forthcoming) describes a subset of ecosystem services—watershed services—that include erosion control, enhanced soil quality, improved water yield, stabilization of streamflows, and sediment reduction.

[2] Such simplified addition is inappropriate because (1) forest ecosystem services are nondivisible and nonexclusive, (2) unit values do not reflect declining marginal willingness to pay (WTP), (3) interdependence among and changes in other ecosystem conditions are not considered, and (4) income and general equilibrium price effects are ignored.

[3] A resource accounting approach is characterized by project evaluation in which intertemporal cash flows are generated using parametric economic values drawn from secondary sources. In the econometric approach, simple production functions are estimated to relate agricultural production to soil erosion. A mathematical programming approach seeks an optimum, given an objective function that is subject to constraints with predetermined parameters.

[4] For example, under *household production theory*, households may combine goods such as water (from the streams in watersheds) and labor to provide a *service* such as drinking or cooking, which enhances *utility*. By comparison, conventional *production theory* would conceptualize streams (raw material) and labor as *inputs* in the production of water as an *output* that could then be sold or consumed. The relationship between the nonmarketed ecosystem service and market commodities falls under one of three general categories: complements, substitutes, or differentiated goods (Freeman 1993).

[5] Pattanayak and Kramer (2001) show that the value of the ecosystem service can be measured by incremental profits that are equivalent to a change in household expenditures. The logic is that complete markets imply that market prices (used to calculate household profits), rather than a household-specific virtual price, reflect the relevant opportunity costs (used to calculate household expenditures). Therefore the increase in producer surplus or profits induced by the greater ecosystem service is equivalent to additional expenditures that the household would be willing to incur to realize the level of welfare associated with higher ecosystem services.

[6] Hotelling's lemma states that the derivative of profits with respect to input price is equal to the input demand (Chambers 1988).

[7] In a systems approach, we would estimate equations for profit, output supplies, and input demands as functions of prices and fixed inputs. Using weak complementarity, we could focus on one essential output supply *or* input demand and estimate it as a function of prices and fixed inputs; we would not need data on all quantities.

[8] While cross-sectional data was sufficient for our purposes, undoubtedly time-series data would have been useful to validate such a model.

[9] Benefits, measured as savings in water collection costs, were found to be insignificant in comparison to the agricultural productivity benefits (Pattanayak [forthcoming]).

[10] If $\beta < -1$ in the log-linear case, an adjusted formula, which is described in Adamowicz et al. (1989), must be applied to compute welfare changes.

[11] If this set of five variables is statistically unrelated to labor demand, it would suggest that production decisions are made independent of consumption decisions, because the labor market is perfect and hired labor can be substituted for family labor.

[11] If this set of five variables is statistically unrelated to labor demand, it would suggest that production decisions are made independent of consumption decisions, because the labor market is perfect and hired labor can be substituted for family labor.

[12] In principle, we could estimate demand for each weak complement and calculate the relevant welfare values. Another advantage of this approach in the production setting, unlike the consumption setting where weak complementarity is a maintained hypothesis, is that the analyst can test for complementarity, because the relationship is a physical/technological association. Our results show that physical complementarity of labor and baseflow holds in Ruteng.